Human Performance in Automated and Autonomous Systems

Human Performance in Automated and Autonomous Systems

Current Theory and Methods

Edited by

Mustapha Mouloua and Peter A. Hancock

Technical Editor

James Ferraro

CRC Press is an imprint of the
Taylor & Francis Group, an **informa** business

CRC Press
Taylor & Francis Group
6000 Broken Sound Parkway NW, Suite 300
Boca Raton, FL 33487-2742

© 2020 by Taylor & Francis Group, LLC
CRC Press is an imprint of Taylor & Francis Group, an Informa business

No claim to original U.S. Government works

Printed on acid-free paper

International Standard Book Number-13 978-1-138-31222-7 (Hardback)

This book contains information obtained from authentic and highly regarded sources. Reasonable efforts have been made to publish reliable data and information, but the author and publisher cannot assume responsibility for the validity of all materials or the consequences of their use. The authors and publishers have attempted to trace the copyright holders of all material reproduced in this publication and apologize to copyright holders if permission to publish in this form has not been obtained. If any copyright material has not been acknowledged, please write and let us know so we may rectify in any future reprint.

Except as permitted under U.S. Copyright Law, no part of this book may be reprinted, reproduced, transmitted, or utilized in any form by any electronic, mechanical, or other means, now known or hereafter invented, including photocopying, microfilming, and recording, or in any information storage or retrieval system, without written permission from the publishers.

For permission to photocopy or use material electronically from this work, please access www.copyright.com (http://www.copyright.com/) or contact the Copyright Clearance Center, Inc. (CCC), 222 Rosewood Drive, Danvers, MA 01923, 978-750-8400. CCC is a not-for-profit organization that provides licenses and registration for a variety of users. For organizations that have been granted a photocopy license by the CCC, a separate system of payment has been arranged.

Trademark Notice: Product or corporate names may be trademarks or registered trademarks, and are used only for identification and explanation without intent to infringe.

Library of Congress Cataloging-in-Publication Data

Names: Mouloua, Mustapha, editors. | Hancock, Peter A., 1953- editors.
Title: Human performance in automated and autonomous systems. Current theory and methods / edited by Mustapha Mouloua and Peter A. Hancock.
Description: Boca Raton, FL : CRC Press/Taylor & Francis Group, 2019. | Includes bibliographical references and index.
Identifiers: LCCN 2019016924 | ISBN 9781138312272 (hardback : acid-free paper) | ISBN 9780429458330 (ebook)
Subjects: LCSH: Human-machine systems—Technological innovations. | Automation—Human factors—Research.
Classification: LCC TA167 .H8679 2019 | DDC 620.8/2—dc23
LC record available at https://lccn.loc.gov/2019016924

Visit the Taylor & Francis Web site at
http://www.taylorandfrancis.com

and the CRC Press Web site at
http://www.crcpress.com

Contents

Preface vii
Remembering Fallen Heroes: A Tribute to Raja Parasuraman,
Joel Warm, & Neville Moray ix
Acknowledgments xv
About the Editors xvii
List of Contributors xix

Chapter 1   Designing Wearable Interfaces for People
Who Have Hard Jobs 1

               Matthew Ward, James Wen, James Head, & William S. Helton

Chapter 2   Humans and Automated Decision Aids:
A Match Made in Heaven? 19

               Kathleen L. Mosier & Dietrich Manzey

Chapter 3   The Quest for Alternatives to "Levels of Automation"
and "Task Allocation" 43

               Robert R. Hoffman & Matthew Johnson

Chapter 4   Why Representations Matter: Designing to Support
Productive Thinking 69

               John M. Flach & Kevin B. Bennett

Chapter 5   Vigilance and Workload in Automated Systems:
Patterns of Association, Dissociation, and Insensitivity 85

               James L. Szalma & Victoria L. Claypoole

Chapter 6   Theoretical Perspectives on Adaptive Automation 103

               Mark W. Scerbo

Chapter 7   Fatigue, Automation, and Autonomy: Challenges
for Operator Attention, Effort, and Trust 127

               *Gerald Matthews, Ryan Wohleber, Jinchao Lin, &
               April Rose Panganiban*

Chapter 8 Human-Automation Interaction and the Challenge of Maintaining Situation Awareness in Future Autonomous Vehicles 151

Mica R. Endsley

Chapter 9 Human Factors Issues Regarding Automation Trust in UAS Operation, Selection, and Training 169

Mustapha Mouloua, James C. Ferraro, Alexandra D. Kaplan, Phillip Mangos, & Peter A. Hancock

Chapter 10 Autonomous Systems Theory and Design and a Paradox of Automation for Safety 191

David Kaber

Chapter 11 Workload and Attention Management in Automated Vehicles 213

Joonbum Lee, Vindhya Venkatraman, John L. Campbell, & Christian M. Richard

Chapter 12 Attention Management in Highly Autonomous Driving 231

Carryl L. Baldwin & Ian McCandliss

Chapter 13 To Autonomy and Beyond 249

Peter A. Hancock

Chapter 14 Teleology for Technology 265

Peter A. Hancock

Chapter 15 The Axial Age of Artificial Autonomy 301

Peter A. Hancock

Author Index 309

Subject Index 319

Preface

To some of us, it hardly seems the blink of an eye since we first convened in Rockville, MD, now nearly a quarter of a century ago to discuss, examine, and project the issues associated with humans and automation. Largely due to the siren call of Raja Parasuraman and Mustapha Mouloua, many of the good and great of our science were present to expound upon their thoughts and ideas for an area that, at that time, was just developing. Such was the enthusiasm for the content that a text (the predecessor of this current edition) was mooted, developed, and published in relatively short order. It proved to be a landmark text and one that exerted an important influence upon the directions that human factors in general was to take in the decades that followed. Times have moved on. What were only broad and imprecise projections in the early 1990s have now become materially present in our world. We have driverless cars, drones that can independently operate for much of their projected missions, an ever-greater interpenetration of automation into ever-wider panoplies of complex technologies. What were once only visions are now realties. Advances in robotics alone could fill several texts, and the sophistication of what is now offered for sale on the open market could only barely be envisaged at that time. Some of what was true then, however, remains true today. As Raja himself noted, automation tends not to simply replace human work but to change the nature of that work. The fact that we have lost Raja, his insights, and wisdom is a tragedy that many of us identify with on a personal level but that this whole area of research must bear on a collective basis. It is right and proper that this work is dedicated to him.

In what follows, we have asked some of the original authors to take and examine their statements of the time and to update them in light of the developments that have occurred. We have also sought the input of yet others, some of whom were perhaps just starting in the field when the first edition of *Automation and Human Performance* was first published. Their new and original take on the present state of developments can inform our community significantly since the burgeoning of research in this area has created such a diaspora of interested parties. We cannot claim that we have been able to present works from all relevant or involved constituencies; after all, that would represent a series of volumes. What we have done is to examine the advances in theory, in modeling, and in applications and have sought to provide representations of each. Any text is always a rock in the river of progress. It freezes at one moment what we think we know and understand. The opportunity to present a second edition of the original text allows us explicitly to look back to ask questions about what we did know, and what we might have known at the time. Like the prognostications of visionaries such as Bartlett, that exercise then allows us a 180-degree turn in order to now look into the present's own future. While it remains true that human begins still see more through tears than they do through telescopes, perhaps we might here just catch a glimpse of the world to come, and even more important, we might just be able to help shape it. Such is our aspiration and our goal. The degree to which the present work is successful in doing so must be left to the deliberations of history—or perhaps to a third edition in another quarter-century's time.

Peter A. Hancock
Mustapha Mouloua
Orlando, FL

Remembering Fallen Heroes
A Tribute to Raja Parasuraman, Joel Warm, & Neville Moray

Peter A. Hancock & Mustapha Mouloua

INTRODUCTION

In the immediate past years, we have lost three individuals whose work and efforts lie at the very heart of the issues described and discussed in the present text. The following, brief overview provides a deserved homage to these scientists and a precis of their contributions and influence. Given that the present book is an updating and revision of the original Parasuraman and Mouloua (1996) text, it is more than reasonable to begin with a tribute to our departed colleague Raja Parasuraman.

RAJA PARASURAMAN

It is important to say that what follows are by no means exhaustive accounts of the lives and contributions of the three specific people featured. Such an exposition would take many papers and several texts to achieve (thus see Hancock, 2019; Hancock, Baldwin, Warm, & Szalma, 2017; Hancock & Szalma, 2019; Hoffman, Boles, Szalma, & Hancock, 2018; Miller, 2015). Rather, the present purpose is to recognize our debt and to convey to the present readers how these, and other luminaries, have guided us on the path to our present state of understanding. Raja's antecedents on the subcontinent of India are widely recognized, but perhaps less well-known was his transition from Indian scholar to English public schoolboy during his teenage years. What he, as the winner of an all-India scholarship, must have made of wet and windy Oakham in the center of England we cannot really envisage. It speaks much to Raja's character and flexibility that he not only adapted but thrived and flourished in the latter environment, moving on to complete a degree at one of the major London universities. The turning point for both Raja and arguably for our field came when he moved on to doctoral studies under the inimitable Roy Davies at the University of Birmingham, also in central England.

It was at Birmingham then that Raja parlayed and exploited his electronic engineering skills into the domain of experimental psychology. As many in our community know, Raja's early published contributions were to the understanding of the area of vigilance, and they proved vital to this aspect of our science. Reasonably, we can say that, without Parasuraman, explanatory studies in the field of sustained attention may have dissolved and even died altogether. Fortunately for us, and especially in relation to automation and autonomy studies, Parasuraman and his mentor Roy Davies, energized the area through the now famous vigilance taxonomy (Davies & Parasuraman, 1982; Parasuraman, 1979). It was the foundation of Raja's lifelong concern for humans in technological settings, especially those involved in monitoring.

A postdoctoral sojourn at UCLA was followed by an appointment at Catholic University in Washington, DC, where Raja elaborated on his earlier interests and skills sets to embrace both a more detailed focus on neuroscience as well as an expanded concern for human performance in technological contexts (e.g., Parasuraman 1987). At Catholic, Raja mentored generations of our current leaders who have themselves now contributed many new and original insights. In the early 1980s, it was becoming more and more evident that automation represented a new and rather special challenge to human factors and ergonomics (HF/E). The confluence of these concerns led members of the HF/E community to attend a foundational meeting, not unexpectedly led by Raja and one of us, his then student (Mustapha Mouloua). It produced a publication (Parasuraman & Mouloua, 1996) that is the direct forebear of the present text. Our science owes greatly to Raja, who passed away in an untimely manner. He is known for publishing a series of profound works (e.g., Parasuraman & Manzey, 2010; Parasuraman & Riley, 1997; Parasuraman, Sheridan, & Wickens, 2000) that have guided HF/E since. His observations on the change in the very nature of work, as opposed to the elimination of certain discrete tasks with progressive automation, remains more than pertinent to our contemporary world and is a foundational and vital pillar of innovations such as resilience engineering (see e.g., Hollnagel, Woods, & Leveson, 2007). Here, we want to publicly and explicitly acknowledge these debts that we owe and to communicate to those that follow us how influential Raja proved in this crucial area. However, Raja was not an island unto himself, and it is to other influential colleagues, now also sadly deceased, that we must also provide relevant acknowledgment.

JOEL WARM

Despite his prominence in the area of sustained attention, we are sure that Raja would defer to Joel Warm the title of "master of vigilance." From his early years at Alabama to a long and distinguished career at Cincinnati, Joel Warm led the field on this aspect of attention for more than 30 years. His association with Bill Dember created a formidable partnership but one that he readily shared with many others. It comes as little surprise that we, like many others, benefitted from this nexus (e.g., Warm, Dember, & Hancock, 1996; Warm, Parasuraman, & Matthews, 2008). Warm's experimental attacks on the vigilance question took many forms. Psychophysical explorations were joined to physiological investigations in order to unlock the nuances of both sensory and cognitive vigilance alike. One of the great lines of investigation Warm consistently pursued was to specify the level of workload associated with vigilance (see e.g., Becker, Warm, Dember, & Hancock, 1995; Hancock & Warm, 1989). His decade's long, systematic empirical sequence of works stands as a benchmark for all of our science in general. In his later years, Warm acted in the capacity as a senior scientist to the U.S. Air Force (USAF) at Wright-Patterson Air Force Base in Dayton, OH. Teaming with HF/E colleagues, as well as his own former students, he brought important scientific understanding to questions critical to the USAF. Here, his understanding of monitoring and the "vigilance decrement" played strongly into the burgeoning wave of their concern with automation and autonomy. We know that it is unwise to place humans into the automation oversight role, especially when interface displays are configured in ways that actually promote this form of performance decrement (Hancock, 2013). It is one of the central contributions of Joel Warm that he taught us where, when, and why such critical signals are likely to be missed. As such, he saved lives through his insights—not simply those serving in the military forces but those working in security, search and rescue, and individuals driving on our highways in either manual, or in the growing numbers of automated, vehicles (Hancock, Nourbakhsh, & Stewart, 2019; Hancock & Szalma, 2019). This achievement is a fitting epitaph for one of the giants of our field. As leaders at the forefront of applied behavioral science, Raja Parasuraman and Joel Warm were joined by another scientist who helped found and elaborate HF/E, especially in relation to automation interaction.

NEVILLE MORAY

Thus, the last, but not least, of the present triumvirate of heroes was Neville Moray. Originally pointed down a road to medicine, Moray like many others, found the applied behavioral realm of overwhelming fascination, and his Oxford dissertation work on divided attention resulted in a classic of our field (Moray, 1967). Potentially, Moray could have sustained his whole career imbued in pure experimental psychology. In this fashion, he would, no doubt, have been very successful. Yet almost in mid-career he was influenced by the likes of Senders and Sheridan while on sabbatical at Massachusetts Institute of Technology (MIT). He turned his formidable analytic abilities to the problems of the real world and made numerous impacts, especially in his pioneering efforts concerning cognitive workload (Moray, 1979). Therein after, Moray's career path was set in championing the utility of experimental psychology to the solution of real-world issues. And Moray contributed to so many of these: the nuclear industry, aviation, hazardous waste disposed, as well as some forensic accident investigation. His bright, engaging, and fecund mind ranged over so many issues, but always the bedrock of empirical attack founded even the most complex and abstruse models and applications. Following a career spent in England, Scotland, Canada, France, and the United States, Moray's later years were devoted to art but also to the larger questions of life that face us all. One of us (Peter Hancock) had the privilege to work with him on one of these latter efforts (Moray & Hancock, 2009); it is a paper that hopefully will exert its rightful influence, though it may take some years to be recognized (there being a slight possibility that we may be biased in this matter). His final opus (Moray, 2014) is one of his most profound and most important works. This particular text, *Science, Cells and Souls*, returns more than great value for the time invested in reading it. It reflects a mature scientist's wisdom and a true vision of the world. Neville truly embodied an involved, enthusiastic, and fun scholar and his assured legacy lies in the tranche of wisdom that he left us (Hancock, Senders, & Lee, 2018).

SALUTATION AND VALEDICTION

Besides their preeminence in HF/E, what ties these three individuals together was an inveterate cursory expressed as an abiding interest in human attention. While we are aware that there are "varieties of attention," (Parasuraman & Davies, 1984) and that Moray pursued selective attention while the others more emphasized sustained attention, it was the application of these faculties in the real-work context that represented their own individual challenges. That these real-world contexts more and more feature automated and now rising autonomous systems makes their science and contributions ever relevant to our present world (Hancock, 2017). Before finishing, we should acknowledge that there is a litany of others, too long to mention, who have also founded our science but now sadly passed on. The featuring of Raja, Joel, and Neville here derives from their special relevance to the current text and the sad fact that we lost all of them in such a short space of time. If we do not understand our history, we cannot comprehend either our place in it or our future. And if we don't acknowledge and respect those to whom we owe so much, wherefore our own legacy?

REFERENCES

Becker, A. B., Warm, J. S., Dember, W. N., & Hancock, P. A. (1995). Effects of jet engine noise and performance feedback on perceived workload in a monitoring task. *International Journal of Aviation Psychology*, 5 (1), 49–62.

Davies, D. R., & Parasuraman, R. (1982). *The Psychology of Vigilance*. New York: Academic Press.

Hancock, P. A. (2013). In search of vigilance: The problem of iatrogenically created psychological phenomena. *American Psychologist*, 68 (2), 97–109.

Hancock, P. A. (2017). Imposing limits on autonomous systems. *Ergonomics*, 60 (2), 284–291.

Hancock P. A. (2019) The life and contributions of Neville Moray. In: S. Bagnara, R. Tartaglia, S. Albolino, T. Alexander, and Y. Fujita (Eds.) *Proceedings of the 20th Congress of the International Ergonomics Association (IEA 2018). IEA 2018. Advances in Intelligent Systems and Computing*, 822, (pp. 621–726), Cham, Switzerland: Springer.

Hancock, P. A., Baldwin, C., Warm, J. S., & Szalma, J. L. (2017). Between two worlds: Discourse on the vigilant and sustained contributions of Raja Parasuraman. *Human Factors*, 59 (1), 28–34.

Hancock, P. A., Nourbakhsh, I., & Stewart, J. (2019). On the future of transportation in an era of automated and autonomous vehicles. *Proceedings of the National Academy of Sciences*, 116 (16), 7684–7691.

Hancock, P. A., Senders, J. W., & Lee, J. (2018). Neville Moray (1935–2017). *American Journal of Psychology*, 131 (3), 381–384.

Hancock, P. A., & Szalma, J. L. (2019). Sustained attention to science: A tribute to the life and scholarship of Joel Warm, *Human Factors*, 61 (3), https://doi.org/10.1177/00187 20819839370

Hancock, P. A., & Warm, J. S. (1989). A dynamic model of stress and sustained attention. *Human Factors*, 31 (5), 519–537.
Hoffman, R. R., Boles, D. B., Szalma, J. L., & Hancock, P. A. (2018). Joel S. Warm (1933–2017). *American Journal of Psychology*, 131 (2), 227–230.
Hollnagel, E., Woods, D. D., & Leveson, N. (2007). *Resilience engineering: Concepts and precepts*. Chichester, UK: Ashgate Publishing.
Miller, C. (2015). Raja Parasuraman, an innovator in human-automation interaction. *IEEE Systems, Man and Cybernetics Magazine*, 1 (2), 41–45.
Moray, N. (1967). Where is attention limited? A survey and a model. *Acta Psychologica*, 27, 84–92.
Moray, N. (Ed.). (1979). *Mental Workload: Its Theory and Measurement*. New York: Springer.
Moray, N. (2014). *Science, Cells and Souls*. Bloomington, IN: Author House.
Moray, N. P., & Hancock, P. A. (2009). Minkowski spaces as models of human-machine communication. *Theoretical Issues in Ergonomic Science*, 10 (4), 315–334.
Parasuraman, R. (1979). Memory load and event rate control sensitivity decrements in sustained attention. *Science*, 205 (4409), 924–927.
Parasuraman, R. (1987). Human-computer monitoring. *Human Factors*, 29 (6), 695–706.
Parasuraman, R., & Davies, D. R. (Eds.). (1984). *Varieties of Attention*. Orlando, FL: Academic Press.
Parasuraman, R., & Manzey, D. H. (2010). Complacency and bias in human use of automation: An attentional integration. *Human Factors*, 52 (3), 381–410.
Parasuraman, R., & Mouloua, M. (Eds.). (1996). *Automation and Human Performance: Theory and Applications*. Hillsdale, NJ: Lawrence Erlbaum Associates, Inc.
Parasuraman, R., & Riley, V. (1997). Humans and automation: Use, misuse, disuse, abuse. *Human Factors*, 39 (2), 230–253
Parasuraman, R., Sheridan, T. B., & Wickens, C. D. (2000). A model for types and levels of human interaction with automation. *IEEE Transactions on Systems, Man, and Cybernetics—Part A: Systems and Humans*, 30 (3), 286–297.
Warm, J. S., Dember, W. N., & Hancock, P. A. (1996). Vigilance and workload in automated systems. In: R. Parasuraman and M. Mouloua. (Eds.). *Automation and Human Performance: Theory and Applications* (pp 183–200). Hillsdale, NJ: Erlbaum.
Warm, J. S., Parasuraman, R., & Matthews, G. (2008). Vigilance requires hard mental work and is stressful. *Human Factors*, 50 (3), 433–441.

Acknowledgments

The production of these two books are made possible thanks to the efforts of several people who have contributed to their planning and execution. First, we are indebted to the commitment and dedication of the present contributors who have written the chapters in these two books. Their insightful words and thorough analyses add immensely to the current state of the literature. Furthermore, they will also serve as a guide for future researchers, educators, and practitioners in the field of human factors and ergonomics.

Chapters included in these two books attest to the necessity of incorporating various theories, methods, approaches, and design principles that will not completely remove the human from the operation of these advanced technological systems. Rather it puts him/her in the center stage of the design process. Additionally, we also wish to extend our deepest gratitude to those who have graciously reviewed some manuscripts as we neared the final phase of the production stage. We feel very indebted to these individuals as well as to other students and staff in our respective departments including Tiffani Marlow, Logan Clark, Salim Mouloua, Dolores Rodriguez, Sandra Montenegro, Elizabeth Merwin, Miguel Ortiz, Eviana Le, and Yazmin Diaz.

We also want to express our sincere thanks to our technical editorial team consisting of Alexandra Kaplan and James Ferraro (University of Central Florida) who have marshalled the various production processes throughout the differing phases of this project. Their substantial efforts have been central and are very much appreciated.

Finally, we also would like to express our sincere appreciation to CRC Press for their assistance throughout various stages of production. These individuals include the anonymous CRC Press reviewers, the Editorial Assistant Ms. Erin Harris, and our CRC Press Executive Editor Ms. Cindy Carelli. We could not imagine a better and more dedicated team of professionals to work with on this present project.

Mustapha Mouloua
Peter A. Hancock

About the Editors

Mustapha Mouloua, Ph.D., is professor of psychology at the University of Central Florida, Orlando, and was the director (2009–2017) and associate director (2006–2009) of the Applied/Experimental and Human Factors Psychology Doctoral Program, director and chief scientist (2001–2014), associate director and senior research scientist (1998–2001) of the Center for Applied Human Factors in Aviation, and associate director of Human Factors Research at the Center for Advanced Transportation Systems Simulation (2001–2003). Dr. Mouloua has over 30 years of experience in the field of human factors and ergonomics. His research interests include human-automation interaction, attention and workload, assessment of older drivers and pilots, UAS operation, selection, training, and simulation technologies across a variety of transportation systems. Dr. Mouloua is the editor or coeditor of several books including *Human Performance in Automated Systems: Current Research and Trends* (1994, Lawrence Erlbaum Associates), *Human-Automation Interaction: Research and Practice* (1997, Lawrence Erlbaum Associates), *Automation and Human Performance: Theory and Applications* (1996, Lawrence Erlbaum Associates), *Automation Technology and Human Performance: Current Research and Trends* (1999, Lawrence Erlbaum Associates), *Human Factors in Simulation and Training* (2009, Taylor & Francis Group), *Automation and Human Performance: Theory, Research, and Practice* (2004, Volumes I & II; Lawrence Erlbaum Associates), and *Proceedings of the 2nd ACM Symposium on Computer Human Interaction for Management of Information Technology* (ACM, CHIMIT, 2008). Dr. Mouloua published over 190 papers and scientific reports and made over 300 presentations at various national and international meetings. Dr. Mouloua was the recipient of the prestigious Jerome Ely (1997) and the Tidewater (1999) Awards from the Human Factors and Ergonomics Society. At UCF, Dr. Mouloua received eight prestigious teaching and research awards including the recent Teaching Incentive Program Award (2015) and Research Incentive Award (2019) for his outstanding teaching and research contributions. Similarly, he was also a recipient of the UCF International Golden Key and Honorary Award (2011) and the UCF "Twenty Years of Service" Award (2014) for his dedicated work and commitment to students. Dr. Mouloua is currently the director of the Neurolinguistics, Aviation, and Driving Research Laboratories at UCF.

Peter A. Hancock, D.Sc., Ph.D., is Provost Distinguished Research Professor in the Department of Psychology and the Institute for Simulation and Training, as well as at the Department of Civil and Environmental Engineering and the Department of Industrial Engineering and Management Systems at UCF. At UCF in 2009, he was created the sixteenth-ever University Pegasus Professor (the Institution's highest honor) and in 2012 was named sixth-ever University Trustee Chair. He directs the MIT2 Research Laboratories. Prior to his current position, he founded and was the director of the Human Factors Research Laboratory (HFRL) at the University of Minnesota where he held appointments as professor in the Departments of Computer Science and Electrical Engineering, Mechanical Engineering, Psychology, and Kinesiology, as well as being a member of the Cognitive Science Center and the Center on Aging Research. He continues to hold an appointment as a clinical adjunct professor in the Department of Psychology at Minnesota. He is also an affiliated scientist of the Humans and Automation Laboratory at Duke University, a research associate of the University of Michigan Transport Research Institute, and a senior research associate at the Institute for Human and Machine Cognition in Pensacola, Florida. He is also a member of the Scientific Advisory Board of the Hawaii Academy.

Professor Hancock is the author of over 1,000 refereed scientific articles, chapters, and reports as well as writing and editing more than 20 books including: the *Human Performance and Ergonomics in the Handbook of Perception and Cognition* series, published by Academic Press in 1999; *Stress, Workload, and Fatigue,* published in 2001 by Lawrence Erlbaum Associates; and *Performance under Stress,* which was published in 2008 by Ashgate Publishing. He is the author of the 1997 book, *Essays on the Future of Human-Machine Systems,* and the 2009 text, *Mind, Machine and Morality,* also from Ashgate Publishers. He has been continuously funded by extramural sources for every one of the 36 years of his professional career. This includes support from NASA, NSF, NIH, NIA, FAA, FHWA, NRC, NHTSA, DARPA, NIMH, and all of the branches of U.S. Armed Forces. He has also been supported by numerous state and industrial agencies. He was the principal investigator on a Multidisciplinary University Research Initiative (MURI), in which he directed $5 million of funded research on stress, workload, and performance. It was the first MURI in behavioral science ever awarded by the U.S. Army. He was also the recipient of the first-ever research grant (as opposed to contract) given by the Federal Aviation Administration. To date, he has secured over $20 million in externally funded research during his career. He has presented, or been an author on, over 1,000 scientific presentations. In 2015, Dr. Hancock became the only two-time recipient of the John Davey Award for Medieval Studies by the Richard III Foundation. Most recently, he has been elected a fellow of the Royal Aeronautical Society (RAeS) and, in 2016, was named the thirtieth honorary member of the Institute of Industrial and Systems Engineers (IISE). In 2017, he was elected a member of the International Academy of Aviation and Space Medicine as well as receiving the Specialist Silver Medal Award from the Royal Aeronautical Society. In 2018, he was the recipient of the International Ergonomics Association (IEA) Outstanding Educators Award, (IEA).

List of Contributors

Carryl L. Baldwin
George Mason University
Fairfax, Virginia, USA

Kevin B. Bennett
Wright State University
Dayton, Ohio, USA

John L. Campbell
Exponent
Bellevue, Washington, USA

Victoria L. Claypoole
US Air Force Research Laboratory
Dayton, Ohio, USA

Mica R. Endsley
SA Technologies
Gold Canyon, Arizona, USA

James C. Ferraro
University of Central Florida
Orlando, FL, USA

John M. Flach
Mile Two, LLC
Dayton, Ohio, USA

Peter A. Hancock
University of Central Florida
Orlando, Florida, USA

James Head
University of Canterbury
Christchurch, New Zealand

William S. Helton
George Mason University
Fairfax, Virginia, USA

Robert R. Hoffman
Institute for Human and Machine Cognition
Pensacola, Florida, USA

Matthew Johnson
Institute for Human and Machine Cognition
Pensacola, Florida, USA

David Kaber
University of Florida
Gainesville, Florida, USA

Alexandra D. Kaplan
University of Central Florida
Orlando, Florida, USA

Joonbum Lee
Battelle Center for Human Performance and Safety
Seattle, Washington, USA

Jinchao Lin
Institute for Simulation and Training
University of Central Florida
Orlando, Florida, USA

Phillip Mangos
Adaptive Immersion Technologies
Tampa, Florida, USA

Dietrich Manzey
Technical University Berlin
Berlin, Germany

Gerald Matthews
Institute for Simulation and Training
University of Central Florida
Orlando, Florida, USA

Ian McCandliss
George Mason University
Fairfax, Virginia, USA

Kathleen L. Mosier
San Francisco State University
San Francisco, California, USA

Mustapha Mouloua
University of Central Florida
Orlando, Florida, USA

April Rose Panganiban
US Air Force Research Laboratory
Dayton, Ohio, USA

Christian M. Richard
Battelle Center for Human
    Performance and Safety
Seattle, Washington, USA

Mark W. Scerbo
Old Dominion University
Norfolk, Virginia, USA

James L. Szalma
University of Central Florida
Orlando, Florida, USA

Vindhya Venkatraman
Battelle Center for Human
    Performance and Safety
Seattle, Washington, USA

Matthew Ward
University of Canterbury
Christchurch, New Zealand

James Wen
University of Canterbury
Christchurch, New Zealand

Ryan Wohleber
Institute for Simulation and Training
University of Central Florida
Orlando, Florida, USA

1 Designing Wearable Interfaces for People Who Have Hard Jobs

Matthew Ward, James Wen, James Head, & William S. Helton

INTRODUCTION

Mary works as a wilderness firefighter in the Pacific Northwest of the United States. Her job involves backbreaking physical labor over harsh terrain in a high-stress environment where one misstep means severe injury or death. Mary could use technological assistance. For example, when Mary is suppressing fire with a hose she needs to know how much water or fire retardant is left in the tank. Moving a fire suppression hose is not physically easy. Fifty feet of charged five-inch hose weighs up to 425 lbs., and moving more than necessary could fatigue Mary. In Mary's work, too much physical fatigue may literally kill her if it contributes to an accident. There is no point in moving the hose if the tank is almost out of suppressant. Mary needs a wearable technology that informs her how much is left in that tank. The problem is almost all the literature on human-computer interaction or human-machine interaction was done on people like Bob instead of Mary. Bob works sitting down in a climate-controlled building. Bob's greatest threat is a paper cut from filing the papers on his desk or a heart attack from the junk food he eats every day from the office vending machine while he sits on his rear end all day. Bob has a physically soft job, and all the information technology so far is mostly built for Bob and his soft job. Mary and folks like Mary who do physically hard jobs have been mostly ignored by researchers, but with the emergence of wearable technologies, Mary has the potential for information system assistance. A challenge is we do not know that much yet about Mary and how this wearable technology will work for her. This chapter is about Mary and the people like Mary who work physically hard jobs.

WEARABLE TECHNOLOGY

Accessing information systems from almost anywhere in the world with small portable devices changes how we work (Pitichat, 2013), how we travel (Wang, 2013), and how we entertain ourselves (Hollister, 2017). Demand for mobile devices is extensive and growing. Of the 2.4 billion computational devices sold in 2016, 80% were smartphones (Keizer, 2017). A major advantage of mobile devices is their ability to be used while people move about untethered like the firefighter Mary. Current mobile computers can be held with a single hand thus leaving the second hand free for other actions. Moving while operating a device opens up a range of new technology interaction possibilities. For example, a theme park system could track the movement of visitors to the park in real time and provide feedback to visitors about where rides are less crowded and to theme park operators enabling them to move employees to areas that are being underserved (Barnes, 2010). These technologies are also useful in occupations where physical movement is an occupational requirement, such as military and emergency response; but in these settings, these technologies would be even more useful if they were wearable where the user could operate them mostly hands free and heads up.

Wearables enable users to undertake tasks where they must access virtual information while manipulating real-world objects. Wearable head-mounted displays (HMDs) with a transparent display, such as Google Glass or Alto Tech's Cool Glass, can overlay virtual information on top of the physical world rather than have the user attend to a screen that entirely occludes the environment. Even HMDs that do partially occlude the environment maintain some of these relative benefits in comparison to handheld displays. Other nonvisual inputs are also possible with wearables including three-dimensional (3D) spatial audio and worn tactile displays. Many people use a vibratory setting with their smartphone to be aware of incoming calls and texts even when an audio signal would be inappropriate, so the concept of tactile signals is not entirely novel.

Wearables are designed to enable multitasking, interacting with the real tangible world while being able to get information from other sources, and this requires the user to allocate and prioritize the allocation of mental resources to both the surrounding physical environment and the information presented to the user via the wearable. In order to design usable wearables, designers should take into consideration what we know about people's attention and performance limitations in dual-tasking or multitasking situations. We will first describe the current literature on models of attention relevant to dual- or multitasking situations. We will then describe how these models may apply to wearable design in general terms. Next, we will provide case studies examining how this may or may not work for people who have physically hard jobs. Finally, we will summarize some basic concerns regarding what we know that is useful for wearable designs for people with these hard jobs.

MODELS OF ATTENTION

Early researchers of divided attention proposed the existence of perceptual or attention bottleneck. In these bottleneck models, multiple stimuli were thought to be sensed by a person, but somewhere in the sequence of processes occurring between perception and response execution there was a limited serial process, which enabled only one stimuli to be processed at a time. Prior to the attention bottleneck, stimuli are processed in parallel but, after this bottleneck, stimuli are processed in serial.

Donald Broadbent's filter model was one of the earliest of this family of models. In his model, multiple incoming stimuli could enter a mental sensory store, also called sensory memory (Broadbent, 1957; Broadbent, 1958). From the sensory store, information was passed through a selective filter that concerned physical properties of the stimulus, such as loudness, brightness, or location of source. Some cognitive control of this filter was possible, with the person being able to selectively filter some, but not all, unwanted information, and this was the stage when the bottleneck occurred. The model stated that only one stimulus could pass through the filter at any given time. Divided attention performance impairment was accounted for by participants alternating the targets of their filter, a process that requires effort and relies on the key components of the stimulus being stored in the sensory memory long enough for the signals to be processed in serial. The next module in the model was one where the incoming stimuli were cognitively processed. At this stage, the stimuli were screened for abstract meaning, coherence, and perceived value for the purposes of deciding which information gets passed to the final module, working memory. From working memory, the processed stimuli could then be encoded into long-term memory or used to select an appropriate response. Evidence in support of Broadbent's model was the finding that when two stimuli are presented simultaneously, often only one is perceived, and when both are noted, the responses to the events are usually given serially (Colavita, 1974).

Broadbent's filter model was unable, however, to provide an explanation for the "cocktail party effect" where a person is able to listen to one conversation, filtering out all others, but can attend to the sudden mention of their name or a high-interest keyword in another audible conversation. Because it is the higher-level, post-bottleneck cognitive process that scans a message for semantic content, this finding meant that at least some kind of cognitive analysis preceded the supposed attention bottleneck. Deutsch and Deutsch (1963) investigated the effects of message importance and concluded that filtering unwanted messages required extensive cognitive processing, implying the cognitive processing module proposed in Broadbent's original model needs to be active before the selective filter.

Instead of focusing on the presence of bottlenecks per se, other researchers began to propose capacity models of attention. In capacity models, dual-task or dual-attention interference is caused by the global cognitive demands of both tasks exceeding the total capacity available in the moment, which is in limited supply. So unlike in bottleneck models, the limitation is not always a strictly either/or limitation of processing but one of relative degree. These capacity models share more in common with economic or resource considerations found in other systems, but they do not rule out potential bottlenecks occurring. Using an analogy of building a house, if both bricks and boards cost a minimum of $5 and the builder is down to their last $5, then the builder has to choose either a brick or a board. Similarly, a resource capacity model can explain the presence of bottlenecks and, so, they are not mutually exclusive theories. Properly detailed capacity or resource models can explain processing bottlenecks.

Kahneman (1973; Kahneman, Peavler, & Onuska, 1968), for example, discovered that the primary determinant of a person's arousal was the difficulty of the task or tasks they were currently engaged in. This correspondence between arousal and task demands has subsequently been supported for arithmetic (Bradshaw, 1968a); short-term memory tasks (Kahneman & Beatty, 1966); pitch discriminations (Kahneman & Beatty, 1967); standard tests of concentration (Bradshaw, 1968b);

sentence comprehension (Wright & Kahneman, 1971); paired-associate learning (Kahneman & Peavler, 1969); and imagery tasks with abstract and with concrete words (Simpson & Paivio, 1968). In Kahneman's model, the relationship between task demand and arousal is causal but is mediated such that increased arousal causes an increase in available capacity (i.e., the total pool of mental resources). As needed capacity increases, so too does the capacity supplied but a slower rate. The disparity between these rate increases causes spare capacity, which is used for other tasks like environment monitoring but becomes constrained and eventually begins to negatively impact performance on the primary task. Kahneman's model also includes an allocation policy that divides mental resources among possible activities. Tasks which are allocated insufficient resources suffer in terms of performance with increased response times and higher error rates. Like the allocation of resources to a task, the resources demanded by tasks can vary from moment to moment. Task demands include the discriminability of target stimuli, the amount of motion of the head or body needed for detection or response execution, and time limits set on responding. In particular, any task requiring information to be held in the short-term working memory naturally imposes such time limits because the participant's rate of activity must now exceed the rate at which items are lost from short-term memory.

Kahneman's model (1973) was successful at predicting effects of dual tasking on the individual component tasks; but in the model, the underlying physiological resource that is actually being divided is not specified. Additionally, over the next few years, evidence was found that the patterns of dual-task interference could not be explained by the combination of task resource demand and the participants' allocation policy alone. Notably, tasks that used separate sensory units or cognitive skills resulted in less interference than in conditions where tasks used overlapping ones (Kantowitz & Knight, 1976; Wickens 1976). Wickens (1980) found evidence that there were separate processing resources as opposed to a single unitary resource. Wickens (2002) proposed a four-dimensional model of attentional resources. The first of the four dimensions is the stage of processing. There are three levels of this dimension, and they occur in order, starting at perception where the incoming stimuli are processed, then cognition, where the processed stimuli are analyzed, and finally response selection/execution, where the appropriate response is decided upon and performed. The second dimension is processing code, which has two levels referring to the type of information the stimuli carries: spatial information or verbal information. This dimension intersects with the stage's dimension at all levels. The third dimension perceptual modalities, or modes, divides the stimuli into categories based on the sensory organ used to detect them. While not included in the model as described by Wickens (2002), other senses, such as tactile signal detection, could hypothetically be included in an expanded model. The perceptual modalities dimension divides stimuli in the perception stage, with the information being combined or fused upon reaching the cognition stage. The fourth and last dimension further divides visual perception into focal and ambient visual channels. Focal visual stimuli come from within the three-degree cone of clear visual perception for human adults, and the remaining ambient comes from the peripheral visual field.

APPLICATION OF ATTENTION MODELS

Employing Wickens' multiple resource model, a designer may estimate where the major conflicts over allocation of mental resources will arise when using wearables. The primary application of a multiple resource perspective that raises minimal controversy would be at the perceptual modality and focal versus ambient visual channels dimensions. In this situation, much of the interference appears obvious and objectively structural. For example, HMDs add visual information to the user's perceptual space, which increases the demand on the focal visual perception components of Wickens' model. If a task being performed in the physical environment demands focal visual attention, then a head-mounted display presenting text may not be the best choice of wearable design. If, alternatively, that information could be presented in another modality, such as audio via headphones or bone conductance headsets, that would be preferable (Barde, Ward, Helton, Billinghurst, & Lee, 2016). A person cannot foveate in two separate locations simultaneously. This is why texting while driving, even with an HMD, is a bad idea (Sawyer, Finomore, Calvo, & Hancock, 2014). Indeed, navigating complex terrain requires visual attention for sighted people, and this will interfere with the use of an HMD in many cases. Woodham, Billinghurst, and Helton (2016) examined the impact of reading text via an HMD on climbing performance and found climbers would slow or entirely stop climbing to read the HMD-presented text.

Where there is more controversy is how multiple resource theory may be employed to better inform wearable designs in regard to the other dimensions of the model, in particular processing code issues. How do real-world movement tasks fit within the existing processing code and stage of processing in the model? Wearables are intended to be used while people interact with the real external world, and this is where there is a considerable research gap. How does human cognition work in real-world settings where extensive body movement is required? This gap is odd given that cognition is presumably an evolutionary product of needing to make decisions by animals on the move, but nevertheless, for expedience and control considerations, psychologists have focused extensively on seated (or laying in the era of functional imaging) behavior. Psychologists have focused more on Bob and his needs than Mary and her needs.

MOVEMENT IN THE REAL WORLD IS INTERFERING

At the cognitive stage in Wickens' model, there is continuing debate whether the cognitive resources are shared in parallel between two concurrently performed tasks, as described in the resource models of Kahneman (1973) and Wickens (2002), or instead is limited to a single task at a time, creating a bottleneck leading to serial responding to the two tasks (Pashler, 1994a; Pashler, 1994b). There are studies that clearly support Wickens' (2002) model; for example, Sims and Hegarty (1997) had participants predict the motion of an object in a diagram after force had been applied to an attached pulley system. As a secondary task, participants were asked to also remember a set of three dot positions on a four-by-four grid and report at a later time if a second set matched the memory set or to remember a string of six consonant letters and at a later time report if the set contained a particular letter. Attempting to hold the spatial information of the dot patterns in memory caused greater interference with the force prediction task than the letter memory task as measured by the proportion of errors in both the primary and secondary tasks. This suggests that two spatial tasks share more resources than a spatial and a verbal task. Similar interference findings have been found in other seated tasks where two spatial tasks or a spatial and a verbal task have been done simultaneously (Helton & Russell, 2011; 2013).

Complex movement tasks, requiring whole body movement over complex terrain, however, are harder to specify in regard to the cognitive resources they actually utilize. For example, a first responder may have to climb a cliffside to assist a victim of a fall or disaster. Is climbing complex terrain spatial, verbal, or both? In a recent series of studies, we have attempted to address this issue by holding a cognitively demanding task constant and comparing across other tasks including physically challenging real-world movement tasks likely to be encountered by users of wearables, like climbing and running (Darling & Helton, 2014; Epling, Blakely, Russell, & Helton, 2016; 2017; Epling, Russell, & Helton, 2016; Green, Draper, & Helton, 2014; Green & Helton, 2011; Head & Helton, 2014; Woodham, Billinghurst, & Helton, 2016). We decided to utilize word recall of a presented word list as the constant task as verbal recall would be representative of the verbally demanding communication tasks for which wearables would often be utilized. Word recall for words only represents a passive, one-way communication process, but performance can be precisely assessed by counting the number of words recalled correctly in the dual-task condition and comparing them to the number of words accurately recalled when the task is performed without another task. We standardized this performance effect using Cohen's d_z.

In addition, we can compare the performance on the other task when done in concert with the verbal recall task and when done alone. For this performance difference, we also calculated Cohen's d_z. This enables us to compare the interference effect on both tasks. The interference is displayed in Figure 1.1.

Climbing produces a significant loss of information in regard to word recall, and word recall significantly disrupts climbing distance. When both interference effects are considered together, they cause greater interference than when word recall is combined with a seated semantic discrimination task (verbal vigil), a seated response inhibition task (SART), or a seated spatial puzzle task (Quadra). Running over flat terrain with no obstacles (grass running track) causes considerably less interference but still some interference. Increasing movement complexity would likely shift the interference toward the amount seen with climbing. Information presented to emergency responders during complex physical tasks, like scrambling over obstacles, is subject to substantial memory loss and interference. This suggests built-in memory augmentation may be a desirable feature of communication technology for emergency responders as much of the information communicated to the responders during complex movement tasks is going to be forgotten. This includes the design of easily navigable memory systems that are built into the communication gear to facilitate the retrieval of previously presented information. When Mary is interrupted inappropriately, unlike Bob, Mary could end up dead.

Designing Wearable Interfaces

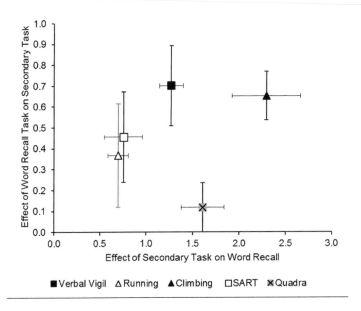

FIGURE 1.1 Standardized effect sizes (Cohen's d_z) for dual-task interference for different experimental tasks on word recall (x-axis) and the interference effect of the word recall task on those same experimental tasks (y-axis). Error bars are standard error of the mean for their respective dimensions. SART = seated response inhibition task. Figure is adapted from Epling, Blakely, Russell, & Helton (2017).

THE ASYMMETRIC CARRYOVER EFFECT OF COGNITIVE WORKLOAD ON PHYSICAL OUTPUT VERSUS PHYSICAL WORKLOAD ON COGNITIVE OUTPUT

Formalized physical education in U.S. public schools was originally adopted post-World War I to address the concerns of mitigating soldier recruits who were physically unprepared for military training (Heaton, Anderson, & Wiltse, 1967). In contemporary time, the inclusion of physical activity in school curriculum for children and adults have shown evidence of enhancing academic performance and suppressing neurodegenerative disease in aging populations (Hogan, Mata, & Carstensen, 2013). These findings do not suggest causation; however, they do imply that a relationship appears to exist. More formalized experimental approaches have provided evidence that physical activity performed prior to a cognitive task shows measurable cognitive enhancement. For example, it has been shown that short bouts (15 minutes) of moderate exercise and high-intensity interval training exercise can significantly enhance subsequent cognitive performance (Hancock & McNaughton, 1986; Hogan et al., 2013; Slusher, Patterson, Schwartz, & Acevedo, 2018).

As briefly outlined earlier, physical activity such as exercise appears to be advantageous to not only general health but also to enhancing cognitive performance. An important question to ask is whether this relationship is reciprocal. In other words, is it advantageous for an individual to engage in a cognitive task prior to physical activity? Interestingly, completing a physical task is similar to completing cognitive task in that they both result in fatigue, albeit different forms. Let's consider both Mary and Bob and what they do for a living. When duty calls, Mary is geared up with heavy equipment and is hiking, climbing, digging, and chopping trees down. Mary's work is grueling and is physically exhausting. Now let's imagine that Mary got lucky and her shift ends after eight hours. Mary returns back to the station to stow her gear and get some rest; however, some of her coworkers ask if she is keen to go rock climbing. Mary loves climbing but begrudgingly declines because every muscle in her body is fatigued. Mary knows if she tries to climb, it would be difficult if not impossible. Now let's consider our keyboard warrior, Bob, who spent eight hours in the office. Sure, the pinnacle of Bob's physical exertion was when he had to climb two flights of stairs to get a candy bar from the vending machine; however, the cognitive demand on Bob today is very high. Bob's boss e-mailed him a large file containing 300,000 lines of data that contains information on technology buying habits of teenagers. Bob is tasked with discovering patterns that will enable his company to focus the most optimal advertising for a future release of a commercial. Bob's boss wants the result by the end of the day in the form of tables, figures, and statistical results. This requires Bob to use all of his mental resources to painstakingly navigate the data with complex coding techniques in R (programming language). Bob has to assess whether data are missing, generate meaningful dependent measures, choose the correct statistical analysis, and generate meaningful tables, figures, and appropriate statistical analysis. Bob's work is very cognitively demanding and mentally difficult. After Bob's eight-hour shift, he returns back home to his roommates who are playing his favorite video game. Bob has to decline joining his friends because he is too fatigued, albeit mentally.

Mental fatigue is a psychobiological state that is generally caused by a prolonged cognitively demanding task (Marcora, Staiano, & Manning, 2009). Mental fatigue is "psychobiological" in that it is both characterized by psychological and physiological indicators. From a psychological perspective, mental fatigue is characterized as a feeling of mental lethargy or burning out (Marcora et al., 2009). Individuals who are mentally fatigued will generally self-report a greater workload after completion of a cognitively intensive task (Head et al., 2017; Head et al., 2016). From a physiological perspective, mental fatigue is characterized by changes in heart rate variability (i.e., parasympathetic stress response) and changes in cerebral oxygenation in the prefrontal cortex (Stevenson, Russell, & Helton, 2011). Research over the past decade has begun to address whether mental and physical fatigue interact. For example, prior investigations have shown that completing a mentally fatiguing task relative to a control prior to aerobic exercise like stationary cycling and treadmill running produces measureable performance impairments. Those that are mentally fatigued cover less distance or give up sooner on the physical task as a result of mental fatigue (Marcora et al., 2009; Pageaux, Lepers, Dietz, & Marcora, 2014; Smith, Marcora, & Coutts, 2015). Mental fatigue also appears to influence more skill-specific physical tasks such as coordinated passes and shots in soccer (Smith et al., 2016).

Additionally, it has been shown that those who complete a mentally fatiguing computer task prior to full-body exercise such as a high-intensity interval training (push-ups, pull-ups, and air squats) show measureable increases in task disengagement

Designing Wearable Interfaces

during the physical task. In other words, those that complete the mental fatigue relative to the control prior to full-body exercise take more rest breaks during the task (Head et al., 2016). Thus, the influence of mental fatigue on subsequent physical performance does not appear to be exercise specific. On the surface, the influence of mental on subsequent physical performance may appear to be benign in that the outcome is not often immediately life or death for causal exercisers. However, consider this phenomena when it is applied to an activity where the consequence can be life or death such as live-fire (Head et al., 2017). For example, Head and colleagues investigated the influence of mental fatigue relative to a control on live-fire marksmanship with trained infantry soldiers. In a repeated measures design, soldiers completed a cognitively fatiguing computer task or viewed a neutral video control prior to shooting a high-shoot/low no-shoot friend-verse-foe shooting task. Soldiers were required to engage E-silhouette targets that would pop up from behind grass berms. The E-silhouette targets would either be a friend or foe as designated by a visual indicator on the target; see Figure 1.2.

Soldiers had to make a fast and accurate decision as to whether to shoot or withhold a shot at the E-silhouette target. Interestingly, there was no significant effect of mental fatigue on whether the soldier hit the target or not. However, the results revealed that when soldiers were mentally fatigued, they had a 16% increase in shooting an E-silhouette target that was designated as a friendly (i.e., friendly fire). Unlike the previously discussed investigations, this finding presents evidence that mental fatigue could have dire consequences for not only the war fighter but also for civilian occupations and sports such as law enforcement and hunting, respectively.

As outlined in the "Application of Attention Models" section, physical activity is not purely a physical endeavor. Real-world activities like climbing or running outside require cognitive engagement such as visual attention, route planning, and communication. As outlined in this section, a cognitive task can result in mental fatigue, which can influence physical performance. The utilization of wearable technology may provide the user with a unique way of mitigating cognitive load while performing their job. Upon speaking with a California wilderness firefighter, they stated that real-time information such as global positioning system (GPS) location, air temperature, humidity, wind speed and direction, navigation, a real-time fire behavior map, and communication is lacking and would be beneficial to have available in an HMD.

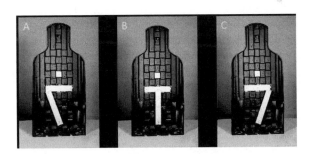

FIGURE 1.2 E-silhouette targets commonly used by soldiers for training with visual indicators. (A) and (C) represent friendly, no-shoot conditions. (B) Represents a foe shoot condition.

However, the use of wearable technology is a double-edged sword in that using it also requires cognitive engagement, which is prone to mental fatigue.

For example, consider Mary is redeployed to a major wilderness fire, but now she is equipped with an HMD that is mounted to her helmet. The HMD provides her with the useful information outlined previously. The first obvious concern is screen real estate, but the major issue is all of the information being presented, which requires attention, processing, and judgment. As Mary begins her hike from base camp to the firebreak two miles away she is not only negotiating tuff terrain with heavy gear but also monitoring her HMD for GPS directions and pertinent information for fire behavior. By the time Mary arrives at the firebreak, she is physically fatigued, which may be compounded by mental fatigue. This presents a real challenge in that physical performance can be further compromised by mental fatigue, which could result in a dire consequence for Mary and her team.

The utilization of wearable technology presents a potentially useful tool that could enhance performance to get a job done more efficiently and safely. However, it is imperative that user-centered research is done to not only understand the user's goals but also their capability to use the technology while performing the actual job. Academic researchers love laboratory-based research because it is controlled and provides reliable data that are less contaminated by third-variable confounds. However, having a college student sitting in the comforts of a research laboratory responding to stimuli on a computer screen not only lacks ecological validity but also fails to capture the behavior of the target user and the complexity of the real world. If wearable technology is going to be used, then, as the saying goes, "know thy user" and "honor thy user" (Wickens, Gordon, Liu, & Lee, 1998). The effects of cognitive workload not only impact physical performance during concurrent activity but also impact later physical performance. The cognitive load of wearables designed for people with physical duties needs to be carefully considered before deployment.

THE DOUBLE-EDGED SWORD OF AUGMENTATION

One example of a task where wearables would be useful augmentation is navigation assistance. People currently use their smartphones and other devices to navigate around instead of using traditional map and compass skills. Mary, in particular, might be assisted by wearable navigational cues to follow when fighting a fire. But what happens if her navigation guidance tool fails due to battery depletion or some other malfunction when she needs to retrace her steps after coordinating additional help and resources? Will the expectation that she can retrace her steps in a timely manner lead to dire consequences if her navigation tool is not available? Mary's dependency on the navigation tool can be a critical consideration in a high-demand environment. This could be a risk factor with the technology. As a firefighter, she does not have the luxury to devote substantial cognitive effort to navigation tools that should simply guide her to the location for doing her job and not distract her from focusing on the upcoming task. Therefore, her use of the tool would be to retrieve quick guidance information periodically while she focuses on her primary duties and activities. However, there could be consequences to a lack of invested effort in a navigation tool that is designed to offer ease of use as a primary feature. How Mary

Designing Wearable Interfaces

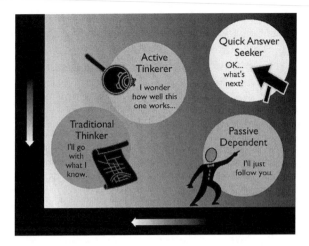

FIGURE 1.3 Graphic display of technology (navigation cues) user types.

uses her navigation technology may impact her overall effectiveness, and there are individual differences in how navigational technologies are utilized.

Wen, Helton, and Billinghurst (2013) classified users of mobile navigation tools into four categories based on collected usage pattern: traditional thinker, active tinkerer, quick answer seeker, and passive dependent, as shown in Figure 1.3. Traditional thinkers exhibit a preference for map-based tools and willingly invest time and effort into building up a mental map so that they do not have to refer to the mobile device frequently. The active tinkerer likes technology and enjoys trying out new tools that may present information in novel ways. The quick answer seeker is less concerned about the nature of the tool as the simplicity and effectiveness of the instructions given and may switch between tools depending on which seems most effective at the moment. Finally, the passive dependent seeks to minimize cognitive effort and prefers a tool that minimizes analysis. Given Mary's primary job of fighting fires and saving lives, she likely uses her navigation tool as a passive dependent, relying on the tool to perform accurately and reliably during the short but frequent spurts when she refers to it, so that she can concentrate on other critical tasks. However, navigation is a fairly complicated task in and of itself, and while Mary would be wise to offload the task to technologies designed to execute such tasks efficiently and effectively, she may not reap the benefits that may arise from an investment of effort into the navigation process, which could prove lifesaving if even seconds are lost to disorientation.

Traditional map reading requires a considerable amount of skill and effort, which is why approximately one in three people profess to have difficulties reading maps (Wen, Helton, & Billinghurst, 2013). This is because maps present cognitive challenges that the user must address. Finding one's self on a map to begin the process of creating a navigation solution is often the initial task when consulting a map. Associating the symbolic representation of real-world landmarks with the actual structures in the surrounding environment can require a degree of concentration that may be easily disrupted by other tasks. Rotating a paper map physically or mentally so that the direction of travel is aligned with the route depicted on the map relies upon a level of spatial ability that is confusing for many people. The effort required to overcome these challenges may do much to improve the user's ability to retain a cognitive map of the area even if the effort itself may not be trivial.

In contrast, directional tools provide simple instructions informing users of the direction they should head and the distance they should traverse. Such tools are simple by nature and provide an immediacy that minimizes the need for intense cognitive processing.

These tools are, effectively, digital compasses that point to desired destinations, such as the nearest cafe, rather than magnetic north. However, like traditional compasses, these tools are not aware of physical obstructions. A building that stands in the way of the destination is not recognized by the tool, which will simply instruct the user to walk through it. The user will, of course, need to find a way around the obstruction, rendering the directional instruction somewhat meaningless. Additionally, the distance to be traversed is potentially a difficult measure to convey since it is often provided numerically, and converting a number to a physical approximation can be difficult for many people.

With the introduction of mobile technology and GPS, tools now exist that combine the advantages of the two types of navigational tools into small, mobile, and wearable technologies that may serve multiple preferences. In particular, maps that depict the surrounding environment around a you-are-here marker that dynamically updates to indicate the user's current position while maintaining correct orientation has made following directions so simple that users have become focused on the navigational information constantly being updated may fail to notice that they are walking into a lamppost or off a curb. With navigational tools integrated into augmented reality eyewear, there is a potential of combining the simplicity of mobile navigational tools with safety concerns.

Other wearable options that address both advantages include tactile cure or spatial audio cues. But while wearable technologies can ameliorate the impact that other complex tasks may have on the retention of mental maps through the use of augmented reality eye gear that project navigation routes to follow, there is evidence that such tools come at a cost.

Specifically, there is data suggesting that the use of augmented reality for navigation guidance results in the formation of weaker cognitive maps. People construct, store, or enact mental representations of the lay of the land or where things are in real space and these representations are referred to as cognitive maps. It is not clear how a cognitive map is represented and so direct measures of cognitive maps cannot be made. How cognitive maps are represented or can best be judged is not clear and they may, in fact, have very few map-like elements but are based, instead, upon a wide and potentially rich set of memory mechanisms.

Wen and colleagues (2013) found that, when augmented reality navigation tools were compared against traditional maps, recall of traversed routes took significantly longer for users of augmented reality tools than traditional map users. This suggests that battery depletion or similar power outages in wearable technologies may have a greater detrimental impact on users who have more fully integrated the advantages of augmented reality wearable tools into their routine. The convenience and efficiency of such tools may create a dependency that may reveal serious consequences in high-stress environments, such as emergency response. Reliance on tools creates, in general, dependencies on those tools that may introduce risk when the tools suddenly malfunction. Preparing for contingencies is therefore an essential factor when considering the overall readiness of an emergency responder.

Traditional power tools may often offer some sort of fallback mechanism that allows the user to manually operate the tool if there is a power failure. With wearable devices, where the functionality of the electronics are binary—on or off, there is often little recourse if power is not available. In order to address this concern, the counterintuitive notion of making interfaces more challenging has been proposed by Wen to train and exercise the user in the event that the electronically based tool itself is unexpectedly put out of commission (Wen, Deneka, Helton, & Billinghurst, 2014; Wen, Deneka, Helton, Dünser et al., 2014). In dual-task studies, it was found that the degree to which cognitive resources are taxed when using augmented reality, as compared to maps, is so low that the possibility of introducing more challenging interfaces can potentially be undertaken without penalty to the efficiency and effectiveness of the system. For example, in the case of an augmented reality navigation tool, the visual cues that provide the guidance can be rendered in a noncontinuous manner. That is, the user is not constantly being provided the guidance and so will need to rely on a combination of the persistence of vision and of visual memory. By withholding constant guidance, the user is forced to retain the visual cues. This may increase the chance that the cue held in visual short-term memory can transition into long-term memory and therefore be retained; this may make it more accessible, even in the absence of the original guidance technology. Sometimes making everything easier is not really making it better. It is possible that increasing the effort in a guided navigation tool may result in better recall in a traversed path since the additional effort may serve as an investment of cognitive effort that leaves a greater cognitive map within the navigator. When the proverbial "shit hits the fan," those too dependent on their technological tools may be unable to cope with reality.

DISCUSSION

Wearables are designed to enable users to utilize augmented technology during movement in the physical world. Psychologists, however, have fixated on self-report questionnaires and computer-based tasks where the overt behavior is finger movements (Baumeister, Vohs, & Funder, 2007). There is a substantial gap in the literature in regard to how cognition works during complex physical movement tasks (Blakely, Kemp, & Helton, 2016), and this gap is a significant issue in the proper design of wearables. Literature is emerging that may be useful for designing wearables for people, like Mary, with hard jobs. As discussed in this chapter, we know three general bits of information so far that are useful for designing wearables for people with hard jobs.

First, complex movement tasks demand considerable attention and cognitive resources. They demand more than mentally engaging and demanding seated tasks. Bob, while seated, is able to attend to more accessory mental tasks than Mary while she is physically negotiating the world. For example, while using an axe to fell a tree, Mary not only must attend to the tree and its state, watching for any indications it may fall, but she must coordinate her own movements to make sure she hits the tree with the axe in the right location and with the appropriate amount of force. She must control her stance so she does not lose balance. She must keep track of her exit plan, where she will move, if the entire tree begins to fall in an unintended direction. More importantly, if Mary makes a mistake, she will not suffer a paper cut but will have tons of wood crush her. Mary's job may be considered a physical job, but complex physical jobs are extremely attention demanding. This may be vastly underestimated by research looking at concurrent physical and cognitive activity while people use treadmills and stationary cycles. The later while requiring physical output do not require significant attention to the environment. They are too artificial to be particularly informative or relevant for Mary and her needs.

Second, cognitive fatigue negatively impacts later physical output. This is different from cognitive output, which can be facilitated by prior physical output. Bob, while doing his demanding cognitive activities, may benefit from getting up occasionally and moving around. The physical activity may help energize him, proving restorative for his cognitive resources. On the other hand, however, Bob may not feel like working out after work because his mental exhaustion is demotivating. He's feels too tired to physically work. While this may not be good for Bob's long-term health and well-being, it will unlikely be immediately fatal. Although if Bob has a cardiovascular episode due to his sedentary lifestyle, then it could be fatal. Mary, on the other hand, must be physically capable. If Mary is given exhausting cognitive work, it will limit her physical output. When she is running from a fire or a tree is falling, running a little slower may be fatal. Mary really does not need unnecessary cognitive workload because in her job this could be immediately fatal.

Third, while relieving Mary of cognitive workload is a good idea, unfortunately, this augmentation can come at a cost. Unlike Bob, who can come back to his work later if there is a technical glitch or information system hiccup, Mary needs to know where she is and what is going on. For Mary, overreliance on technology could mean life or death, if the technology fails at the wrong time. Mary may need back-up skills and some built-in redundancy.

In conclusion, wearables for people with hard physical jobs like Mary's will require further research on how people perform in these challenging settings. Modern psychology has become too fixated on easy-to-conduct research on seated or lying people doing finger movements. Psychologists need to examine cognition during physically challenging tasks. They have to start doing hard research, research where a participant is not stationary, for hard people—like Mary—and not become excessively fixated on people, like Bob, doing physically soft jobs.

REFERENCES

Barde, A., Ward, M., Helton, W. S., Billinghurst, M., & Lee, G. (2016). Attention redirection using binaurally spatialised cues delivered over a bone conduction headset. *Proceedings of the Human Factors and Ergonomics Society*, 60: 1534–1538.

Barnes, B. (2010). Disney command center aims to keep lines moving. Nytimes.com. Retrieved 26 January 2017, from http://www.nytimes.com/2010/12/28/business/media/28disney.html

Baumeister, R. F., Vohs, K. D., & Funder, D. C. (2007). Psychology as the science of self-reports and finger movements: Whatever happened to actual behavior? *Perspectives on Psychological Science*, 2(4): 396–403.

Blakely, M. J., Kemp, S., & Helton, W. S. (2016). Volitional running and tone counting: The impact of cognitive load on running over natural terrain. *IIE Transactions on Occupational Ergonomics and Human Factors*, 4(2–3): 104–114.

Bradshaw, J. L. (1968a). Pupil size and problem solving. *Quarterly Journal of Experimental Psychology*, 20: 116–122.

Bradshaw, J. L. (1968b). Load and pupillary changes in continuous processing tasks. *British Journal of Psychology*, 59: 265–271.

Broadbent, D. E. (1957). A mechanical model for human attention and immediate memory. *Psychological Review*, 64(3): 205–215.

Broadbent, D. E. (1958) *Perception and communication*. London, UK: Pergamon Press Ltd.

Colavita, F. B. (1974). Human sensory dominance. *Perception and Psychophysics*, 16: 409–412.

Darling, K. A. & Helton, W. S. (2014). Dual-task interference between climbing and a simulated communication task. *Experimental Brain Research*, 232(4): 1367–1377.

Deutsch, J. A. & Deutsch, D. (1963). Attention: Some theoretical considerations. *Psychological Review* 70: 80–90.

Epling, S. L., Blakely, M. J., Russell, P. N., & Helton, W. S. (2016). Free recall and outdoor running: Cognitive and physical demand interference. *Experimental Brain Research*, 234(10): 2979–2987.

Epling, S. L., Blakely, M. J., Russell, P. N., & Helton, W. S. (2017). Interference between a fast-paced spatial puzzle task and verbal memory demands. *Experimental Brain Research*, 235(6): 1899–1907.

Epling, S. L., Russell, P. N., & Helton, W. S. (2016). A new semantic vigilance task: Vigilance decrement, workload, and sensitivity to dual-task costs. *Experimental Brain Research*, 234(1): 133–139.

Green, A. L., Draper, N., & Helton, W. S. (2014). The impact of fear words in a secondary task on complex motor performance: A dual-task climbing study. *Psychological Research*, 78(4): 557–565.

Green, A. L. & Helton, W. S. (2011). Dual-task performance during a climbing traverse. *Experimental Brain Research*, 215(3–4): 307–13.

Hancock, S. & McNaughton, L. (1986). Effects of fatigue on ability to process visual information by experienced orienteers. *Perceptual and Motor Skills*, 62(2): 491–498.

Head, J. & Helton, W. S. (2014). Sustained attention failures are primarily due to sustained cognitive load not task monotony. *Acta Psychologica*, 153: 87–94.

Head, J., Tenan, M. S., Tweedell, A. J., LaFiandra, M. E., Morelli, F., Wilson, K. M., … Helton, W. S. (2017). Prior mental fatigue impairs marksmanship decision performance. *Frontiers in Physiology*, 8: 680. doi:10.3389/fphys.2017.00680

Head, J., Tenan, M. S., Tweedell, A. J., Price, T. F., LaFiandra, M. E., & Helton, W. S. (2016). Cognitive fatigue influences time-on-task during bodyweight resistance training exercise. *Frontiers in Physiology*, 7: 373. doi:10.3389/fphys.2016.00373

Heaton, L. D., Anderson, R. S., & Wiltse, C. M. (1967). *Physical standards in World War II*. Washington, DC: Office of the Surgeon General, Department of the Army.

Helton, W. S. & Russell, P. N. (2011). Working memory load and the vigilance decrement. *Experimental Brain Research*, 212: 429–437.

Helton, W. S. & Russell, P. N. (2013). Visuospatial and verbal working memory load: Effects on visuospatial vigilance. *Experimental Brain Research*, 224: 429–436.

Hogan, C. L., Mata, J., & Carstensen, L. L. (2013). Exercise holds immediate benefits for affect and cognition in younger and older adults. *Psychology and Aging*, 28(2): 587.

Hollister, S. (2017, July 6). The rise and not-quite-fall of Pokemon Go. *CNET*. Retrieved 26 January 2017, from https://www.cnet.com/news/pokemon-go-million-dollar-monthly-active-users/

Kahneman. D. (1973). *Attention and effort*. Englewood Cliffs, NJ: Prentice-Hall Inc.

Kahneman D. and Beatty, J. (1966). Pupil diameter and load on memory. *Science*, 154: 1583–1585.

Kahneman D. and Beatty, J. (1967). Pupillary responses in a pitch-discrimination task. *Perception and Psychophysics*, 2: 101–105.

Kahneman D. and Peavler, W. S. (1969) Incentive effects and pupillary changes in association learning. *Journal of Experimental Psychology*, 79: 312–318.

Kahneman, D., Peavler, W. S., and Onuska, L. (1968). Effects of verbalization and incentive on the pupil response to mental activity. *Canadian Journal of Psychology*, 22: 186–196.

Kantowitz, B. H. and Knight, J. L. (1976). Testing tapping timesharing, II: Auditory secondary task. *Acta Psychologica*, 40: 343–362.

Keizer, G. (2017). Windows comes up third in OS clash two years early. *Computerworld*. Retrieved 25 January 2017, from http://www.computerworld.com/article/3050931/microsoft-windows/windows-comes-up-third-in-os-clash-two-years-early.html

Marcora, S. M., Staiano, W., & Manning, V. (2009). Mental fatigue impairs physical performance in humans. *Journal of Applied Physiology*, 106(3): 857–864. doi:10.1152/japplphysiol.91324.2008

Pageaux, B., Lepers, R., Dietz, K. C., & Marcora, S. M. (2014). Response inhibition impairs subsequent self-paced endurance performance. *European Journal of Applied Physiology*, 114(5): 1095–1105. doi:10.1007/s00421-014-2838-5

Pashler, H. (1994a). Dual-task interference in simple tasks: Data and theory. *Psychological Bulletin*, 116(2), 220–244. doi:10.1037/0033-2909.116.2.220

Pashler, H. (1994b). Graded capacity-sharing in dual-task interference? *Journal of Experimental Psychology: Human Perception and Performance*, 20(2), 330–342. doi:10.1037/0096-1523.20.2.330

Pitichat, T. (2013). Smartphones in the workplace: Changing organizational behaviour, transforming the future. *LUX: A Journal of Transdisciplinary Writing and Research from Claremont Graduate University*, 3(1): 13.

Sawyer, B. D., Finomore, V. S., Calvo, A. A., & Hancock, P. A. (2014). Google Glass: A driver distraction cause or cure? *Human Factors*, 56: 1307–1321.

Simpson H. M. and Paivio, A. (1968) Effects on pupil size of manual and verbal indicators of cognitive task fulfillment. *Perception and Psychophysics*, 3: 185–196.

Sims, V., & Hegarty, M. (1997). Mental animation in the visuospatial sketchpad: Evidence from dual-task studies. *Memory & Cognition*, 25(3), 321–332. doi:10.3758/bf03211288

Slusher, A. L., Patterson, V. T., Schwartz, C. S., & Acevedo, E. O. (2018). Impact of high intensity interval exercise on executive function and brain derived neurotrophic factor in healthy college aged males. *Physiology & Behavior*, 191: 116–122.

Smith, M. R., Coutts, A. J., Merlini, M., Deprez, D., Lenoir, M., & Marcora, S. M. (2016). Mental fatigue impairs soccer-specific physical and technical performance. *Medicine & Science in Sports & Exercise*, 48(2): 267–276. doi:10.1249/MSS.0000000000000762

Smith, M. R., Marcora, S. M., & Coutts, A. J. (2015). Mental fatigue impairs intermittent running performance. *Medicine & Science in Sports & Exercise*, 47(8): 1682–1690. doi:10.1249/MSS.0000000000000592

Stevenson, H., Russell, P. N., & Helton, W. S. (2011). Search asymmetry, sustained attention, and response inhibition. *Brain and Cognition*, 77(2): 215–222. doi:10.1016/j.bandc.2011.08.007

Wang, D. (2013). *A framework of smartphone use for travel*. Temple University. Available at https://login.ezproxy.net.ucf.edu/login?url=https://search.proquest.com/docview/1294910187?accountid=10003

Wen, J., Deneka, A., Helton, W., & Billinghurst, M. (2014, April). Really, it's for your own good... making augmented reality navigation tools harder to use. In *CHI'14 Extended Abstracts on Human Factors in Computing Systems* (pp. 1297–1302). ACM.

Wen, J., Deneka, A., Helton, W. S., Dünser, A., & Billinghurst, M. (2014, June). Fighting technology dumb down: Our cognitive capacity for effortful AR navigation tools. In *International Conference on Human-Computer Interaction* (pp. 525–536). Springer, Cham.

Wen, J., Helton, W. S., & Billinghurst, M. (2013). Classifying users of mobile pedestrian navigation tools. In *Proceedings of the 25th Australian Computer-Human Interaction Conference: Augmentation, Application, Innovation, Collaboration* (pp. 13–16). ACM.

Wen, J., Helton, W. S., & Billinghurst, M. (2015). If reality bites, bite back virtually: Simulating perfection in augmented reality tracking. In *Proceedings of the 14th Annual ACM SIGCHI_NZ conference on Computer-Human Interaction* (p. 3). ACM.

Wickens, C. D. (1976). The effects of divided attention on information processing in tracking. *Journal of Experimental Psychology: Human Perception and Performance*, 2: 1–13.

Wickens, C. D. (1980). The structure of attentional resources. In R. Nickerson (Ed.). *Attention and Performance VIII* (Hillsdale, NJ: Lawrence Erlbaum): 239–257.

Wickens, C. D. (2002). Multiple resources and performance prediction. *Theoretical Issues in Ergonomics Science*, 3(2): 159–177.

Wickens, C. D., Gordon, S. E., Liu, Y., & Lee, J. (1998). *An introduction to human factors engineering*. New York, NY: Longman.

Woodham, A., Billinghurst, M., & Helton, W. S. (2016). Climbing with a head-mounted display dual-task costs. *Human Factors: The Journal of the Human Factors and Ergonomics Society*, 58: 452–461.

Wright, P. and Kahneman, D. (1971) Evidence for alternative strategies of sentence retention. *Quarterly Journal of Experimental Psychology*, 23: 197–213.

2 Humans and Automated Decision Aids: A Match Made in Heaven?

Kathleen L. Mosier & Dietrich Manzey

In the tiny villages and towns around Winchester, UK, motorists have long had problems with GPS navigation. They follow the route their devices suggest and repeatedly wind up stuck — the roads are simply too narrow for some wider vehicles. The problem got so bad in the town of Exton that a bright yellow sign was erected to warn drivers NOT to follow their GPS and to, instead, rely on common sense. These are the times we live in, folks (https://www.ranker.com/list/9-car-accidents-caused-by-google-maps-and-gps/robert-wabash).

AUTOMATED DECISION SUPPORT: WHAT AND WHY?

For the last four decades, computer-based automated or even autonomous systems have changed our life considerably. Specific sorts of automation are the so-called automated decision support systems (DSSs), which are ubiquitous in today's high-technology world. Designed as a support tool for humans, they provide — at discrete points in time, either automatically or on demand — certain information about the state of the world that can improve informed decision making. Beyond that, they also may provide recommendations on actions and/or predictions of outcomes in the near future in order to help the human to arrive at proper decisions about actions. Their main goal is to improve the reliability and quality of human decision making at work or in everyday life.

Two basic aspects of human decision making have been referred to as front-end and back-end processes (Mosier & Fischer, 2010). Front-end processes include all cognitive judgment processes needed for a proper situation assessment and situation awareness (SA) (Endsley, 1995, 2000), for example, information search, information analysis, risk assessment, or identification of constraints for possible actions. These usually provide the basis for the back-end processes, which include the actual decision making, that is, the selection and planning of proper actions. Corresponding to this basic distinction, DSSs can be categorized, albeit relatively coarsely, as to whether they support the front-end or back-end of decision making. Typical examples of front-end DSSs are alarm systems, such as alarm lights that illuminate on cockpit or other control displays to alert the operator about a problem, or the collision prevention assist in an automobile, a sensor that monitors a vehicle's distance from other vehicles in front and beeps if the driver accidentally gets too close to the car ahead. More complex front-end assistance systems include information systems, such as those that provide traffic information on radar displays or moving maps or provide surgeons with information about the location of specific landmarks to help them navigate through a patient's anatomy. In contrast, back-end DSSs include all systems that provide automated support for decision choice and action selection, such as the GPS systems in cars, which analyze a given situation and then suggest the quickest route to a location or instruct the individual exactly on what to do. Even more complex systems may dynamically incorporate both front-end and back-end processes. Examples of this latter type are sophisticated DSSs in aircraft cockpit like the Ground Proximity Warning System (GPWS) or the traffic collision avoidance system (TCAS) that start as pure front-end systems alerting pilots of a possibly critical situation (e.g., "terrain, terrain" or "traffic, traffic") but switch to back-end directives instructing the pilot exactly what to do if the situation becomes more critical (e.g., "pull up" or "climb/descend"). Note that our distinction of front-end and back-end DSSs also fits to models describing different types and levels or degrees of automation (e.g., Endsley & Kaber, 1999; Onnasch, Wickens, Li, & Manzey, 2014; Parasuraman, Sheridan, & Wickens, 2000). For example, Parasuraman et al. (2000) distinguish different types and levels of automation with types of automation referring to the stage of information processing that is supported. Automation support of the first two stages — information acquisition and analysis — would correspond to front-end support, while automation support of later stages of information processing, that is, decision and response selection, would correspond to back-end support. Because back-end DSSs often implicitly also include automated situation assessment, they can be considered to represent a higher degree of automation than front-end support, that is, they support the human user in both situation assessment and selection of actions (Onnasch et al., 2014).

The intent of all of these DSSs is to compensate for the inadequacies of the human decision-maker and to make decisions more rational, compared to the often-observed issues of biases and heuristics in human decision making (Gigerenzer & Todd, 1999; Kahneman, 2011). If this would work, providing a DSS as a support tool for a human decision maker would result in a well-suited pairing with an almost perfect outcome — like a match made in heaven! However, before treating DSSs in such a romantic way, some caveats are necessary. Evidence over the past 22 years since this volume was first published suggests that the relationship between human decision makers and DSSs is not as ideal as expected. Although much of the time DSSs successfully enhance human performance, reliance on imperfect automation has the unintended consequence of worsening performance when automated information or advice is not correct. In fact, human tendencies toward the use of heuristics, or shortcuts, in judgment and decision making, which DSSs should help to avoid, can be triggered by DSSs, eroding the quality of the human-automation match and causing it to be more earth-like, with weaknesses as well as strengths, than a heavenly ideal. In particular, the presence of DSSs may foster automation bias, "the tendency to use automated cues as a heuristic replacement for vigilant information seeking and processing..." (Mosier & Skitka, 1996, p. 205).

This chapter particularly focuses on the performance consequences of DSSs with a particular emphasis on the issue of automation bias as demonstrated in research across many domains. We examine the factors that impact automation bias including operator characteristics such as trust in the system and level of perceived accountability; system characteristics such as reliability, transparency, and understandability, and the level and degree of automation involved; and the characteristics of the task-context including time pressure, consequences of errors, workload, or working in redundant work teams. We also discuss the challenge of balancing the pros and cons of using automation as a heuristic in decision making as well as current trends and future advances in automated systems.

AUTOMATION BIAS: STILL AROUND AFTER ALL THESE YEARS

Designers responded to "pilot error" and the increasing cockpit workload by attempting to remove the error at its source, that is, to replace human functioning with device functioning — in their view, to automate human error out of the system. But there were two flaws in this reasoning: (1) The devices themselves had to be operated and monitored by the very humans whose caprice they were designed to avoid; human error was not eliminated but relocated. (2) The devices themselves had the potential for generating errors that could result in accidents (Wiener, 1985, p. 78).

Automation bias was first identified in cockpit crews and was characterized in the tradition of classical decision heuristics and biases (Kahneman, Slovic, & Tversky, 1982) as a decision shortcut. It manifests in two types of performance consequences: *omission errors* result when decision makers do not take appropriate action because they are not informed about a situation, problem, or imminent system failure by automated decision aids; and *commission errors* occur when decision makers take inappropriate action because they over-attend to information or directions from an automated decision aid (Mosier & Skitka, 1996; Mosier, Skitka, Heers, & Burdick, 1998). The distinction between two manifestations of automation bias is related to the distinction of reliance and compliance in human-automation interaction proposed by Meyer (2004). Whereas reliance describes how much an individual trusts the alert function of a DSS in the case of critical events and, thus, reduces his or her own monitoring of the environment, compliance describes the behavioral tendency to trust and follow a DSS's recommendation. Given this, automation bias in terms of omission errors can be understood as a consequence of overreliance on a DSS, whereas commission errors reflect a sort of "over compliance" with a DSS.

Several possible mechanisms underlie these errors. First, automated aids present powerful, salient decisional cues and are widely believed to be accurate. The "cognitive miser" perspective suggests that people tend to take the road of least cognitive effort and may diffuse responsibility for decision tasks to automated aids, resulting in incomplete cross-checking of available information. This is related to automation-induced complacency, which has been documented as a factor in automation-related incidents and accidents (see Parasuraman & Manzey, 2010, for a review). Second, operators may inattentively process information that contradicts automated directives, analogous to a looking but not "seeing" or inattentional blindness effect (Simons & Chabris, 1999), or may "see" information that confirms the automation even though it is not actually present, a phenomenon labeled "phantom memory" (Mosier et al., 1998). A third possibility is that people see but actively discount information from less automated sources because they believe the automated decision support system is more likely correct (Skitka, Mosier, & Burdick, 1999). It is likely that the mechanism varies along with the task context, for example, whether automated information is made more salient than other information or whether time pressure is present. Additionally, the mechanism may depend on whether the DSS targets the front or back end of decision making.

Since the early research in the 1990s, automation bias has been identified not only in aviation but also in several other domains in which technology and automated decision support are commonplace.

Evidence for Automation Bias from Different Domains
Aviation

A series of studies of human-automation interaction in aviation (and aviation-like laboratory tasks) by Mosier and colleagues provided the cradle of automation bias research (Mosier, Palmer, & Degani, 1992; Mosier et al., 1998; Skitka et al., 1999). In the first of this series, Mosier et al. (1992) studied the effects of an automated electronic checklist on decision making in an ambiguous simulated engine fire situation, that is, there were conflicting cues concerning which engine was on fire. The electronic checklist for ENGINE FIRE prompted the pilots to shut down the wrong engine (i.e., the one not affected by fire). Although available information from other cockpit instruments contradicted the electronic device, a total of 75% of pilots followed the wrong recommendation of the decision aid. This was in stark contrast to a control group of pilots who used a traditional checklist for the same task and committed this error to a considerably lesser extent (25%).

In another study, a sample of experienced commercial pilots performed simulated flights in a low-fidelity flight trainer equipped with various advanced cockpit automation systems (Mosier et al., 1998). During the flight, they were exposed to automation failures creating situations for omission and commission errors, including an altitude clearance that did not get loaded correctly, a heading change that was not executed properly by the flight system, and a false engine-fire alert on the electronic EICAS (engine-indicating and crew-alerting system) that was not supported by any other instrument readings. While the first two automation errors provided opportunities for an error of omission if not detected by the pilots, the latter one provided an opportunity to commit a commission error. Basically, all of these automation failures were easily detectable by the use of relevant information available from other cockpit instruments. Yet, the frequency of omission errors was 55% — and 100% of the pilots decided to shut down the engine in response to the erroneous EICAS alert. Interestingly, 67% of the pilots reported that they had seen at least one other indication supporting the EICAS fire alert, which in fact was not there.

Further evidence for automation bias was also found in a group of nonpilots performing a low-level flight simulation task with and without automation support (Skitka et al., 1999). During the task, several automation failures occurred — six failures where the automation failed to prompt the participants of a critical event and another six where the aid gave a wrong recommendation. Again, the omission error rate in response to the first type of automation failure was 41%. In contrast, only

3% of the critical events were missed by a control group performing the task without automation support. A similar picture also emerged for commission errors. A total of 65% of participants followed the wrong advice of the decision aid, although other available information directly contradicted the advice and the participants were informed that the other information would be perfectly valid. The common characteristic of all these findings was that the presence of an automated support for decisions made pilots less likely to seek and assess information from other sources. Since this early work, many other examples of automation bias in interaction with decision aids in advanced cockpits and also air traffic control have been reported (e.g., Metzger & Parasuraman, 2005; Rovira & Parasuraman, 2010; Sarter & Schroeder, 2001) also as contributing factors to fatal accidents (e.g., Dutch Safety Board, 2010). Many of these cases suggest that automation bias might not necessarily result in manifest errors of omission (or commission) but rather a delayed detection and response to contradictory information leaving insufficient time to respond to a critical situation (de Boer, Heems, & Hurts, 2014).

Health Care

Automation bias in interactions with automated decision aids has also become a concern in the health care domain (Goddard, Roudsari, & Wyatt, 2012, 2014; Lyell et al., 2017). One set of studies addressed the performance consequences of so-called clinical decision-support systems (CDSSs) on the quality of clinical judgment and decision making (Goddard et al., 2014; Lyell et al., 2017; McKibbon & Fridsma, 2006; Tsai, Fridsma, & Gatti, 2003; Westbrook, Coiery, & Gosling, 2005). CDSSs are computer-based systems that provide physicians with information or advice, including clinical knowledge, warnings, and recommendations, in order to improve their clinical judgment and decision making. The researchers included a variety of systems ranging from electronic information resources like PubMed or Google, which loosely can be considered CDSSs — to specific diagnostic systems or electronic prescribing systems. Overall, consulting these systems was shown to improve the quality of clinical decisions and diagnoses (Garg et al., 2005). However, this general benefit can be offset at least to some extent by consequences of too much reliance on the CDSS when the suggestions or recommendations are wrong or suboptimal, reflecting automation bias. This negative effect has been shown by a number of studies comparing the correctness of clinical diagnoses before and after consulting a CDSS (Friedman et al., 1999; McKibbon & Fridsma 2006; Westbrook et al., 2005). Correct CDSS advice increased the number of correct diagnoses after consultation of the aid by 2%–21%. However, at the same time, incorrect advice led clinicians in 6%–11% of cases to commit commission errors, that is, to change their initially correct diagnoses into wrong ones after consulting the false advice of the aid, which reduced the net gain effects of the CDSS significantly.

While these instances of automation bias involved only commission errors, more recent studies investigating the effects of electronic prescribing systems have revealed increased risks for both omission and commission errors caused by using this sort of CDSS (Lyell et al., 2017). The main function of electronic prescribing systems is to alert and warn a physician of possible risks due to an inappropriate or dangerous medication or unwanted interaction effects among different prescriptions.

Compared to a control group that made prescription decisions without the support of the CDSS, correct alerts by the system successfully decreased the rate of prescription errors by about 43%. However, in the case of missed alerts or false alarms of the CDSS, prescription errors due to omission errors (e.g., overlooking a wrong dose or a risky interaction effect that was not indicated by the system) and commission errors (e.g., abstaining from prescribing an actually safe medicine that was falsely indicated as risky by the system) increased by 28.7% and 56.9%, respectively. Participants who made omission errors in this study reported lower cognitive load compared with those who did not, suggesting that the allocation of insufficient cognitive effort to the prescription task was a factor in these automation bias errors (Lyell, Magrabi, & Coiera, 2018).

Another set of studies investigated possible risks of automation bias in interaction with computer-aided detection aids (CADs) in radiology (Alberdi, Povyakalo, Strigini, & Ayton, 2004; Alberdi, Poviakalo, Strigini, Ayton, & Given-Wilson, 2008). Experienced radiologists examined a set of mammograms either with or without the support of a detection aid that suggested areas containing lesions. The first study mainly focused on the risk of omission errors in cases where the CAD failed to prompt critical areas, that is, left a mammogram unmarked although there were signs of cancer or misplaced a prompt away from an area that was critical (Alberdi et al., 2004). Compared to a film reading without the aid, such false prompts decreased the detection rates considerably from 46% to 21% and 66% to 53%, respectively. Obviously, the film readers in this study tended to take the absence of a computer prompt as evidence for the absence of cancer. The authors interpreted this finding as evidence for automation bias, possibly due to a reduced vigilance when supported by a CAD. The second study looked specifically at performance consequences of false positive prompts, that is, prompts that were placed although actually no signs of cancer were present (Alberdi et al., 2008). Again, some evidence for automation bias was found. Falsely placed prompts significantly raised the probability by 12.3% that the prompted areas actually were marked as malignancy compared to an unaided condition. Compared to the performance consequences of missed prompts, however, this effect was relatively weak, suggesting that with such systems' false positive prompts providing opportunities for commission errors might be more tolerable than false negative prompts possibly provoking errors of omission.

Process Control

Some even stronger effects of automation bias have been reported from studies investigating performance consequences of automated decision aids in simulated process control tasks (Bahner, Hueper, & Manzey, 2008; Manzey, Reichenbach, & Onnasch, 2012; Reichenbach, Onnasch, & Manzey, 2011; Sauer, Chavaillaz, & Wastell, 2016; Wickens, Gutzwiller, & Santamaria, 2015). In all of these studies, a process-control microworld that simulates an autonomously running life-support system in a remote space habitat was used (AutoCAMS; Manzey et al., 2008). As a rule, students with a background in engineering served as operators in these studies. Their task was to monitor the system and to intervene manually in the case of failures. While performing this supervisory-control task, they were supported by an automated fault identification and repair agent (AFIRA) that, in the case of failures in the life-support systems, provided the operators with an alert accompanied by an automatically generated failure diagnosis and/or further hints for proper failure management. In the case of correct diagnoses and advice provided by AFIRA, operators' performance in terms of failure identification and fault management improved and became close to perfect, compared to when they had to perform the task without the decision aid. However, in the case of an incorrect diagnosis provided by the aid, converging evidence for automation bias was found in all studies referred to earlier. This was reflected in up to 80% of operators committing a typical commission error, that is, following the AFIRA advice although inspecting other available information would have proved it wrong. The effect was particularly evident for the first occurrence of such a failure of the decision aid ("first-failure effect"; Wickens, Clegg, Vieane, & Sebok, 2015). More specific analyses revealed that this sort of automation bias often occurred despite the fact that the operators had inspected other relevant data in order to cross-check the diagnosis of the aid at hand. That is, they either discounted this contradictory information or did not process it properly (Manzey et al., 2012).

Command and Control

Additionally, issues of automation bias have also been reported from interactions with intelligent decision-support systems for command-and-control operations in the military domain (e.g., Crocoll & Coury, 1990; Cummings, 2004; Rovira, McGarry, & Parasuraman, 2007). Decision-support systems in this domain include, among others, automated target identification and engagement systems. As reported by Cummings (2004), erroneously following such automated aids has already contributed to fatal military decisions, including friendly fires during the Iraq War. Systematic research in this domain has addressed the effects of front-end and back-end support on military decision making and the risks if the decision aids provided inaccurate advice (e.g., Crocoll & Coury, 1990; Rovira et al., 2007). For example, Rovira et al. (2007), investigated how different target-identification systems would impact the quality of military decisions under time pressure. The DSSs used in this study differed with respect to their reliability and to what extent they just provided front-end information analysis support or also back-end decision support. In line with the results from other domains, all of the aids led to performance improvement in terms of reduced decision times and higher number of correct decisions made compared to unaided control conditions. However, in the case of incorrect advice of the aids, a number of commission errors occurred and decision accuracy declined from 89% in the unaided condition to a mean of only 70% in the supported conditions. The most pronounced performance decrements were found for aids that provided a high level of back-end support (i.e., made a specific recommendation for one particular response option) with an overall high level of reliability.

Factors Impacting Automation Bias

> It does little good to remind human operators that automation is not always reliable or trustworthy when their own experience tells them it can be trusted to perform correctly over long periods of time (Billings, 1996, p. 97).

A number of factors have been suggested to impact the management of errors in interaction with automation and, thus, also the risk of automation bias when using DSSs. Major factors include individual characteristics and attitudes of the human operator, characteristics of system design, and different aspects of the task context (see also McBride, Rogers, & Fisk, 2013 for an elaborated discussion of error management in human-automation interaction).

INDIVIDUAL FACTORS

Trust and Overtrust. A key component in human-automation interaction is the extent to which the operator trusts the system (e.g., Chen & Barnes, 2014; Lee & See, 2004). Bailey and Scerbo (2007), for example, found an inverse relationship between trust and monitoring performance. In general, findings across many studies suggest that automation reliability is strongly associated with trust development and maintenance, and that operators adjust their trust in automation in line with its performance (Hoff & Bashir, 2015). Experience with reliable automated systems increases trust; negative experiences with automation can reduce it, with the negative feedback loop causing considerably stronger and longer-lasting effects (Manzey et al., 2012). It seems reasonable to posit that the level of individual trust in a certain DSS also impacts automation bias. This might also explain in part why professionals working with highly reliable systems are equally as susceptible to automation bias as novices (e.g., Mosier et al., 1998) — extensive experience with highly reliable automation induces trust and reduces the odds that experts will conduct a thorough information search to verify automated information and advice. However, clear empirical evidence that explicitly supports a link between individual trust and the occurrence of omission and commission errors is still lacking.

Furthermore, there is evidence that individuals differ in their proneness to rely too much on automated systems. This individual tendency has been referred to as complacency potential (Singh, Molloy, & Parasuraman, 1993) and is assumed to be dependent on perceived system properties (e.g., reliability) as well as individual characteristics of the human operator such as general attitudes toward technology or personality characteristics such as self-efficacy (Parasuraman & Manzey, 2010; Prinzel, 2002). More recent work even suggests that the basis of these individual differences is a genetic variance in the dopamine beta hydroxylase gene (Parasuraman, de Visser, Lin, & Greenwood (2012). However, more research is certainly needed before any decisive conclusions about systematic individual differences in proneness toward automation bias can be made.

Accountability for Decisions. Early attempts to mitigate automation bias drew from the debiasing techniques of the social psychology literature. Mosier, Skitka, and colleagues demonstrated that the imposition of predecisional accountability (before performing or making a decision; Hagafors & Brehmer, 1983) for accuracy and overall performance resulted in lower rates of omission and commission errors compared to nonaccountable student participants (Skitka, Mosier, & Burdick, 1999). Experimentally manipulated accountability demands did not have the same impact for professional pilots in flight scenarios; however, pilots who reported an internalized perception of "accountability" for their performance and strategies of interaction with automation were significantly more likely to double-check automated cues against other information and less likely to commit errors than those who did not share this perception (Mosier et al., 1998). The researchers suggested that "debiasing" decision making through externally imposed accountability does not show much promise for eliminating automation bias in professionals; it may be better to train behaviors that can mitigate its impact, such as verifying automated information against other available data. Other interventions, such as providing practical experiences with automation errors during training, may be more effective (Bahner et al., 2008).

SYSTEM PROPERTIES

Besides factors related to the human user of DSS, characteristics of the DSSs themselves, such as their reliability, whether they provide front-end or back-end support, their transparency and understandability, and how easy it is to cross-check or verify their outputs, have been suggested to impact the risk of automation bias.

> Reliability. Automation bias is a direct reflection of what has been referred to as ironies of automation (Bainbridge, 1983), that is, that highly reliable automation improves performance if it works correctly but may make things even worse compared to no automation support in the case of automation errors. More specifically, the likelihood of automation bias seems to be directly linked to the reliability of DSSs. Direct supporting evidence for this relationship has been provided by Bailey and Scerbo (2007), using a complex system monitoring task supported by different alarm systems as examples of simple front-end DSSs. They found that omission errors as a consequence of automation errors (misses of detecting a critical state) increased from 32.4% to 48.3% when the reliability of the DSSs increased from 0.87 to 0.98. Similar results were also reported from Rovira et al. (2007). Using an automation monitoring task supported by different DSSs providing front-end or back-end support (see also the next section), they found overall higher levels of omission errors with more reliable (0.75) than unreliable (0.50) DSSs. Furthermore, the predictability of reliability in terms of its dynamic variability seems to play a role with respect to omission errors. This is suggested by the study by Parasuraman, Molloy, & Singh (1993), which showed that omission errors in interaction with imperfect alarm systems were more likely when the reliability of the alarm system remained constant across experimental blocks than when it varied between low and high in different blocks. However, less and somewhat inconsistent data are available for direct links between DSS

reliability and the occurrence of commission errors in the case of wrong DSS recommendations. As reported earlier, Rovira et al. (2007) found higher rates of commission errors with a reliable (0.80) than with an unreliable (0.60) back-end DSS in a command and control task, which corresponds to the findings for omission errors described previously. Other studies also reported higher compliance and smaller rates of cross-checking the automation with higher reliability alarm systems, but this seemed to reflect more of a well-adapted behavior than a form of over compliance and automation bias (Manzey, Gérard, & Wiczorek, 2014). Studies that investigate compliance rates with complex DSSs or that vary reliability and possible performance consequences in terms of automation bias are still lacking.

Front-End versus Back-End Support. Some of the evidence for the possible significance of front-end versus back-end support in command and control tasks has been mentioned previously. When studying risks of automation bias in interaction with automated target identification systems, Rovira et al. (2007) found that particularly systems providing back-end support reduced the overall reliability of decisions made in cases of wrong advice, which confirmed earlier findings of Crocoll and Coury (1990). Similar effects have also been reported from studies in aviation. For example, Sarter and Schroeder (2001) investigated the performance consequences of inflight-icing alerting systems in airplane cockpits. Two systems were compared in this study. The first one included a status display providing pilots (front end) with information about the icing condition (wing icing or tailplane icing) but leaving the decision about a proper action with the pilots. The second system included a command display, directly providing (back-end) specific advice to the pilots about what to do. The results showed that both sorts of the DSS considerably improved the overall speed and number of correct decisions made by the pilots. However, in the case of inaccurate advice, the systems caused more decision errors than in an unaided control condition, and this effect was stronger for the back-end than the front-end aid. Laboratory studies addressing the impact of different types of DSSs in process control did not find differences between a front-end and back-end DSS in terms of manifest commission and omission errors. However, individuals spent less effort in cross-checking information against other available sources, that is, they showed monitoring behavior indicative of automation bias, when supported by a back-end rather than a front-end DSS (Manzey et al., 2012). Furthermore, there is evidence that the extent to which back-end support is provided might make a difference with respect to risks of automation bias in the case of incorrect advice (Layton, Smith, & McCoy, 1994). In a study addressing the possible performance consequences of a DSS supporting flight planning of pilots in aviation, the authors compared electronic flight planning tools that provided pilots with automatically generated suggestions for an optimum flight plan. They found that these tools do not always improve the quality of flight planning. Specifically, they found that pilots tended to accept the electronic flight plans even though they were suboptimal. Most important in this context, this tendency was stronger for tools recommending a specific flight plan (high-level back-end support) than for tools that created different options but left the final choice to the pilots (low-level back-end support).

Altogether, the different studies referred to here suggest that front-end DSSs might be less conducive to automation bias in the case of incorrect advice than more directive back-end DSSs, and that the more definitive the advice provided by the latter, the more conducive to automation bias they are. This fits to the more general "lumberjack" hypothesis stated by Onnasch et al. (2014) that higher degrees of automation provide higher benefits than lower levels in cases of reliable routine operation but increase risks of errors and return-to-manual performance decrements in cases of automation failures. However, the effects with respect to automation bias found in the different studies were usually relatively small, and other design factors of DSSs might be even more important for affecting risks of automation bias.

Automation Visibility. Such other design factors might include the transparency of DSSs, the understandability of their functioning, and the effort it takes to cross-check and verify their outputs. Specifically, system understandability and transparency have been considered to be key antecedents to the development and calibration of system trust (e.g., Chen, Barnes, & Harper-Sciarini, 2011; Lyons et al, 2016; Sheridan, 1989). A recent study in the railway sector, for example, identified understanding of the automation, rather than system reliability, as the key component in developing trust in a DSS (Balfe, Sharples, & Wilson, 2018). Evidence for the importance of visibility in mitigating automation bias has been provided by McGuirl and Sarter (2006). Capitalizing on the earlier research on effects of inflight-icing alerts (Sarter & Schroeder, 2001), they investigated how pilots' decision making in response to inflight icing encounters would be affected by DSSs that provided system confidence information together with the automatically generated front-end or back-end information/advice. The confidence information was updated on a trial-by-trial basis and was presented to the participants in a separate trend display. Independent of whether front-end or back-end support was provided, the pilots receiving this additional information were less prone to automation bias in the case of inaccurate information, apparently because they were better able to assess the validity of the aids' recommendations. Analogous effects have recently been reported from so-called likelihood alarm systems (front-end DSS) that provide better options to estimate the likelihood of critical events in the case of alerts than binary alarm systems and that were found to improve human decision making particularly in response to system errors, that is, false or missed alerts (Wiczorek & Manzey, 2014). However, independent of whether or not likelihood information was provided, the fewest number of commission and omission errors were made when information to cross-check and verify the output of the alarm systems was available and easily accessible. This suggests that also the accessibility of automation verification information can impact the risk of automation bias. Thus, it can be expected that design decisions concerning the display of verification data (e.g., surface display versus layered or hidden display) will impact whether or not decision makers will trust DSSs or make the effort to verify their output. Overall, automation visibility, that is, its transparency in terms of its sources and functioning (Dorneich et al., 2015) and the ease of accessibility of automation verification information, seems to be a key determinant in whether decision makers will cross-check or just (over-)trust system information. This directly fits to recent findings from a meta-analysis of automation bias in interaction with DSSs in different domains, which also suggests that risks of automation bias are directly related to the effort needed to verify the output of a DSS (Lyell & Coiera, 2017).

TASK CONTEXT

DSSs are introduced to support individuals in accomplishing certain tasks. Thus, it seems reasonable to assume that the task itself or the specific situational circumstances under which a task has to be performed, that is the task context, also impacts how individuals interact with DSSs and make use of them. For example, it can be assumed that factors like time pressure, workload, the expected performance consequences of decision errors, or the social context in which a task has to be performed might have an impact on the risk of automation bias. However, only a few studies have addressed these sorts of factors thus far.

> Time Pressure. It has been known for some time that time pressure can have a considerable impact on human judgment and decision making (Edland & Svenson, 1993). More specifically, humans have been shown to compensate for time pressure in judgment and decision making by abandoning a time-consuming rational analytic strategy, based on comprehensive information search, in favor of a faster heuristic strategy based on considering only the most important (valid) cues (Rieskamp & Hoffrage, 2008). Often, this use of heuristics instead of a full information search and weighting allows for making quicker decisions without compromising quality to a significant extent (Gigerenzer & Gaissmaier, 2011). Analogous to these findings, one might assume that time pressure also would lead humans to depend more on DSSs, that is, to use their particularly salient and usually (but not always) correct advice directly as a basis for fast judgments and decision making without fully evaluating the available information from the environment. This assumption has been addressed in a series of experiments by Rice and colleagues (Rice, Hughes, McCarley, & Keller, 2008; Rice & Keller, 2009; Tunstall, Rice, Mehta, Dunbar, & Oyman, 2014). Specifically, they investigated to what extent time pressure would affect dependence on front-end DSSs in situations where humans had to make binary decisions about the presence of critical targets based on the visual inspection of complex images. After getting the advice of the DSS (i.e., critical target present or absent), participants in their studies had either two or eight seconds for their

own inspection of the visual images before they had to commit to a decision. The DSSs used in these studies did not miss any targets but provided a different number of false alarms affording the opportunity for automation bias in terms of commission errors. What Rice et al. consistently found was a higher compliance rate with the DSS under the high compared to low time-pressure conditions. This effect emerged largely independent of the reliability of the DSSs ranging from highly reliable (0.95) to fairly reliable (0.65) systems. At least part of this higher compliance with DSS advice under time pressure seems to reflect fairly rational behavior, given that the time for a comprehensive information search is limited and the objective dependence on a DSS gets higher. This was also reflected in the performance effects. The higher compliance rates under time pressure did not lead to overall performance decrements when the reliability of the DSS was high. Compared to an unsupported control condition, even more correct decisions were produced under time pressure with support of the highly reliable DSS.

It seems that, in this case, the performance costs of automation bias effects in terms of commission errors when erroneously responding to false alarms were less than the performance benefits gained by also following the DSS more often when it was correct. As a consequence, Rice et al. recommend moderate time pressure as a possible means to improve compliance with highly reliable DSSs. However, clear overall performance costs due to automation bias effects emerged with the least reliable DSS. Here, the same strategy of higher dependence on the DSS under time pressure caused performance to drop even below the performance achieved when no DSS was available (Rice et al., 2008; Rice & Keller, 2009). These results directly indicate the mixed blessing of heuristic use of a DSS under conditions of time pressure. It can improve performance if the reliability of the DSS is high but can lead to severe automation bias errors in the case of comparatively low DSS reliability. Surprisingly, the empirical evidence for the impact of time pressure is largely limited to front-end DSSs. A generalizability to back-end DSSs seems highly plausible but requires further evidence.

Workload. Similarly, also the overall workload of an individual has repeatedly been shown to affect an individual's reliance and compliance in interactions with DSSs. An example of the former is an early finding of Parasuraman et al. (1993) suggesting that complacency effects and resulting omission errors in response to unreliable automation particularly arise in multitasking contexts, that is, when the task supported by a DSS is one of several tasks to be performed concurrently. It seems that when multitasking, individuals tend to lower the priority of the task supported by a DSS and the effort to be invested in cross-checking the automation in favor of the other tasks. Whereas such behavior, on the one hand, certainly reflects a reasonable (and intended) consequence of introducing a DSS (Moray, 2003), it also can increase the risk of automation bias if it is not balanced properly.

Other findings have provided converging evidence of the same processes occurring for errors of commission. For example, Manzey et al. (2014; Exp. 2 and 4) compared compliance rates with front-end DSSs (alarms systems) under conditions where participants had to perform one or two tasks concurrently to the task supported by the DSS. The raised workload in the two-task condition increased the direct compliance rate, that is, the frequency of responses to the alarm without consulting other available information, for a relatively unreliable (0.70) DSS from about 20% to almost 60%, resulting in a considerably higher rate of commission errors when the workload was high. Even more direct support for workload impacting the dependence on DSSs has been provided by other studies (Biros, Daly, & Gunsch, 2004; Dorneich et al., 2015). Biros et al. (2004) studied to what extent workload would change the link between trust in DSSs and reliance on their advice in a command and control scenario. More specifically, they varied trust in a DSS supporting tactical decisions by means of different instructions and investigated to what extent this variation would have an impact on DSS use depending on the overall task load. Task load was operationally defined by the complexity of information to be considered in the decision task. While differences in trust directly predicted the use of automation in the low task-load condition, participants tended to rely on the automation independent of their trust in the DSS when the task load was high. The higher task load obviously led to less skeptical overreliance on the DSS, although the participants were aware that the advice might be wrong and, thus, accepted the higher risk of committing commission errors when using the aid. Similarly, Dorneich et al. (2015) found that air transport pilots tended to over-trust an information automation system when they were under a high workload and chose the top plan suggested by the system, even though information was missing and the plan was not the best one. Note that these findings parallel the findings of time-pressure effects on human dependence on DSSs. Particularly multitasking situations, but also high workload situations in general, can be considered as situations where the time for judgment and decision making is lowered due to other task demands, implicitly inducing a sort of time pressure for humans.

However, workload that is too low ("underload") might also be detrimental for human-automation performance and can increase automation bias effects. This is suggested by the recent work of Lyell and colleagues, who found that participants who made omission errors reported lower cognitive load compared with those who did not, suggesting that low cognitive load led to the allocation of insufficient cognitive effort to the experimental task, which was a factor in these automation bias errors (Lyell et al., 2018). Yet, it is not clear what cause and effect is in these findings, and clearly more research on effects of underload will be needed before any decisive conclusions about effects of underload on automation bias can be drawn.

Consequences of Errors. Evidence that the consequences of potential errors committed by an operator might affect verification behavior and the risk of automation bias was provided by two of the first studies of automation bias (Mosier et al., 1998; Mosier, Skitka, Dunbar, & McDonnell, 2001). In this research, pilots were required to perform different flight tasks during a simulated flight from San Francisco to Los Angeles (Mosier et al., 1998) or different approaches to San Francisco (Mosier et al., 2001) in a low-fidelity part-task flight simulator. During the flight, different automation failures occurred, which provided opportunities to commit omission and commission errors as manifestations of overreliance or over compliance with the automated systems. In both studies, the criticality of errors turned out to be an important predictor of omission errors. For example, automation failures related to altitude control (e.g., misloading an altitude clearance in the autopilot) were more often detected than failures related to the communication system (e.g., misloading a frequency change). Thus, it seems that pilots still monitored and verified actions of DSSs related to flight critical tasks but were less vigilant in monitoring subsystems less critical for flight safety, and this effect occurred independent of whether the simulations were accomplished by single pilots (Mosier et al., 1998) or two-person crews (Mosier et al., 2001).

Individuals versus Teams. Most of the research on the automation bias phenomenon has been conducted in a single-person performance configuration. However, some highly automated decision environments, such as aircraft cockpits, involve teams rather than solo decision makers. Whether or not this condition of shared responsibility makes a difference with respect to automation bias and risks of committing omission and commission errors when DSSs err was investigated in a pair of studies by Mosier and colleagues. Specifically, they examined individuals versus teams of commercial class cockpit pilots (Mosier et al., 2001) and students (Skitka, Mosier, Burdick, & Rosenblatt, 2000) performing simulated flight tasks under varying instruction conditions. Results demonstrated the persistence of automation bias in teams for both samples.

CONCLUSION, PRESENT TRENDS, AND FUTURE POSSIBILITIES

Clearly, automation bias is still a ubiquitous phenomenon in the interaction of humans and DSSs, which is particularly relevant with highly, but not perfectly reliable, systems. It thus directly reflects an irony of automation (Bainbridge, 1983), namely that highly reliable DSSs on the one hand certainly improve the human judgment and decision making when working properly but, at the same time, lead to automation bias in the case of automation errors.

Even more important, automation bias does not necessarily result from human weaknesses in automation monitoring but can directly result from quite rational behavior in interactions with automated systems (Moray, 2003). Thus, directly corresponding to the assumption that using heuristics in human decision making is not necessarily bad but can "make us smart" (Gigerenzer & Todd, 1999), using highly reliable DSSs as a heuristic most times leads to positive effects like reduced cognitive effort and better decisions. And most organizations want their decision makers to trust — and not constantly second-guess — the automated support they have provided! Lüdtke and Möbus (2005) have proposed that interactions with highly reliable DSSs over time might lead to an effect of learned carelessness. That is, making the repeated experience that the automation works properly even without close monitoring, the tendency to directly use the advice of DSSs as a heuristic for decision making, is amplified by a sort of positive feedback loop (see also Parasuraman & Manzey, 2010). This might also explain why human-focused approaches to improve automation verification and mitigating automation bias, such as making individuals accountable for their interactions with automation or aware of the nonperfect reliability of systems and the related risk of automation bias, have had minimal success (Skitka, Mosier, & Burdick, 2000) unless human users have the direct experience of automation errors (Bahner et al., 2008). Thus, the main challenge of mitigating automation bias lays with a proper design of DSSs, specifically in finding a delicate balance between designing automation that facilitates decision making and creating an environment that fosters automation bias. Achieving this balance is essential to creating a human-automation match that improves performance considerably without introducing new risks, even though it still may not be as perfect as a match "made in heaven." A few current design trends and future possibilities are geared toward accomplishing this.

BALANCING THE TRADE-OFFS

Several design factors that might help to mitigate the risk of automation bias in interaction with DSS can be identified from the research presented previously. Beside the tendency for front-end DSSs to be a bit more resilient toward automation bias than more directive back-end DSSs, a more general take-away message from the research discussed here concerns the role of transparency, including understandability and predictability. Particularly, proper system feedback during use — especially cognitive feedback that provides information on how a DSS functions and the relationships among information and DSS output — is a form of transparency that can facilitate understanding of DSS functioning and accurate calibration for its use (Seong & Bisantz, 2008). Many early DSSs were opaque in their functioning and did not provide the decision maker with any map of their situation (mental) model, rationale for their recommendations, or justification for their recommendations. Early in the evolution of aviation automation, Billings (1996) already emphasized the need to train pilots how systems operate rather than simply how to operate systems: "If a pilot does not have an adequate internal model about how the computer works when it is functioning properly, it will be far more difficult for him or her to detect a subtle failure. We cannot always predict failure modes in these more complex digital systems, so we must provide pilots with adequate understanding of how and why aircraft automation functions as it does" (p. 96). Operators must have sufficient knowledge of what DSSs can do, what they "know," and how they function within the context of other systems as well as knowledge of their limitations, in order to utilize them efficiently and avoid errors of automation bias. System feedback that seems to be particularly important in this respect is the provision of confidence information along with advice and recommendations provided by a DSS. Examples include information about the quality of the database used by a DSS (e.g., the accuracy of positions identified by a GPS), likelihood information about the presence of a critical situation indicated by a front-end DSS (Wiczorek & Manzey, 2014), or feedback about the strength of evidence supporting the output of a back-end DSS (McGuirl & Sarter, 2006).

DSS AS A TEAM MEMBER

An overarching design principle is to look at DSSs as members of a human-automation team. The team-member metaphor as a possible guiding principle has been introduced to automation research and design by Christoffersen and Woods (2002). With respect to DSSs this suggests designing a system to function like a true team member, under the assumption that components of successful human-automation teams mirror those of effective human-human teams (Mosier, Fischer, Burian, & Kochan, 2017). This perspective is consistent with current tendencies to adapt human-human concepts — for example, trust — to human-automation interaction and to evaluate automated DSSs according to these human characteristics. The ideal DSS then would incorporate not only the most needed abilities of automated decision support — such as the ability to sense, synthesize, and integrate large amounts of data and information, to perform calculations quickly, and to detect malfunctions and failures — but also incorporate into the design characteristics that are desired in a human crew member and that are known to make a difference in human teamwork. On the one hand, this would include the sort of transparency and understandability mentioned previously as ideal human crewmembers are observable when acting, are predictable with respect to their next actions, and provide a justification of their intentions and actions. Beyond that, ideal DSSs would also resemble human crew members in other aspects, including responsiveness to direction, flexibility in providing advice and recommendations on a level (front-end versus back-end) needed by the human, and the ability to adapt to contextual factors like workload and time pressure that current DSSs may not be able to consider. Finally, an ideal DSS would be able to monitor the human as the human monitors the DSS.

Concrete examples of more advanced DSSs in this respect include DSSs that are adaptable or even adaptive to the needs of human users, ensuring that the best type of support (front end and/or back end; low level versus high level) is easily available and verifiable at the right time, especially when decisions need to be made under time pressure (Mosier et al., 2017; Burian, Mosier, Fischer, & Kochan, in press). Adaptable DSSs are adjusted by the operator, who maintains control over automation and is able to designate whether the human or automation will do all or part of tasks. Thus, individual operators can choose the level of support they prefer. Examples how this might be implemented and used by individual operators are provided by Sauer and Chavaillaz (2018). In contrast, with adaptive systems, the automation controls the division of tasks and may automatically change the sort of support provided in response to individual states of the human user (e.g., fatigue, stress) or task-context factors (e.g., time pressure, workload; Sheridan & Parasuraman, 2006). An example of the latter is the TCAS in modern airplanes, which dynamically changes its basic characteristic from front-end to back-end support with increasing time pressure. As long as the situation allows, TCAS provides front-end informational support with accompanying lower risks of automation bias and changes to back-end action directives only when time pressure becomes so high (i.e., imminent collision) that immediate compliance with the highly reliable conflict resolution advisory — despite associated risks of automation bias — is the safer option. Another conceptualization of adaptive automation is context-sensitive information automation, which would sense and take into account the situation or context, tailoring information support to characteristics such as specific task demands, environmental factors, system status, and human cognitive and performance variables (Mosier et al., 2017). This notion, however, is only at the conceptual stage, and the potential automation bias risks in this type of system are unknown.

A recent and even more far-reaching approach, enabled by current technological advances, indeed treats human and "automated agents" as interdependent team members who share a common mental model of situations and goals and can coordinate and collaborate activities, monitor each other, provide feedback to each other, and adapt dynamically to contextual demands. This approach, characterized as Coactive Design (e.g., Bradshaw, Dignum, Jonker, & Sierhuis, 2012; Johnson et al., 2011), adheres to the design principles of observability, predictability, and directability (Johnson et al., 2014), which is intended to facilitate teamwork behaviors such as monitoring progress and providing back-up behavior and make coordinated action possible. The implementation of Coactive Design is still in its infancy, although it has already been applied to the development of a humanoid robot to assist a human operator during disaster relief (Johnson et al., 2014). However, the concept of human-DSS interdependence and automation autonomy will most likely figure in the design of future automated decision support. One caveat for the notion of a more human-like automated team member — because "humanizing" automation (particularly anthropomorphism) enhances trust — designers will have to ensure that it does not inadvertently promote automation bias (e.g., de Visser et al., 2016; Pak, Fink, Price, Bass, & Sturre, 2012)

In summary, it seems that the match between humans and DSSs has many advantages but is not perfect. Actually, it seems that humans and DSSs represent more a match made on earth than a match made in heaven! As with all other earth-made matches, continuous attention is needed from both sides, the system design and the human user, to keep the match working smoothly and to achieve the best balance between maximizing the benefits of technological advances and, at the same time, minimizing the risks of automation bias. The present review suggests that this is a complex undertaking that needs consideration of system characteristics, characteristics of the human user, and characteristics of the situational context in which DSSs are used.

REFERENCES

Alberdi, E., Povyakalo, A., Strigini, L., & Ayton, P. (2004). Effects of incorrect computer-aided detection (CAD) output on human decision-making in mammography. *Academic radiology*, 11(8): 909–918.

Alberdi, E., Poviakalo, A. A., Strigini, L., Ayton, P., & Given-Wilson, R. (2008). CAD in mammography: Lesion-level versus case-level analysis of the effects of prompts on human decisions. *International Journal of Computer-Assisted Radiology and Surgery*, 3: 115–122.

Bahner, E., Hueper, A.-D., & Manzey, D. (2008). Misuse of automated decision aids: Complacency, automation bias and the impact of training experience. *International Journal of Human-Computer Studies*, 66: 688–699.

Bailey, N. R., & Scerbo, M. W. (2007). Automation-induced complacency for monitoring highly reliable systems: The role of task complexity, system experience, and operator trust. *Theoretical Issues in Ergonomics Science*, 8(4): 321–348.

Bainbridge, L. (1983). Ironies of automation. *Automatica*, 19(6): 775–779.

Balfe, N., Sharples, S., & Wilson, J. R. (2018). Understanding is key: An analysis of factors pertaining to trust in a real-world automation system. *Human Factors*, 60: 477–495.

Billings, C. E. (1996). *Human-Centered Aviation Automation: Principles and Guidelines*. NASA TM 110381. Moffett Field, CA: NASA Ames Research Center.

Biros, D. P., Daly, M., & Gunsch, G. (2004). The influence of task load and automation trust on deception detection. *Group Decision and Negotiation*, 13: 173–189.

Bradshaw, J. M., Dignum, V., Jonker, C., & Sierhuis, M. (2012). Human-agent-robot teamwork. *IEEE Intelligent Systems*, 27(2): 8–13.

Burian, B. K., Mosier, K. L., Fischer, U. M., & Kochan, J. A. (in press). New teams on the flight deck: Humans and context-sensitive information automation. In E. Salas & M. Patankar (Eds.) *Human Factors in Aviation* (3rd edition). New York: Academic Press.

Chen, J. Y., & Barnes, M. J. (2014). Human–agent teaming for multirobot control: A review of human factors issues. *IEEE Transactions on Human-Machine Systems*, 44(1): 13-29.

Chen, J. Y., Barnes, M. J., & Harper-Sciarini, M. (2011). Supervisory control of multiple robots: Human-performance issues and user-interface design. *Systems, Man, and Cybernetics, Part C: Applications and Reviews, IEEE Transactions on*, 41(4): 435–454.

Christoffersen, K., & Woods, D. D. (2002). How to make automated systems team players. In E. Salas (Ed.), *Advances in human performance and cognitive engineering research*. Vol. 2. Automation (pp. 1–12). Emerald Group Publishing Limited, pp. 1–12. https://www.emeraldinsight.com/doi/abs/10.1016/S1479-3601(02)02003-9 accessed 1 June, 2019.

Crocoll, W. M., & Coury, B. G. (1990). Status or recommendation: Selecting the type of information for decision aiding. In *Proceedings of the Human Factors Society 34th Annual Meeting* (pp. 1525–1528). Santa Monica, CA: Human Factors and Ergonomics Society.

Cummings, M. L. (2004). Automation bias in intelligent time critical decision support systems. Paper presented to the AIAA 1st Intelligent Systems Technical Conference, September 2004, p. 6313. [Available from: web.mit.edu/aeroastro/labs/halab/papers/CummingsAIAAbias.pdf; accessed 28 Feb 2010.]

de Boer, R. J., Heems, W., & Hurts, K. (2014). The duration of automation bias in a realistic setting. *International Journal of Aviation Psychology*, 24: 287–299.

de Visser E. J., Monfort, S. S., McKendrick, R., Smith, M. A. B., McKnight, P. E., Krueger, F., & Parasuraman, R. (2016). Almost human: Anthropomorphism increases trust resilience in cognitive agents. *Journal of Experimental Psychology: Applied*, 22(3): 331–349.

Dorneich, M. C., Dudley, R., Rogers, W., Letsu-Dake, E., Whitlow, S. D., Dillard, M., & Nelson, E. (2015). Evaluation of information quality and automation visibility in Information automation on the flight deck. In *Proceedings of the Human Factors and Ergonomics Society Annual Meeting* (Vol. 59, No. 1, pp. 284–288). Los Angeles, CA: SAGE Publications.

Dutch Safety Board. (2010). Crashed during approach, Boeing 737-800, near Amsterdam Schiphol airport, 25 February 2009. The Hague, The Netherlands: Dutch Safety Board.

Edland, A., & Svenson, O. (1993). Judgment and decision making under time pressure. In O. Svenson & A. J. Maule (Eds.), *Time Pressure and Stress in Human Judgment and Decision Making* (pp. 27–40). Springer, Boston, MA.

Endsley, M. R. (1995). Toward a Theory of Situation Awareness in Dynamic Systems. *Human Factors Journal*, 37: 32–64.

Endsley, M. R. (2000). Theoretical underpinnings of situation awareness: A critical review. In M.R. Endsley & D. Garland (eds.), *Situation Awareness Analysis and Measurement*. Mahwah, NJ: Lawrence Erlbaum Associates.

Endsley, M. R., & Kaber, D. B. (1999). Level of automation effects on telerobot performance and human operator situation awareness and subjective workload. In M.W. Scerbo & M. Mouloua (eds.), *Automation Technology and Human Performance: Current Research and Trends* (pp. 165–170). Mahwah, N.J.; Lawrence Erlbaum.

Friedman, C. P., Elstein, A. S., Wolf, F. M., Murphy, G. C., Franz, T. M., Heckerling, P. S., ... & Abraham, V. (1999). Enhancement of clinicians' diagnostic reasoning by computer-based consultation: A multisite study of 2 systems. *Jama*, 282(19): 1851–1856.

Garg, A. X., Adhikari, N. K., McDonald, H., Rosas-Arellano, M. P., Devereaux, P. J., Beyene, J., ... &Haynes, R. B. (2005). Effects of computerized clinical decision support systems on practitioner performance and patient outcomes: A systematic review. *Jama*, 293(10): 1223–1238.

Gigerenzer, G., & Gaissmaier, W. (2011). Heuristic decision making. *Annual Review of Psychology*, 62: 451–482.

Gigerenzer, G., & Todd, P. M. (1999). *Simple Heuristics that Make Us Smart*. New York: Oxford University Press.

Goddard, K. L., Roudsari, A., & Wyatt, J. C. (2012). Automation bias: A systematic review of frequency, effect mediators, and mitigators. *Journal of the American Medical Informatics Association*, 19: 121–127.

Goddard, K., Roudsari, A., & Wyatt, J. C. (2014). Automation bias: Empirical results assessing influencing factors. *International Journal of Medical Informatics*, 83: 368–375.

Hagafors, R., & Brehmer, B. (1983). Does having to justify one's decisions change the nature of the decision process? *Organizational Behavior and Human Performance*, 31: 223–232.

Hoff, K. A., & Bashir, M. (2015). Trust in automation integrating empirical evidence on factors that influence trust. *Human Factors*, 57(3), 407–434.

Johnson, M., Bradshaw, J. M., Feltovich, P. J., Jonker, C. M., van Riemsdijk, M. B., & Sierhuis, M. (2011). The fundamental principle of coactive design: Interdependence must shape autonomy. In M. De Vos, N. Fornara, J. Pitt, & G. Vouros (Eds.), *Coordination, Organizations, Institutions, and Norms in Agent Systems VI* (Vol. 6541, pp. 172–191). Berlin, Heidelberg: Springer.

Johnson, M., Bradshaw, J. M., Feltovich, P. J., Jonker, C. M., Van Riemsdijk, M. B., & Sierhuis, M. (2014). Coactive design: Designing support for interdependence in joint activity. *Journal of Human-Robot Interaction*, 3(1), 43–69.

Kahneman, D. (2011). *Thinking, Fast and Slow*. New York: Farrar, Straus and Giroux.

Kahneman, D., Slovic, P., & Tversky, A. (Eds.). (1982). *Judgment under Uncertainty: Heuristics and Biases*. Cambridge, UK: Cambridge University Press.

Layton, C., Smith, P. J., & McCoy, C. E. (1994). Design of a cooperative problem-solving system for en-route flight planning: An empirical evaluation. *Human Factors*, 36(1): 94–119.

Lee, J. D., & See, K. A. (2004). Trust in automation: Designing for appropriate reliance. *Human Factors*, 46(1): 50–80.

Lüdtke, A., & Möbus, C. (2005). A case study for using a cognitive model of learned carelessness in cognitive engineering. In *Proceedings of the 14th International Conference of Human-Computer Interaction*. Mahwah, NJ: Retrieved from https://www.uni-oldenburg.de/fileadmin/user_upload/informatik/ag/lks/download/Publikationen/2005/Luedtke_Moebus_crv.pdf Accessed 26 June 2018.

Lyell, D., & Coiera, E. (2017). Automation bias and verification complexity: A systematic review. *Journal of the American Medical Informatics Association*, 24(2): 423–431.

Lyell, D., Magrabi, F., & Coiera, E. (2018). The effect of cognitive load and task complexity on automation bias in electronic prescribing. *Human Factors*, 60(7): 1008–1021.

Lyell, D., Magrabi, F., Raban, M. Z., Pont, L. G., Baysari, M. T., Day, R. O., & Coiera, E. (2017). Automation bias in electronic prescribing. *BMC Medical Informatics and Decision Making*, 17, 28. https://doi.org/10.1186/s12911-017-0425-5

Lyons, J. B., Koltai, K. S., Ho, N. T., Johnson, W. B., Smith, D. E., & Shively, R. J. (2016). Engineering trust in complex automated systems. *Ergonomics in Design*, 24(1): 13–17.

Manzey, D., Bleil, M., Bahner-Heyne, J. E., Klostermann, A., Knnasch, L., Rmichånbach, J., & Vöttger, S. (2008). Auto-CAMS 2.0 manual. Retrieved from https://www.aio.tu-berlin.de/fileadmin/a3532/Berichte/Manual_AutoCAMS_2.0_190908.pdf Accessed 01 June, 2019.

Manzey, D., Gérard, N., & Wiczorek, R. (2014). Decision-making and response strategies in interaction with alarms: The impact of alarm reliability, availability of alarm validity information and workload. *Ergonomics*, 57(12): 1833–1855.

Manzey, D., Reichenbach, J., & Onnasch, L. (2012). Human performance consequences of automated decision aids: The impact of degree of automation and system experience. *Journal of Cognitive Engineering and Decision Making*, 6(1): 57–87.

McBride, S. E., Rogers, W. A., & Fisk, A. D. (2013). Understanding human management of automation errors. *Theoretical Issues in Ergonomics Science*, 15: 545–577.

McGuirl, J. M., & Sarter, N. B. (2006). Supporting trust calibration and the effective use of decision aids by presenting dynamic system confidence information. *Human Factors*, 48(4): 656–665.

McKibbon, K. A., & Fridsma, D. B. (2006). Effectiveness of clinician-selected electronic information resources for answering primary care physicians' information needs. *Journal of the American Medical Informatics Association*, 13(6): 653–659.

Metzger, U., & Parasuraman, R. (2005). Automation in future air traffic management: Effects of decision aid reliability on controller performance and mental workload. *Human Factors*, 47(1): 35–49.

Meyer, J. (2004) Conceptual issues in the study of dynamic hazard warnings. *Human Factors*, 46: 196–204.

Moray, N. (2003). Monitoring, complacency, scepticism and eutactic behaviour. *International Journal of Industrial Ergonomics*, 31(3): 175–178.

Mosier, K. L., & Fischer, U. M. (2010). Judgment and decision making by individuals and teams: Issues, models, and applications. *Reviews of Human Factors and Ergonomics*, 6(1): 198–256.

Mosier, K., Fischer, U., Burian, B., & Kochan, J. (2017). *Autonomous, Context-Sensitive, Task Management Systems and Decision Support Tools I: Contextual Constraints and Information Sources*. NASA/TM-2017-219565. NASA Ames Research Center.

Mosier, K. L., Palmer, E. A., & Degani, A. (1992). Electronic Checklists: Implications for Decision Making. *Proceedings of the 36th Annual Meeting of the Human Factors Society*, Atlanta, GA, October 12-16, pp. 7–12.

Mosier, K. L., & Skitka, L. J. (1996). Human decision makers and automated decision aids: Made for each other? In R. Parasuraman & M. Mouloua (eds.), *Automation and Human Performance: Theory and Applications* (pp. 201-218). Hillsdale, NJ, US: Lawrence Erlbaum Associates, Inc.

Mosier, K. L., Skitka, L. J., Dunbar, M., & McDonnell, L. (2001). Aircrews and automation bias: The advantages of teamwork? *The International Journal of Aviation Psychology*, 11(1): 1–14.

Mosier, K. L., Skitka, L. J., Heers, S., & Burdick, I. (1998). Automation bias: Decision-making and performance on high-tech cockpits. *The International Journal of Aviation Psychology*, 8: 47–63.

Onnasch, L., Wickens, C. D., Li, H., & Manzey, D. (2014). Human performance consequences of stages and levels of automation: An integrated meta-analysis. *Human Factors*, 56(3): 476–488.

Pak, R., Fink, N., Price, M., Bass, B., & Sturre, L. (2012). Decision support aids with anthropomorphic characteristics influence trust and performance in younger and older adults. *Ergonomics*, 55(9): 1059–1072.

Parasuraman, R., de Visser, E., Lin, M. K., & Greenwood, P. M. (2012). Dopamine beta hydroxylase genotype identifies individuals less susceptible to bias in computer-assisted decision making. *PloS One*, 7(6): e39675.

Parasuraman, R., & Manzey, D. H. (2010). Complacency and bias in human use of automation: An attentional integration. *Human Factors*, 52(3): 381–410.

Parasuraman, R., Molloy, R., & Singh, I. L. (1993). Performance consequences of automation-induced "complacency." *The International Journal of Aviation Psychology*, 3(1): 1–23.

Parasuraman, R., Sheridan, T. B., & Wickens, C. D. (2000). A model for types and levels of human interaction with automation. *IEEE Transactions on Systems, Man, and Cybernetics–Part A: Systems and Humans*, 30(3): 286–297.

Prinzel, L. J. III (2002). The relationship of self-efficacy and complacency in pilot-automation interaction. NAA/TM-2002-211925. Hampton, VA: NASA Langley Research Center.

Reichenbach. J., Onnasch, L., & Manzey, D. (2011). Human performance consequences of automated decision aids in states of sleep loss. *Human Factors*, 53: 717–728.

Rice, S., Hughes, J., McCarley, J. S., & Keller, D. (2008). Automation dependency and performance gains under time pressure. In *Proceedings of the Human Factors and Ergonomics Society Annual Meeting* (Vol. 52, No. 19, pp. 1326–1329). Los Angeles, CA: SAGE Publications.

Rice, S., & Keller, D. (2009). Automation reliance under time pressure. *Cognitive Technology*, 14: 36–44.

Rieskamp, J., & Hoffrage, U. (2008). Inferences under time pressure: How opportunity costs affect strategy selection. *Acta Psychologica*, 127(2): 258–276.

Rovira, E., McGarry, K., & Parasuraman, R. (2007). Effects of imperfect automation on decision making in a simulated command and control task. *Human Factors*, 49(1): 76–87.

Rovira, E., & Parasuraman, R. (2010). Transitioning to future air traffic management: Effects of imperfect automation on controller attention and performance. *Human Factors*, 52(3): 411–425.

Sarter, N. B., & Schroeder, B. (2001). Supporting decision making and action selection under time pressure and uncertainty: The case of in-flight icing. *Human Factors*, 43: 573–583.

Sauer, J., & Chavaillaz, A. (2018). How operators make use of wide-choice adaptable automation: Observations from a series of experimental studies. *Theoretical Issues in Ergonomics Science*, 19(2): 135–155.

Sauer, J., Chavaillaz, A., & Wastell, D. (2016). Experience of automation failures in training: Effects on trust, automation bias, complacency and performance. *Ergonomics*, 59(6): 767–780.

Seong, Y. & Bisantz, A. M. (2008). The impact of cognitive feedback on judgment performance and trust with decision aids. *International Journal of Industrial Ergonomics*, 38(7–8): 608–625.

Sheridan, T. B. (1989). Trustworthiness of command and control systems. In Analysis, Design and Evaluation of Man–Machine Systems 1988: Selected Papers from the Third IFAC/IFIP/IEA/IFORS Conference, Oulu, Finland, 14–16 June 1988. (pp. 427–431).

Sheridan, T.B. & Parasuraman, R. (2005). Human-automation interaction. *Reviews of Human Factors and Ergonomics*, 1: 89–129.

Simons, D. J., & Chabris, C. F. (1999). Gorillas in our midst: Sustained inattentional blindness for dynamic events. *Perception*, 28(9): 1059–1074.

Singh, I. L., Molloy, R., & Parasuraman, R. (1993). Automation-induced "complacency": Development of the complacency-potential rating scale. *The International Journal of Aviation Psychology*, 3(2): 111–122.

Skitka, L. J., Mosier, K. L., & Burdick, M. (1999). Does automation bias decision-making? *International Journal of Human-Computer Studies*, 51(5): 991–1006.

Skitka, L. J., Mosier, K. L., & Burdick, M. (2000). Accountability and automation bias. *International Journal of Human-Computer Studies*, 52: 701–717.

Skitka, L. J., Mosier, K. L., Burdick, M., & Rosenblatt, B. (2000). Automation bias and errors: Are crews better than individuals? *The International Journal of Aviation Psychology*, 10(1): 85–97.

Tsai, T. L., Fridsma, D. B., & Gatti, G. (2003). Computer decision support as a source of interpretation error: The case of electrocardiograms. *Journal of the American Medical Informatics Association*, 10(5), 478–483.

Tunstall, C., Rice, S., Mehta, R., Dunbar, V., & Oyman, K. (2014). Time pressure has limited benefits for human-automation performance. In *Proceedings of the Human Factors and Ergonomics Society Annual Meeting* (Vol. 58, No. 1, pp. 1043–1046). Los Angeles, CA: SAGE Publications.

Westbrook, J. I., Coiera, E. W., & Gosling, A. S. (2005). Do online information retrieval systems help experienced clinicians answer clinical questions? *Journal of the American Medical Informatics Association*, 12(3), 315–321.

Wickens, C. D., Clegg, B. A., Vieane, A. Z., & Sebok, A. L. (2015). Complacency and automation bias in the use of imperfect automation. *Human Factors*, 57(5), 728–739.

Wickens, C. D., Gutzwiller, R. S., & Santamaria, A. (2015). Discrete task switching in overload: A meta-analyses and a model. *International Journal of Human-Computer Studies*, 79, 79–84.

Wiczorek, R., & Manzey, D. (2014). Supporting attention allocation in multitask environments: Effects of likelihood alarm systems on trust, behavior and performance. *Human Factors*, 56, 1209–1221

Wiener, E. L. (1985). Beyond the sterile cockpit. *Human Factors*, 27, 75–90.

3 The Quest for Alternatives to "Levels of Automation" and "Task Allocation"

Robert R. Hoffman & Matthew Johnson

> There are encouraging signs within academe and government that a new respect for subjectivity is emerging… our subjective sense of what is right, beautiful, and consistent with a just and sustainable society, and what contributes most to human fulfillment, ought to dictate our use of [computers] with their enormous potential (Sheridan, 1980, pp. 71–72).

INTRODUCTION

This chapter is a historical review of a variety of conceptual approaches that system developers have adduced to guide the design of human-machine systems. The goal of this chapter is to convince cognitive systems engineers, human factors engineers, and systems developers more broadly that macrocognitive work systems must not be designed around methods of task allocation and schemes based on levels of automation. An alternative approach is proposed that regards the human and the machine as operating in a number of different interdependence relations.

BACKGROUND: THE CONCEPT OF "ALLOCATION"

The classic "Fitts' List" appeared in a 1951 report to the National Research Council. The purpose of the list was to encapsulate "a long-range integrated plan for human engineering research to parallel and support long-range planning for equipment and systems design" (Fitts et al., 1951, p. iii). The list had been developed during and just after World War II by human factors psychologist Paul Fitts and others who were designing cockpits, radar devices, navigation displays, and air traffic control displays for the U.S. Army Air Force (Fitts et al., 1951). Fitts' List specified some of the things that humans can do and that machines cannot do at all (perceive, reason, judge, adapt), but the focus was on the ways in which machines surpass humans. Machines (i.e., computers and control systems) can routinize tasks, perform tasks faster, and conduct tasks in parallel. They can calculate better. They can mitigate human error. They do not suffer from emotionality or memory limitations.

Fitts et al. advocated for an approach in which "machines are made for men, not men forcibly adapted to machines" (p. iv). The intent was to avoid the creation of "mechanical monstrosities which tax the capabilities of human operators and hinder the integration of man and machine into a system designed for most effective accomplishment of designated tasks" (p. iv). While this is manifestly a worthwhile goal, and was premised on a work systems viewpoint, the fine print in the Fitts et al. report reveals a stance of man versus machine. The path to effectiveness was for the machine to compensate for human limitations. The making of machines hinges on an awareness of the human's "physiological and sensory handicaps" and "human limitations in the integration of complex responses" (p. iv). Indeed, it was suggested "... that great caution be exercised in assuming that men can successfully monitor complex automatic machines and take over if machines break down" (p. x).

While rostering the capacities that only humans possess (perception, reasoning, etc.), the design focus entailed in the Fitts et al. approach is on leveraging the things machines do well in order to compensate for the human's limitations. In other words, what tasks can be taken from the human and handed to or allocated to the machine. The premise underlying the "allocation" concept is that tasks that are defined in terms of the work system's primary goals can be cleanly divided into subtasks (or subgoals) and then those can be portioned out to either the human or machine (not both), and that such portioning is sufficient as a method for the design of work systems.

It is certainly the case that there are situations in which an agent exercises authorities, and possibly obligations as well, to command another agent to conduct certain activities, sometimes highly specified or proceduralized activities, to achieve some particular goal. But if subtasks can be allocated to either the human or the machine, one can imagine that the more subtasks that are allocated to the machine, the more autonomous the machine will be. This led to the notion of levels of automation. Thus, if more tasks are allocated to the machine, the work falls at a "higher level" of automation, meaning there is more of it, or what there is of it plays more of a role in the work. This might seem to be what levels of automation is all about (Sheridan, 1980), but the original idea was subtler than this since it depended on a concept of classes of types of subtasks. We therefore expand the discussion of the levels concept.

LEVELS OF AUTOMATION

The initial presentation of the concept of levels of automation was in a 1978 Massachusetts Institute of Technology technical report by Thomas Sheridan and William Verplank (see also Verplank, 1978). This report was about the remote control of underwater vehicles. The report was seminal in the field of teleoperation and was more about the design of submersible robots and control models for teleoperations than it was about the concept of autonomous machines. The report is laden with William Verplank's beautiful hand-drawn diagrams and the authors' insightful discussions spanning cost-risk-benefit analysis motivating the creation and employment of teleoperated effectors, a survey of tasks appropriate for teleoperation, and technical constraints on vehicles, sensors, effectors, and

TABLE 3.1
A Summary of the Sheridan-Verplank Levels of Autonomy

LOW
1	The computer offers no assistance; the human must take all decisions and actions.
2	The computer offers a complete set of decision/action alternatives.
3	The computer narrows the selection down to a few.
4	The computer suggests one alternative.
5	The computer executes if the human approves.
6	The computer allows the human only a limited time to veto before automatic execution.
7	The computer executes autonomously, then necessarily informs the human.
8	The computer informs the human only if asked.
9	The computer informs the human only if the computer decides to.
10	The computer decides everything and acts autonomously, ignoring the human.

HIGH

communication systems. The report concludes with recommendations for research involving humans in simulated teleoperations. A main focus of the report is theories of control and "taxonomic models of man-machine interactions" (Sheridan & Verplank, 1978, pp. 1–11). It is in the discussion of this latter topic that the notion of levels is introduced.

The levels appeared in a table that bore the title, "Levels of Automation in Man-Computer Decision Making for a Single Elemental Decisive Step" (Sheridan & Verplank, 1978, Table 8.2, pages 8–17 through 8–19). Each level involved one or more functions or behaviors that were referred to as operators in the original work. Either the human or the machine could REQUEST something, FETCH something, SELECT an optional course of action, APPROVE a course of action, START an action, or a TELL the other what had been done. Figure 3.1 is a reproduction of the original table.

Table 3.1 presents a summary description of what seems to have been the intent of the Sheridan-Verplank levels. The levels were framed around the following question: Which agent (human or machine) has to make a decision and which agent has to act upon that decision?

The levels are not defined in this table by a dimension, but that is how the levels are described in the text of the report (p. 8–15). It is difficult to see how a concept of levels applies since the scheme is clearly multidimensional. That is, the row entries are categories and the numbering is not ordinal. Its mixture of activities and responsibilities as specified in Table 3.2.

The common understanding of the levels assumes that significantly different kinds of work can be handled equivalently (e.g., taskwork and teamwork; reasoning, communication, decisions, and actions). For option generation, option selection, and action implementation (columns 3, 4, and 7 in Table 3.2), either the human can do it or the machine can do it. This reinforces the automation myth that "automation activities simply can be substituted for human activities without otherwise affecting the operation of the system" (Christoffersen & Woods, 2002, p. 3).

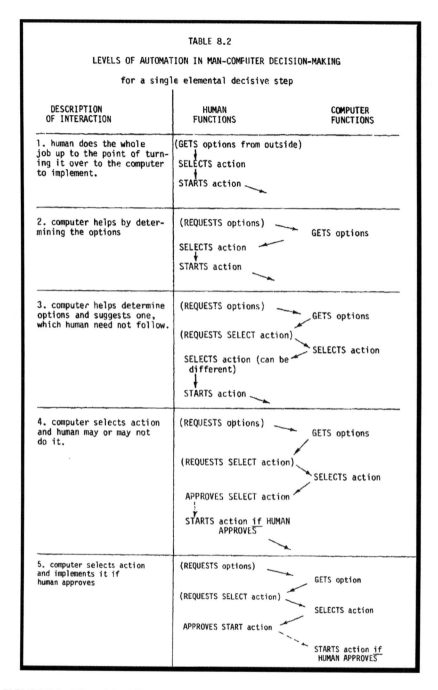

FIGURE 3.1 The original Sheridan-Verplank Table 8-2 (reproduced with permission).

Alternatives to "Levels of Automation" and "Task Allocation"

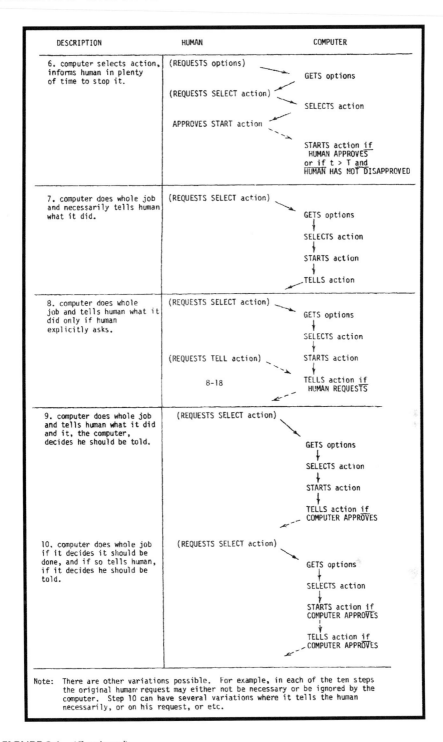

FIGURE 3.1 (*Continued*)

TABLE 3.2
The Sheridan-Verplank Levels of Automation are a Mixture of Responsibilities and Activities

	Who Requests the Options?	Who Generates the Options?	Who Selects the Option?	Who Requests an Action?	Who Approves the Action?	Who Starts the Action?	Who Stops the Action?	Who Communicates What to Whom?
1.		Human	Human		Human	Human		
2.	Human	Computer	Human		Human	Human		
3.	Human	Human	Computer or Human		Human	Human		
4.	Human	Computer	Computer		Human	Human		
5.	Human	Computer	Computer		Human	Computer		
6.	Human	Computer	Computer		Human	Computer starts the action if the human does not disapprove.	Human	
7.		Computer	Computer	Human	(Human approval not needed)	Computer		Computer tells human what it did.
8.		Computer	Computer	Human	(Human approval not needed)	Computer		Computer tells human what it did only if the human asks.
9.		Computer	Computer	Human	(Human approval not needed)	Computer		Computer decides whether to tell the human.
10.		Computer	Computer	Human	Computer	Computer		Computer decides whether the job should be done and whether to tell the human.

Alternatives to "Levels of Automation" and "Task Allocation"

The levels notion piggybacks on the Fitts' List scheme that the computer compensates for the human. Only the computer does the communicating (column 9 in Table 3.2). For option generation, option selection, and action implementation (columns 3, 4, and 7 in Table 3.2), the machine does it all at Levels 7, 8, 9, and 10. At Level 1, the machine is completely controlled, as in the teleoperation of submersible robots. Between the "extremes" are such cases as the detection of an off-nominal state requiring the human to make sense of it and institute corrective action. Yet on close inspection, the scheme is decidedly one-sided. For all the Levels 2 through 9, the computer controls the flow of events. The computer selects actions, even if the human has to approve them.

Contrary to what one might suppose, Level 1 is not "the human does everything." The human implements the decision, but that decision has the machine doing something. The human does the whole job up to the point of turning it over to the computer to implement.

In a paper published the following year (1980), Sheridan included a significantly truncated description of the levels, anchored in Level 1 (human considers alternatives, makes and implements decisions) and Level 10 (computer makes and implements decisions and informs the human only if it feels this is warranted). This is the unidimensional interpretation that formed the levels scheme as it has persisted in the human factors literature to this day (e.g., Parasuraman, Sheridan, & Wickens, 2000; Sheridan, 1997; 2011). Unfortunately, the levels scheme has been further reinterpreted as an algorithm for any enterprise that seeks to solve human-machine work system complexity (and cost) issues by simply injecting more and more automation. This is not how the original levels notion was intended to be used (Sheridan, personal communication).

In reviewing the notion of autonomous systems, the U.S. Defense Science Board (Murphy & Shields, 2012) asserted that:

> [The] taxonomy is organized into levels, and is often incorrectly interpreted as implying that autonomy is simply a delegation of a complete task to a computer, that a vehicle operates at a single level of autonomy and that these levels are discrete and represent scaffolds of increasing difficulty. Though attractive, the conceptualization of levels of autonomy as a scientific grounding for a developmental roadmap has been unproductive… The levels served as a tool to capture what was occurring in a system to make it autonomous; these linguistic descriptions are not suitable to describe specific milestones of an autonomous system… Research shows that a mission consists of dynamically changing functions, many of which can be executing concurrently as well as sequentially. Each of these functions can have a different allocation scheme to the human or computer at a given time (pp. 23–24).

As we will argue in the following section, this reference to "dynamically changing functions" is crucial. But to continue this portrayal of the historical development of the current concept of levels, subsequent attempts to refine or extend the levels approach involved adopting its stance (compensation and substitution notions) and explicating the different activities that were scattered among the Sheridan-Verplank levels.

ADAPTIVE ALLOCATION: LEVELS + STAGES

Parasuraman et al. (2000, see also Li, Burns, & Hu, 2016) proposed a taxonomy that was premised on the unidimensional interpretation of the levels, but seeing a limitation to that, they included "stages" of human-automation activity. The stages were (1) information acquisition, (2) analysis of information, (3) decision or choice of action, and (4) execution of action. The Parasuraman et al. stages — which are obviously an information-theoretic, input-output sort of model — were motivated by the idea that the level of automation (LOA) might differ for each of the four stages, with each task involving all four stages in sequence.

The stages could be very easily understood as a rebranding of the Sheridan-Verplank "operators" listed previously, which were also engaged in a sequence (e.g., request options, generate options, select decision, approve action, engage action). Sheridan understood this, of course (see Sheridan, 1998). "Most of the literature on adaptive automation has focused on the criteria for allocating control fully to the human or fully to the automation. However, there seems to be no reason why adaptive allocation cannot be made at one of the intermediate levels" (Sheridan, 2011; p. 663). (Here, by "intermediate levels," Sheridan likely meant stages.) Indeed, in the original Sheridan-Verplank table, each level was accompanied by a little diagram showing a sequence of operators. Unfortunately, subsequent presentations of the levels, especially that by Sheridan (1980), left out those little diagrams (see Figure 3.1).

The operators in the Sheridan-Verplank scheme and the stages in the "levels+stages" scheme are categories of mental operations that can be dated to the earliest days of task analysis in a prebehaviorist era when activities such as perceiving (information acquisition), deciding, and action implementation were all considered in task analysis (see Hoffman & Militello, 2008). However, in the levels+stages formulation, some confusion emerges regarding levels, stages, and indeed the very concept of a task. The four stages together essentially define what a task is, but each of the stages can just as well be thought of as a subtask (e.g., acquiring information).

The levels+stages concept suffers from the same fundamental flaw as other stage models of cognitive processes. Specifically, the stages assume that the mental events (and the "final" physical activity of implementing a decision) are sequential. Take for example one of the three possible pairs in the stages scheme: the leap between deciding and implementing. After the human has made a decision and an action has been implemented, that does not mean that the process of deciding has ended. Decisions are usually contingent, especially in complex work systems, meaning that consequences are investigated for possible surprises that have to be mitigated, sometimes well after implementation (Hoffman & Yates, 2005). Similarly, for the sequential stages of information acquisition and information analysis, acquisition of information does not halt when information analysis commences, and conclusions from the analysis are contingent on additional information. In other words, the mental events that are labeled as stages are only partly stage-like. In fact, they are only stage-like when we look retrospectively at particular instances of cognitive work or narratives of case studies. Fundamentally, when cognition is "in the moment," the mental activities are parallel and interactive (Hoffman & McNeese, 2009; Klein et al., 2003).

Another limitation of the levels+stages scheme is the failure to distinguish self-sufficiency from self-directedness (see Johnson et al., 2011; Johnson et al., 2014). At any of the levels, a given machine capability operating in a specific situation may simultaneously be low on self-sufficiency while being high on self-directedness. At Level 6, the "computer selects an action and informs the human in plenty of time to stop it." This is about not doing something.

The motivation for the levels+stages scheme was the idea that allocation could be adaptive, that is, which activities were engaged by human and machine could be contingent. But some agent would have to make the decisions about when to exercise a contingency. The process of allocation would require a separate authority to monitor the conditional dependencies and alter the allocations, and thus another loop to the control theoretic model was required, in which a human or a computer would make changes to "adaptive parameters." Sheridan (2011) listed the following adaptation contingencies, or parameters: changes to a control law, changes to the interface, changes to the goal state, new inputs, the task dynamics, and the state vector (or task progress). This list covers quite a lot. Arguably, changes to the control law, the interface, and the goal could be thought of as changes to the task. Changes to the vector and the inputs are, arguably, changes to the task context.

From a control theoretic point of view, the addition of an "outer loop" of supervisory control might be regarded as a solution to the directedness problem. Sheridan's reference was one in which the human-machine system is performative, that is, the human is operating on the "world" that is being controlled. The prototypes are industrial process control and airplane piloting. And in that context, adaptive allocation can be discussed in terms of parameter selection based on loads in a plant, changes to altitude, etc. This all speaks to how the controller would operate. But much of the hard work remains undone: Specifying how the level of automation would be determined at each of the stages, and then apportioning the subtasks to human and machine, and doing this in a way that allows for parameter context dependence.

A number of additional variants of the levels+stages and adaptive allocation schemes have been proposed that are intended to make up for perceived limitations of those approaches.

DEGREE OF AUTOMATION

Onnasch, Wickens, and Manzey (2014) proposed an extension of the scaling of levels by introducing a notion of degree of automation (DOA). They retained the assumption that the levels are a single dimension: "[T]he amount of automation autonomy and responsibility (highest at the highest level) and he amount of human physical and cognitive activity (highest at the lowest level)" (p. 477). The authors next assumed that one, two, three, or all four of the Parasuraman et al. stages within each level could be automated. Based on these assumptions, the levels and stages can be combined into a single ordinal scale. Thus, for example, the allocation of three stages to a machine at Level 9 would have a higher DOA number than the allocation of two stages to a machine at Level 8. The allocation of all four stages at Level 10 would have the highest DOA, and the allocation of none of the stages at Level 1 would have the lowest DOA.

While this scheme embeds many conceptual and measurement issues, of which the authors were aware, it is not entirely unreasonable when utilized at a coarse grain. Onnasch et al. (2014) conducted a meta-analysis of 15 studies involving allocations that could have DOA numbers assigned to them. What they found confirms the irony of automation: More automation yields better human-machine performance when situations are routine but more problematic performance when the work system is stretched and there is an increase in mental workload and also a risk associated with the sudden return to manual control (see for instance, Wright, Samuel, Borowsky, Zilberstein, & Fisher, 2016). Thus, as DOA increases, performance increases until there is some failure, at which point performance declines. This interaction was clearest at a coarse grain, that is, when the reviewed studies were combined in a comparison in which allocation of Stages 1 and 2 (information acquisition and analysis) were compared with allocation of Stages 3 and 4 (selection and execution of actions). When a work system crosses this boundary, one sees the greatest effect of DOA. (Somewhat in contrast with this result, there was no clear effect of DOA on situational awareness, that is, loss of situational awareness being associated with higher DOA.)

It is important here to add the following footnote: Of the 18 studies that met the criteria for the meta-analysis, 14 of them involved novices (college students) in simplistic, unfamiliar tasks conducted in the academic laboratory. These features are characteristic of many human factors experiments on function allocation. This aspect of the methodology used in the area should always be kept in mind if the primary reference is intended to be "real-world" cognitive work systems. The primary reference and focus should be on experienced professionals, working on their familiar, realistic tasks, not on college students working tasks that are simple (relative to the "real world") and novel as far as the participants' experience is concerned.

DYNAMIC OR GRADUAL ALLOCATION

In the adaptive allocation schemes, levels are thought of as spanning complete to partial allocation. A number of researchers have drawn a distinction between static and dynamic allocation (e.g., de Brunélis & Le Blaye, 2008; Hancock & Scallen, 1996). In such schemes, allocations are changed as a function of context or other parameters — a contingency scheme not unlike levels+stages but more dynamically context- or situation-sensitive.

A form of dynamic (or adaptive) allocation proposed by Yoo (2012) is called "gradual allocation." Setting aside complete allocation (Levels 1 and 10), allocation at the intermediate levels could be gradual in that the level would be increased or decreased "until demands are sufficiently satisfied" (p. 2134), that is, human performance is optimal. This would be in recognition of the fact that numerous situational variables could influence the operator's state. What is troubling about this approach is that it takes the substitution view to an extreme — not that the machine can substitute for the human but that the human can be regarded as a type of machine, controlled by limits and thresholds, a machine for which all that matters is raw performance and performance optimization. It comes as little surprise that in this approach, emphasis is placed on measuring workload in terms of physiological responses and using physiological data to control the level of automation. Short-term improvements in "system performance" may be possible in the short-term laboratory experiments on gradual allocation, but real-world cognitive work would suffer if the work system were designed in this philosophy, and it would suffer in numerous ways.

ASYMMETRICAL, ADAPTABLE, AND ADAPTIVE ALLOCATION

Calhoun, Ruff, Spriggs, and Murray (2016) proposed that the level of automation could be increased when the worker is overloaded and decreased when the participant is underloaded but argued further that the adaptation cannot be asymmetrical since high performance at high levels of automation would make reductions in the level of automation less likely than increases. The researchers found that an asymmetrical allocation scheme improved performance speed and accuracy in a task involving image analysis (change detection) for vehicle routing. They argued that an asymmetrical algorithm would keep people at a lower level of automation and thus make automation-induced problems less likely.

Miller (2012) and Kidwell, Calhoun, Ruff, and Parasuraman (2012) distinguished adaptable automation and adaptive automation. Miller conceived of adaptability in terms of delegation of responsibilities by a supervisor (Sheridan's "outer control loop"?) that relies on something like a playbook. The supervisor would monitor the execution of the play and make adjustments to the plan as it is implemented. In the similar scheme of Kidwell et al., the LOA is manipulated by the machine based on human performance (apparently using the word "automation" to mean allocation). Somewhat counterintuitive is the claim that, "Adaptive automation has been theorized to alleviate some of the drawbacks attributable to static automation, including subjective feelings of automation surprise, [loss of] mode awareness, and [loss of] situational awareness" (Kidwell et al., 2012, p. 428). It was argued that adaptive automation results in reduced human workload due to the decrease in the human's responsibility. It also was argued that adaptable automation results in increased workload due to increased human responsibility, but the direction of attention to system monitoring lessens the human's tendency to become complacent.

These ideas were presented as if they were empirical facts. Their experiment (also apparently using college students) involved a task of controlling multiple autonomous air vehicles using a simulation testbed. Subtasks that could be allocated were image analysis, change detection, and monitoring of the system's status. The researchers found that adaptive automation led to increased perceived workload but better performance. Adaptable automation resulted in higher ratings of confidence, at the image interpretation (change detection) task, but the "statistically significant" difference was between 92% and 100% correct identifications — hardly of any practical significance. Two things their findings do make clear is that people prefer to be in control; and being in control can reduce or eliminate complacency.

The distinction between adaptive and adaptable automation is interesting, but only marginally because it does not go far enough: In "real-world" jobs, humans are always sensemaking with their technology and flexecuting with their technology (work-arounds, kluges, etc.) (Hoffman, Best, & Klein, 2015; Koopman & Hoffman, 2003).

ALLOCATION + RASMUSSEN + EXPERTISE

Cummings (2014) proposed that roles can be allocated, essentially equating roles with functions. She recapitulated Fitts' List and then adopted the unidimensional levels concept: "Levels of automation can range from a fully manual system with no computer intervention to a fully automated system where the human is kept completely out of the loop" (p. 63). Adhering to the compensation notion, Cummings asked how humans can improve on the sensor and reasoning limitations of machines and how automation can help the human reduce uncertainty.

What Cummings presents that seems new is to add expertise onto Rasmussen's distinction between tasks that can be performed based on skills versus tasks that can be performed based on rules versus tasks that can be performed based on knowledge (Rasmussen, 1983). In the spirit of Fitts' List, Cummings sees expertise as a capability that humans have and that machines do not. It is not clear how expertise takes us beyond Rasmussen's concept of knowledge. On Cummings' model, the exercise of knowledge and expertise both involve collaboration in cognitive work, whereas tasks that depend on skills and rules are ones that can be conducted entirely or primarily by machines. The escape clause to this analysis is that at the level of rule-based tasks, uncertainty can enter into the picture and thus tasks that are rule-based cannot be entirely automated. At the knowledge-based and expertise-based levels of tasks, uncertainty becomes even more of a consideration: "Human judgment and intuition become critical" (Cummings p. 67). Cummings refers to collaborative systems, but seems focused on allocating functions. Even the examples provided, such as the superiority of airplanes landing themselves, implies that the goal is to obviate the need for the human.

While this allocation + Rasmussen + expertise scheme may be a "principled framework" that might be used to "train traditional engineers and computer scientists to consider the human early in the design process" (p. 64), there remains the crucial gap of going from the analytical scheme to a specific method for designing tasks. It seems odd that Cummings keys off of Dekker and Woods (2002), who made a clear case than any task substitution approach is insufficient. Cummings correctly states that "past methodologies for balancing the roles and functionalities between humans and computers in complex systems have been based on heuristics with little formalization" (p. 62) but then does little more than add expertise and uncertainty as "heuristics," while neither is defined in such a way as to make it expressible as an actionable heuristic. The formalization or methodology that is implied is not provided. All it says is: Here are the very general functions that are most likely suited to be automated. But we suspect that most engineers, traditional or otherwise, already know what machines can do and what they cannot do.

SAE INTERNATIONAL LEVELS

With the increasing commercial interest in self-driving cars, a new levels model specific to that domain was developed by SAE International (Shladover, 2016). This levels model provides some insightful improvements to the Sheridan-Verplank levels. Similar to Parasuraman et al. (2000), this model distinguishes different types of automation. However, like the Sheridan-Verplank levels, the SAE model divides

Alternatives to "Levels of Automation" and "Task Allocation"   55

work along functional boundaries, that is, in terms of responsibility for particular functions. Additionally, the SAE model speaks to off-nominal situations (i.e., who takes control when something goes wrong). Importantly, it is the first levels model to acknowledge that work assignment is not simply binary — either human or machine.

Rather than being composed as a list, the SAE model is a matrix. Top to bottom seems to be "less human control" to "more human control," and left to right seems to be "less machine control" to "more machine control." Inspection of what is said about the individual cells clarifies these dimensions somewhat. Individual cells refer to such actions as "which agent accelerates" and "which agent monitors the environment." The Sheridan-Verplank Level 10 would be the bottom-right cell: The machine retains full control of everything, in all conditions. The Sheridan-Verplank Level 1 would be the upper left cell: The human has no machine assistance, not even cruise control.

Overall, the SAE model is an advancement by shifting the focus to responsibility, by considering off-nominal cases, and by representing both the human and machine working together and not just separately. However, the levels themselves are still better described as alternatives than ordinal levels. For example, minimal machine control is when "something goes wrong." Clearly, this is an empirical fact about all human-machine work systems and is not a design choice or option. As another example, the top-to-bottom axis is categorical, referencing both specifics (which agent decelerates) and abstractions (which agent monitors the environment). And interpreted as a dimension, there seems to be a confusion: At Level 3, the car can drive everywhere but might need backup. But this is "lower" than Level 4, at which the car can only drive in limited circumstances but will not require backup in those circumstances.

The top to bottom of the SAE matrix focuses on the control of steering and acceleration. The left to right of the matrix focuses on monitoring the environment. But if the driver is steering, do they need to monitor anything? Level 0 in the top-to-bottom axis (how much driving is assisted) is "no assistance" and Level 5 is "some assistance." Clearly, the middle levels are where human-machine systems will actually operate. Everything except the Levels 0 and 5 are joint activities. We can and must add much to meaningfully distinguish those middle levels. The distinctions among the alternatives in the cells raise excellent questions and draw attention to design needs.

But does considering these options as ordinal or dimensional add any value to what should be an interesting discussion of alternatives? More importantly does the ordering provide predictive capabilities beyond a simple list of alternatives? More likely, it is the description of the functional requirements and responsibility distribution that serve to explain design challenges and performance outcomes.

MEASUREMENT CHALLENGES AND GAPS

The entrenched nature of the allocation model — the assumption that function allocation is a necessary perspective and seemingly the only perspective — is typified and reinforced by a recent investigation on the following question: Given that allocation is the design model, how can allocations be evaluated to determine their effectiveness? Metrics that are proposed include: clarity of role designations and responsibilities, appropriate workload, support for team collaboration, mutual predictability of human and machine agents, avoidance of interruptions, degree to which the automation is pushed to operate beyond its boundary conditions, and support for human adaptivity (see Pritchett, Kim, & Feigh, 2014). These are all worthwhile, if not necessary, things to measure in the evaluation of macrocognitive work systems (e.g., Hoffman & Woods, 2011; Klein, Woods, Bradshaw, Hoffman, & Feltovich, 2004). The contribution of Pritchett, Kim, and Feigh (2014) is a demonstration of how actual measures and measurements can be entrained in the context of flight-deck automation (e.g., task load is measured in terms of the number of actions and their durations).

This raises the question of what allocation scheme was utilized in the design of today's cockpit automation. Arguably, it was not really allocation at all in the sense that decisions were made about who does what given what each agent (human or machine) can or cannot do. Rather, the procurement thrust, generally, is to build machines to do all the things that machines can possibly do and let the human: (1) supervise the machine, (2) handle the things that machines cannot do, and (3) end up being responsible when things go badly. The consequence of this shows in the results of Pritchett, Kim, and Feigh's analysis (2014). On some measures, performance is consistently good whether handled by human or by machine (e.g., duration of vertical deviations). Other measures seem particularly well suited to the evaluation of how automation induces states in which the human has to monitor the automation in a "cognitive control mode."

Especially interesting is their analysis of the mismatches between authorities and responsibilities. These are cases in which the automation is authorized to perform an action but the human is responsible for the consequences, or vice versa. The analysis assumes that mismatches are uniformly a bad thing since they require the human to monitor the actions taken by the automation. We think this is counter to the notion that human-machine interdependence requires mutual observability; there are situations where one wants a mismatch of authority and responsibility. Interruptions are also regarded as bad (i.e., interruptions do not help), but if a machine tells you what to look out for, that is a good interruption. Teamwork is not a penalty, it also adds value that is not captured in the measures. Allocation and levels of automation cannot describe the interleaving between functions.

Overall, the reference to allocation is tangential except that the measures are made so as to enable a comparison of what the automation does with what the human does. Thus, the measurement focus is on counting taskwork (see Tables 1, 2, 3, and Figure 6 in Pritchett et al., 2014); teamwork functions are just more functions to allocate, but coordination and adaptability to accommodate human variety suggest that allocation is not the whole story and may not even be the right story. For example, measuring coherence of human and machine activity in terms of allocation seems at odds with teamwork and interdependence.

DEFENDING THE LEVELS + ALLOCATION SCHEMES

To summarize the presentation to this point, the main concept that connects all of the ideas is that of function allocation. It is based on the premise that work can be divided into independent chunks that can be cleanly distributed to man or machine. The legacy of human factors is that this is not the case. Yet, the concepts of substitution, compensation, levels, and allocation have been preserved in many varieties and most often go unquestioned. Moreover, there are rampant confusions of terminology, such as, for example, the confusion of task allocation and function allocation, the confusion of functions and roles, and the use of the word "automation" to mean "allocation." All of the recent attempts to "correct" the original Sheridan and Verplank levels scheme are flourishes that maintain the view that more automation means reduced mental workload because all the human has to do is supervise (see Breton & Bossé, 2002; Sellers, Fincannon, & Jentsch, 2012), and that the purpose of automation is to compensate for human limitations (see Yoo, 2012).

It has recently been argued that the levels and allocation schemes should persist (de Winter & Dodou, 2014). Specifically, it has been claimed that the Fitts' List approach meets all the criteria for a good scientific theory: plausibility, explanatory adequacy, interpretability, simplicity, descriptive adequacy, and generalizability.

We will address these, but first we need to point out something crucial: Fitts' List is not a theory. At most, it is a bag of hypotheses (assertions), and it would take formidable semantic and grammatical gymnastics to turn each of the entries in Fitts' List into something that actually reads like a testable hypothesis. The gap in the understanding of the rigors of scientific theorizing characterizes quite a lot of the work in both human factors and cognitive systems engineering (see Koltko-Rivera & Hancock, 2005). A theory must have a metatheory, a subject matter (ontology), a set of laws (empirical nomological generalizations), and a methodology (including postulates about testability). Fitts' List is only a partial ontology and is best thought of as just a stance. (Unfortunately, it is a stance that has grown into a paradigm.)

Fitts' List is indeed interpretable and it is indeed simple. But this is part of the problem. The List is a reductive understanding of something that is very complex. The criterion of simplicity is often invoked in naive philosophy of science and seems to have gotten into the genome of psychologists (e.g., Jacobs & Grainger, 1994). It can be dated to a misinterpretation of Occam's Razor, that was current in the days when cognitive psychology was first emerging and the issue on the table at that time was whether mental events play a causal role in behavior. *Essentia praeter necessitatem not sunt multiplicanda* says that a theory should contain all the necessary concepts, not that it should be made to be as simple as possible. Thus, the other edge of the razor is *Essentia praeter fidem not sunt subtrahenda*, or "Don't pretend to do without concepts you cannot really do without" (Hoffman, 1979). For theories of complex systems, including complex cognitive systems, the goal is elegance, not simplicity.

Returning to the list of features that de Winter and Dodou consider:

- Fitts' List is not plausible. Much of the discussion we have presented here attests to this. Fitts' List is woefully lacking in explanatory adequacy. All it does is list some high-level functions that are "best" performed by machines. It does not explain, for example, why automation surprises occur (see Sarter & Woods, 1995; Sarter, Woods, & Billings, 1997).
- Fitts' List is woefully lacking in descriptive adequacy. It does not say anything, for example, about interdependence relations, collaboration, and so forth.
- Fitts' List is generalizable because it is overly general. As we have suggested, and as Fuld (1993) argued, it lists high-level functions and does not reach to the level of particular tasks. Hence, it is ineffectual as guidance for design.

- Fitts' List's tradition gets the compensation backward. Take, for instance, the notion that humans cannot be trusted to be able to monitor machines and take over when things go badly. Is it not usually the case that human reasoning, adapatability, and flexibility (all three capacities being listed in the "plus" column in Fitts' List) are precisely those human capacities that are in play when things do go badly, and it is *always* the human who saves the day by compensating for the machine?

Apart from the active defense of Fitts' List approach, there is another and major reason for the persistence of the Sheridan-Verplank levels and the associated allocation schemes. This reason cannot be ignored. It resides in forces that lie outside of the scientific community proper.

FUNDING FORCES

According to reports from the Human-Systems Community of Interest Technology Interchange (HSI COI, 2015), in the United States, about half a billion dollars of Department of Defense (DoD) funding is targeted to human-systems technologies (see also Bornstein, 2015). Focus areas run the gamut of intelligent systems technologies, including training [e.g., intelligent tutoring, accelerated learning, adaptive training, virtual reality (VR)-based training], performance (e.g., automated readiness assessment, cognitive models), integration (e.g., human-robot integration, human-machine teaming), and interfaces (e.g., intuitive interaction, natural language interfaces). In fact, there is not much in the broad U.S. DoD portfolio that is not aimed at generating more automation.

Frustratingly, the U.S. DoD (and certainly other major organizations around the world) thinks with a split mind on the topic. One report from the U.S. Department of Defense Science Board questions the utility of the levels of automation approach:

> Instead of viewing autonomy as an intrinsic property of an unmanned vehicle in isolation, the design and operation of autonomous systems needs to be considered in terms of human-system collaboration… During the design of an autonomous system, a significant number of decisions are made to allocate specific cognitive functions to either the computer or the human operator. These decisions reflect system-level trade-offs between performance factors, such as computationally efficient, optimal solutions for expected scenarios versus susceptibility to failures or the need for increased manpower when variations in the scenarios or new situations occur. In many cases, these design decisions have been made implicitly without an examination of the consequences to the ultimate system users or to overall acquisition, maintenance, or manpower costs (Defense Science Board, 2012, pp. 1–3).

Yet, the same report adopts a concept of dynamic allocation:

> Multiple concurrent functions may be needed to evince a desired capability, and subsets of functions may require a human in the loop, while other functions can be delegated at the same time. Thus, at any stage of a mission, it is possible for a system to be in more than one discrete level simultaneously… The Task Force recommends that the DoD abandon the use of "levels of autonomy" and replace them with an autonomous systems reference framework that explicitly focuses design decisions on the explicit allocation of cognitive functions

and responsibilities between the human and computer to achieve specific capabilities [and that] recognizes that these allocations may vary by mission phase as well as echelon (p. 4).

Another Defense Science Board report (2016) buys into the view that more automation means savings (faster, better, cheaper): "Autonomy delivers significant military value… the DoD must accelerate its exploitation of autonomy" (p. 1).

We emphasize the following point: The main use of levels and allocation schemes seems to actually be as an independent variable in human factors laboratory experiments. Levels and allocations play little role in the actual design of human-machine work systems. The funding forces do the heavy lifting in design, by pushing for the automation of everything that can be automated. What remains is not allocation but delegation. The things that cannot be automated are handed over to the human. And what are often called "decision support" systems become, in effect, process control systems in which the machine controls the workflow and the human's work process.

We remain hopeful that funding programs will soon escape the levels and allocations schemes. But there has to be some positive paradigm to swap in.

INTERDEPENDENCE ANALYSIS

If all the levels and allocation schemes and variations have not delivered on their promise — because they have a restricted view of the human-machine work system — what are the alternatives? What if we were to take interdependence as fundamental, rather than separable, tasks and acknowledge that the "levels" are really just a sampling of possible interdependence relations? Might this approach help us forge a different means for designing human-machine systems?

The most important human-machine work is work in which the two are communicating, collaborating, and acting together (Roth, 2012; Scott, 2012; Smith, 2012). The execution of an allocated task (or subtask) by one agent in a work system that does not mean that other agents are no longer dependent on it while they work on their own allocated tasks. The fact that some activities can be "allocated" hides the persistent, necessary, and inescapable interdependencies that make human-machine coactivity possible. The injection of automation into the workplace means there is a change to the nature of the workload and change to the ways human and machine have to depend on one another.

Because machines rely upon a limited ontology, at any given point in time, their sensitivity to context is low, their capacity to recognize change is low, and their sensitivity to anomaly is low. Thus, the machines depend upon operators to keep their model of the world aligned to the context, keep their model of the world stable given the variability and change inherent in the world, and update and repair their ontologies (Hoffman et al., 2002). Human and machine must both be able to monitor progress toward goals, remain sensitive to surprise, be able to "flexecute" (replan) on the fly, and be able to adapt their roles and responsibilities (the "common ground") (Klein et al., 2004).

These forms of interdependence are invisible to levels and allocation schemes, which focus only on the separated taskwork elements and miss the interdependencies that enable the human and machine to work together to accomplish the entire activity. To illustrate this, let us look back at the original levels of automation and evaluate them in terms of how the scheme expresses (or does not express) interdependence

TABLE 3.3
Primary Acts Listed in the Levels of Automation and the Adaptive Allocation Schemes

Observe	Can be states in the world or in the controlled process.
Reveal	Can be of states in the world, of the controlled process, or in an agent's status.
Act	Can be to act in the world or act on a process.
Be Directed	Can be to engage any of the other primary acts.
Direct	Includes transfer of control and veto power (negative obligation).
Decide	Can be to engage any of the other primary acts.
Permit	Can be to engage any of the other primary acts.
Adapt	Capability to change rules or procedures.

relations. First, we can abstract from the levels and allocation schemes a list of primary activities: observe, reveal, act, be directed, direct, decide, permit, and adapt (i.e., change the rules or procedures). This list is presented in Table 3.3. An analysis need not be constrained to this list, but it is as good as any place to start given its roots going back to even the earliest forms of task analysis (see Hoffman & Militello, 2008).

We can take this list and apply it to the Sheridan-Verplank levels in such a way as to make clear the shortcomings of that scheme. This analysis is presented in Table 3.4. Empty cells are ones where there was no explicit reference in the original Sheridan-Verplank Table 8-2.

It is striking that nothing appears anywhere in the rightmost column in Table 3.4. There is no interdependence in this scheme at all. The levels are basically a sampling of some of the many possible admixtures of self-directedness and self-sufficiency. The empty cells reveal how the levels leave out interdependence relations, which are a necessary part of the work but are not part of any of the immediate subtasks that are explicitly referenced at the given level (Johnson et al., 2011).

TABLE 3.4
Application of the Primary Acts to the Sheridan-Verplank Levels of Automation

Level	Machine has the Capability of...	Machine Depends on the Human to...	Human has the Capability of...	Human Depends on the Machine to...
10	Decide			
9	Direct	Act		
8	Reveal		Direct	
7	Permit/Reveal			
6	Be Directed	Permit	Direct	
5	Be Directed	Direct		
4	Reveal		Decide	
3	Decide		Decide/Permit	
2	Decide/Reveal		Decide/Permit	
1			Decide/Permit	

Alternatives to "Levels of Automation" and "Task Allocation"

The levels-allocation mythos has persisted by reinvention. But there has also been at least one explicit attempt to salvage the scheme by reference back to the original Fitts' List.

Thus, rather than thinking of work design in terms of levels and allocations, it can be understood as manifesting a number of different interdependence relationships. These are described in Table 3.5. Reference to the table is to "agents" because this scheme applies to human-human interdependence as well as to human-machine interdependence.

The relationships are not mutually exclusive. A human and machine might be coordinating on some activity and simultaneously collaborating on another. Some of the relations entail something that might be regarded as a "division of labor," specifically, tight interdependence ("To achieve our shared goal, you do this and I do that."). However, this is not allocation in the sense in which that concept is used today. Even while each agent is conducting its activity, the other agent is not disengaged from it. The interdependence requirements are capacities that must be stable and continuous, whatever relations are engaged. Any of the relations can be engaged at a given time, and multiple relations can be in effect at the same time for various subtasks or goals. Returning to the example of tight interdependence, even

TABLE 3.5
Some of the Interdependence Relationships

COOPERATION is noninterference.	
Cooperation can be reciprocal but need not be.	 **FIGURE 3.2**
Soft Cooperation	An agent does not act so as to impede another agent's progress toward their goals.
Hard Cooperation	An agent does not act so as to prevent another agent from making progress toward their goals.
COORDINATION is reciprocal knowledge sharing.	
• Depends upon cooperation. • "Here is where I am with respect to my goals, and here are the implications for your progress toward your goals." • Individuals have individual goals, although they are in service of higher-level goals (e.g., team or organization goals). • To achieve the higher-level goals, individual goals have to be achieved. • Coordination is communication and activity intended to inform each team member about the progress other team members are making toward their individual goals.	 **FIGURE 3.3**
Loose Coordination	Individual goals can be achieved in any time frame and in any order.
Tight Coordination	Individual goals have to be achieved in certain time frames and in certain orders.

(Continued)

TABLE 3.5 (*Continued*)
Some of the Interdependence Relationships

COLLABORATION is reciprocal facilitation.	
• Depends on cooperation and coordination. • "I'll actively help you if you help me." • Collaboration depends on coordination. • For any one individual to achieve their goals they require input or assistance from one or more other team members.	 FIGURE 3.4
Loose Collaboration	Individual goals can be achieved in any time frame and in any order.
Tight Collaboration	Individual goals have to be achieved in certain time frames and in certain orders.
INTERDEPENDENCE is coactivity.	
• Depends on coordination, collaboration, and interdependence. • "I need you to help me, and if you do, you'll be able to do what you have to do." • Individuals have individual goals, although they are in service of higher-level goals (e.g., team or organization goals). • Individuals also have at least one shared goal. • Individuals work together to achieve their individual and shared goal(s). • Progress toward the goals of one team member make possible progress toward the goals of other team members.	 FIGURE 3.5
Loose Interdependence	This is reciprocal reliance. Progress toward the goals of one team member make possible progress toward the goals of other team members.
Tight Interdependence	This is coactivity: "For us to accomplish this shared goal, you have to this part of it and I have to do that part of it."

while "you do this and I do that," the agents must be able to recognize anomalies, correct each other's models of the world, adapt to change, and so forth.

CONCLUSION

The work and ideas of Thomas Sheridan and William Verplank opened a door. It is necessary, however, to move beyond simple single scales of degrees of autonomy. Schemes developed from the 1940s through 1970s are no longer adequate. Fitts' List and its descendants (levels of automation, adaptive allocation, levels+stages, etc.) were premised on what was regarded as a fundamental asymmetry: Humans are limited, and there are some things that machines can do far better than humans. But this can be recast as a duality rather than as an asymmetry. The brightest machine agents will be limited in the generality, if not the depth, of their inferential, adaptive, social, and sensory capabilities. Humans, though fallible, are functionally rich in reasoning strategies and their powers of observation, learning, and sensitivity to context. These are the things that make adaptability and resilience of work systems possible. Adapting to appropriate mutual, interdependent roles that take advantage of the respective strengths of humans and strengths of machines, and crafting natural and effective modes of interaction, are key challenges for technology. Not the creation of more widgets.

In the 2015 report from the U.S. DoD's Office of Technical Intelligence, it was argued that more automation is needed especially to deal with "long-duration and continuous-operation missions, which tend to require large numbers of people to execute" (p. 8). Specifically, the report refers to data overload resulting from the introduction of more and more image and data collection systems:

> The volume of information produced by intelligence collection systems over time can be enormous, requiring a large number of analysts to process effectively, and the high cost of training, salaries, and benefits translates this into a large burden for DoD. Autonomous analytic systems have the potential to sift through large amounts of data to cue analysts to important information, decreasing the number of humans required, especially for low-level tasks (p. 8).

What we see here is the argument that we need more automation to compensate for the more automation that we had introduced previously. The first injection (more sensors and image processing systems are needed because we need more information) was based on the promise of increased capability. Following the unintended negative consequence (data overload), the second injection is premised on the promise of more technology. Realization that the reality of "more automation" does not match the promise might encourage researchers and system developers to try and escape the myth that the promise of automation can be achieved only if we develop better (computer-based) methods of validation, verification, and certification. Of course, it is very difficult to validate, verify, and certify complex computerized systems (e.g., Office of Technical Intelligence, 2015). But it is doubtful that more automation, by itself, will solve the emergent problems triggered by more automation.

For a recent project on creating an autonomous vehicle, the various competitors' vehicles integrated numerous sensor packages, algorithms, planning software, fault management software, environment models, and so forth, and yet in the performance test, the vehicles did not perform well at all.

> Extending the performance of this vehicle will, according to the presenters, be addressed by even more sensors, more algorithms, and more computation. There appears to be no limit to the complexity of interacting and interdependent computational elements… Closing the gap between the demonstration and the real thing requires the development of new methods to manage creeping complexity (Woods, 2016, p. 132).

Much has yet to be done for the myths of automation to be overcome (see Hoffman, Cullen, & Hawley, 2016; Hoffman & Hancock, 2014; Hoffman, Hawley, & Bradshaw, 2014; Woods, 2016).

CODICIL: SHERIDAN-VERPLANK REVISITED

While the "ten levels" have been in the crosshairs for over a decade, and have been mortally wounded a number of times, they have also been treated somewhat unfairly. Consider, for example, the details in the original Sheridan-Verplank Table 8-2 (see Figure 3.1) and then Sheridan's own lament:

> The public (and unfortunately, too many political and industrial decision makers) have been slow to realize that task (function) allocation does not necessarily mean allocation of a whole task to either human or machine, exclusive of the other (1998, p. 21).

A closer look at the original Sheridan-Verplank report reveals their appreciation of the importance of interdependence. Their Figure 4-2, which is reproduced in Figure 3.6, is a cartoon expressing the fact that the human and the machine can share tasks and back up each other (Sheridan & Verplank, 1978, pp. 4–6), even though there can also be elements of a task in which the computer replaces the human.

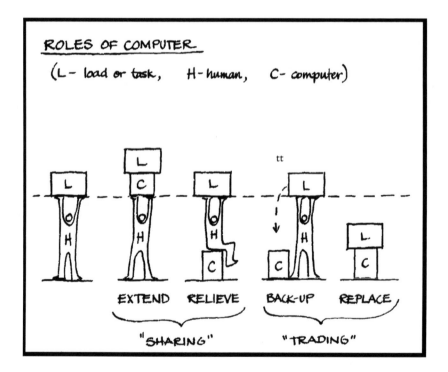

FIGURE 3.6 The original Figure 4-2 from Sheridan and Verplank (1978).

The accompanying text reads:

The computer can extend the human's capabilities beyond what he can achieve alone, it can partially relieve the human, making his job easier, it can back-up the operator in case he falters, and it can replace him completely. In the case where both computer and human are working on the same task at the same time, we call this sharing control. When they work on the same task at different times this is trading control (p. 4–6).

REFERENCES

Bornstein, J. (2015). "DoD Autonomy Roadmap." Presentation at the National Defense Industrial Association 16th Annual Science and Engineering Technology Conference, Springfield, VA.

Calhoun, G. L., Ruff, H. A., Spriggs, S., & Murray, C. (2016). Tailored performance-based adaptive levels of automation. In *Proceedings of the Human Factors and Ergonomics Society 2016 Annual Meeting* (pp. 413–418). Santa Monica, CA: Human Factors and Ergonomics Society.

Christoffersen, K., & Woods, D. D. (2002). How to make automated system team players. In E. Salas (Ed.), *Advances in human performance and cognitive engineering research*, 2: pp. 1–12. Amsterdam: Elsevier.

Cummings, M. (2014, September/October). Man versus machine or man + machine? *IEEE Intelligent Systems*, 29(5): 62–69.

de Brunélis, T. C., Le Blaye, P., Bonnet, P., & Maille, N. (2008). Distant mission management and dynamic allocation of functions1. In *Humans Operating Unmanned Systems (HUMOUS) conference, Brest, France. September* (pp. 3–4).

de Winter, J. C. F., & Dodou, D. (2014). Why the Fitts' list has persisted throughout the history of function allocation. *Cognition, Technology and Work*, 16: 1–11.

Defense Science Board (2012). "The Role of Autonomy in DoD Systems." Office of the Undersecretary for Defense for Acquisition, Technology and Logistics, Department of Defense, Washington, DC.

Defense Science Board (2016). "Summer Study on Autonomy." Office of the Undersecretary for Defense for Acquisition, Technology and Logistics, Department of Defense, Washington, DC.

Dekker, S. W. A., & Woods, D. D. (2002). MABA-MABA or Abracadabra? Progress on human-automation co-ordination. *Cognition, Technology and Work*, 4: 240–244.

Feltovich, P. J., Hoffman, R. R., & Woods, D. D. (2004, May/June). Keeping it too simple: How the reductive tendency affects cognitive engineering. *IEEE Intelligent Systems*, 19 (3): 90–95.

Fitts, P. M., Chapanis, A., Frick, F. C., Garner, W. R., Gebbard, J. W., Grether, W. F., Henneman, R. H., Kappauf, W. E., Newman, E. B., & Williams, A. C. (1951). "Human Engineering for an Effective Air Navigation and Traffic Control System." Report, National Research Council, Washington, DC.

Flach, J. M., & Dominguez, C. O. (1995, July). Use-centered design. *Ergonomics in Design*, 3 (3): 19–24.

Fuld, R. (1993, January). The fiction of function allocation. *Ergonomics in Design*, 1 (1): 20–24.

Hancock, P. A., & Scallen, S. F. (1996, October). The future of function allocation. *Ergonomics in Design*, 4 (4): 24–29.

Hoffman, R. R. (1979). On metaphors, myths, and mind. *The Psychological Record*, 29: 175–178.

Hoffman, R. R., Best, B., & Klein, G. (2015, June). Trust and reliance as emergent phenomena in macrocognitive work: An integrated model. *Proceedings of the 12th International Meeting on Naturalistic Decision Making*. McLean, VA: MITRE Corp.

Hoffman, R. R., Cullen, T. M., & Hawley, J. K. (2016). Rhetoric and reality of autonomous weapons: Getting a grip on the myths and costs of automation. *Bulletin of the Atomic Scientists, 72*, (4): 247-255. [DOI:10.1080/00963402.2016.1194619]

Hoffman, R. R., Feltovich, P. J., Ford, K. M., Woods, D. D., Klein, G., & Feltovich, A. (2002, July/August). A rose by any other name... would probably be given an acronym. *IEEE: Intelligent Systems, 17*:72–80.

Hoffman, R. R., & Hancock, P. A. (2014). Words matter. *Bulletin of the Human Factors and Ergonomics Society, 57*(8): 1. [http://www.hfes.org/web/HFESBulletin/aug2014wordsmatter.html]

Hoffman, R. R., Hawley, J. K., & Bradshaw, J. M. (2014, March/April). Myths of automation Part 2: Some very human consequences. *IEEE Intelligent Systems, 29*, (2): 82–85.

Hoffman, R. R., & McNeese, M. (2009). A history for macrocognition. *Journal of Cognitive Engineering and Decision Making, 3*: 97–110.

Hoffman, R. R., & Militello, L. G. (2008). *Perspectives on cognitive task analysis: Historical origins and modern communities of practice.* Boca Raton, FL: CRC Press/Taylor and Francis.

Hoffman, R. R., & Woods, D. D. (2011, November/December). Beyond Simon's slice: Five fundamental tradeoffs that bound the performance of macrocognitive work systems. *IEEE Intelligent Systems, 26* (6): 67–71.

Hoffman, R. R., & Yates, J. F. (2005, July/August). Decision(?)Making(?). *IEEE: Intelligent Systems, 20* (4): 22–29.

Human-Systems Integration Community of Interest (HSI COI) (2015). 16th Annual SandT Conference. [Downloaded 10 September 2016 at http://www.defenseinnovationmarketplace.mil/hs.html]

Jacobs, A. M., & Grainger, J. (1994). Models of visual word recognition: Sampling the state of the art. *Journal of Experimental Psychology: Human Perception and Performance, 29*: 1311–1334.

Johnson, M. (2014). "Coactive design: Designing support for interdependence in human-robot teamwork." Proefschrift, Technical University Delft, Delft, The Netherlands. [http://repository.tudelft.nl/islandora/object/uuid%3A6925c772-fb7f-4791-955d-27884f037da0]

Johnson, M., Bradshaw, J. M., Feltovich, P. J., Hoffman, R. R., Jonker, C., van Riemsdijk, B., & Sierhuis, M. (2011, May/June). Beyond cooperative robotics: The central role of interdependence in coactive design. *IEEE: Intelligent Systems, 26* (3): 81–88.

Johnson, M., Bradshaw, J. M., Feltovich, P. J., Catholijn, M., Jonker, C. M., van Riemsdijk, B., & Sierhuis, M. (2014). Coactive design: Designing support for interdependence in joint activity. *Human Robot- Interaction, 3*: 43–69.

Kidwell, B., Calhoun, G. L., Ruff, H. A., & Parasuraman, R. (2012). Adaptable and adaptive automation for supervisory control of multiple autonomous vehicles. In *Proceedings of the Human Factors and Ergonomics Society 56th Annual Meeting* (pp. 428–433). Santa Monica, CA: Human Factors and Ergonomics Society.

Klein, G., Ross, K. G., Moon, B. M., Klein, D. E., Hoffman, R. R., & Hollnagel, E. (2003, May/June). Macrocognition. *IEEE: Intelligent Systems, 18* (3): 81–85.

Klein, G., Woods, D. D., Bradshaw, J. D., Hoffman, R. R., & Feltovich, P. J. (2004, November/December). Ten challenges for making automation a "team player" in joint human-agent activity. *IEEE: Intelligent Systems, 19*: 91–95.

Koltko-Rivera, M.E., & Hancock, P.A. (2005). Why and how HFE professionals can better use theory: Meta-theory included; some assembly required. *Proceedings of the Human Factors and Ergonomics Society, 49*, 881–885.

Koopman, P., & Hoffman, R. R. (2003, November/December). Work-arounds, make-work, and kludges. *IEEE: Intelligent Systems, 18* (6): 70–75.

Li, Y., Burns, C., & Hu, R. (2016). Representing stages and levels of automation in a Decision Ladder: The case of automated financial trading. In *Proceedings of the Human Factors and Ergonomics Society 2016 Annual Meeting* (pp. 328–333). Santa Monica, CA: Human Factors and Ergonomics Society.

Miller, C. A. (2012). Collaboration through playbooks. In *Proceedings of the Human Factors and Ergonomics Society 56th Annual Meeting* (pp. 214–215). Santa Monica, CA: Human Factors and Ergonomics Society.

Murphy, R. & Shields, J. *The Role of Autonomy in DoD Systems.* Available online: https://fas.org/irp/agency/dod/dsb/autonomy.pdf accessed June 29, 2019

Office of Technical Intelligence. (2015). "Technical Assessment: Autonomy." Report from the Office of the Assistant Secretary of Defense for Research and Engineering. Washington, DC: Department of Defense.

Onnasch, L., Wickens, C. D., Li, H., & Manzey, D. (2014). Human performance consequences of stages and levels of automation: An integrated meta-analysis. *Human Factors*, *56*: 476–488.

Parasuraman, R., Sheridan, T. B., & Wickens, C. D. (2000). A model for types and levels of human interaction with automation. *IEEE Transactions on Systems, Man and Cybernetics-Part A: Systems and Humans, 30*: 286–297.

Pritchett, A. R., Kim, S. Y., & Feigh, K. M. (2014). Modeling human-automation function allocation. *Journal of Cognitive Engineering and Decision Making, 8*(1): 52–77.

Rasmussen, J. (1983). Skills, rules, and knowledge: Signals, signs and symbols, and other distinctions in human performance models. *IEEE Transactions on Systems, Man and Cybernetics, 13*: 257–266.

Roth, E. M. (Chair) (2012). Discussion panel: Collaborative automation across varying time scales of interaction: What's the same, what's different? In *Proceedings of the Human Factors and Ergonomics Society 56th Annual Meeting* (pp. 213–217). Santa Monica, CA: Human Factors and Ergonomics Society.

Sarter, N. B., and Woods, D. D. (1995). "How in the world did we get into that mode?" Mode error and awareness in supervisory control. *Human Factors, 37*: 5–19.

Sarter, N. B., & Woods, D. D. (1997). Team play with a powerful and independent agent: A corpus of operational experiences and automation surprises on an airbus A-320. *Human Factors, 39*: 553–569.

Sarter, N. B., & Woods, D. D. (2000). Team play with a powerful and independent agent: A full mission simulation. *Human Factors, 42*: 390–402.

Sarter, N., Woods, D. D., & Billings, C. E. (1997). Automation surprises. In G. Salvendy (Ed.), *Handbook of human factors/ergonomics* (2nd ed.) (pp. 1926–1943). New York, NY: Wiley.

Scott, R. (2012). Collaborative automation for dynamic replanning. In *Proceedings of the Human Factors and Ergonomics Society 56th Annual Meeting* (pp. 213–217). Santa Monica, CA: Human Factors and Ergonomics Society.

Sellers, B. C., Fincannon, T., & Jentsch, F. (2012). The effects of autonomy and cognitive ability on workload and supervisory control of unmanned systems. In *Proceedings of the Human Factors and Ergonomics Society 56th Annual Meeting* (pp. 1039–1044). Santa Monica, CA: Human Factors and Ergonomics Society.

Sheridan, T. B. (1980, October). Computer control and human alienation. *Technology Review, 10*: 60–72.

Sheridan, T. B. (1997). Task analysis, task allocation, and supervisory control. In M. G. Helander, T. K. Landauer, P. Prabhu (Eds.), *Handbook of Human-Computer Interaction* (2nd ed.) (pp. 87–105). Amsterdam, The Netherlands: Elsevier Science.

Sheridan, T. B. (1998, July). Allocating functions rationally. *Ergonomics in Design, 6* (3): 20–25.

Sheridan, T. B. (2011). Adaptive automation, level of automation, allocation authority, supervisory control, and adaptive control: Distinctions and modes of adaptation. *IEEE Transactions on Systems, Man and Cybernetics-Part A: Systems and Humans, 41*: 662–667.

Sheridan, T. B., & Verplank, W. L. (1978). "Human and Computer Control of Undersea Teleoperations." Technical Report on Contract N00014-77-0256, Engineering Psychology Program, Office of Naval Research, Arlington, VA.

Shladover, S. E. (2016, June). The truth about "self driving" cars. *Scientific American*, 3–57.

Smith, P. J. (2012). Collaborative automation in a temporally and spatially distributed decision-making environment. In *Proceedings of the Human Factors and Ergonomics Society 56th Annual Meeting* (p. 215). Santa Monica, CA: Human Factors and Ergonomics Society.

Verplank, W. L. (1978). Display aids for remote control of untethered undersea vehicles. In *Proceedings of the Fourth Annual Combined Conference of the IEEE and the MTS*.

Woods, D. D. (2016). The risks of autonomy: Doyle's catch. *Journal of Cognitive Engineering and Decision Making, 10*: 131–133.

Wright, T. J., Samuel, S., Borowsky, A., Zilberstein, S., & Fisher, D. L. K. (2016). Experienced drivers are quicker to achieve situation awareness that inexperienced drivers in situations of transfer of control within a Level-3 autonomous environment. In *Proceedings of the Human Factors and Ergonomics Society 2015 Annual Meeting* (pp. 270–275). Santa Monica, CA: Human Factors and Ergonomics Society.

Yoo, H. -S. (2012). Framework for designing adaptive automation. In *Proceedings of the Human Factors and Ergonomics Society 56th Annual Meeting* (pp. 2133–2136). Santa Moinica, CA: Human Factors and Ergonomics Society.

4 Why Representations Matter: Designing to Support Productive Thinking

John M. Flach & Kevin B. Bennett

INTRODUCTION

The guiding premise of this chapter is that a primary function of interfaces in complex systems is to couple intention, perception, and action. That is, the goal of interface design is to specify the field of possible actions that can be performed and the potential consequences of those actions relative to the functional goals and values of an organization. In this context, the interface is seen as a component of a triadic semiotic (or sensemaking) system as illustrated in Figure 4.1. This chapter describes a holistic approach to this system, consistent with a belief that the function of the interface can only be fully understood in the context of emergent dynamics of the full sensemaking system. Figure 4.1 illustrates three different perspectives on the dynamics of this system: affording, specifying, and satisfying.

Affording reflects the constraints on action that bound the possible behaviors and the achievable consequences (both positive/objectives and negative/risks). In control theoretic terms, this dimension reflects the controllability of the system. That is, it determines what paths are possible through the state space and which states are reachable. This dynamic reflects what *can be done or what can happen for good or ill* (Gibson, 1979).

Specifying reflects the information mapping among the action possibilities, consequences, and functional values. This determines a system's ability to perceive the affordances relative to the organization's values. In control theoretic terms, this dimension reflects the observability of the system. In essence, the interface reflects the feedback and comparator functions in a control system, and the quality of the interface representation will determine how well the affordances are specified relative to the purposes of an organization. In other words, the specifying dimensions reflect the directness of the mapping between intentions and actions. Thus, specifying provides a framework for evaluating what can be skillfully controlled. In other words, specifying is the lynchpin that determines a system's ability *to do what ought to be done.*

Satisfying reflects the values, purposes, or functions of the semiotic system. In particular, this addresses the reason WHY a work domain or work organization

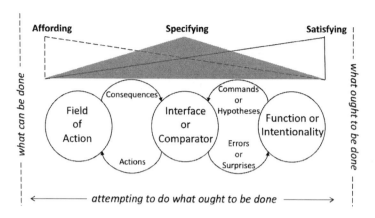

FIGURE 4.1 An illustration of a triadic semiotic or sensemaking system, emphasizing the parallels to control systems and showing three emergent dynamics: affording (controllability), specifying (observability), and satisfying (cost/benefit function).

exists and the qualities that determine whether performance in that domain results in successful (satisfying) or unsuccessful (unsatisfying) consequences. In control theoretic terms, this dimension reflects the goals and the cost functions or pay-off matrices for evaluating performance quality (e.g., as used in the formulation of optimal control or ideal observer models). Satisfying provides a framework for understanding *what ought to be done* and for comparing alternative solutions in terms of their quality.

It is important to note the parallels between the sensemaking system and control systems — to a significant extent the quality of sensemaking is analogous to the quality of a control system in terms of controllability, observability, and performance relative to appropriate cost/payoff functions. However, the sensemaking system is far more sophisticated than the simplistic servomechanisms that were used to introduce the cybernetic hypothesis to social scientists (e.g., Miller, Galanter, & Pribram, 1960). In contrast, the sensemaking system is an adaptive, self-organizing system that cannot depend on an extrinsic source (e.g., a designer) to specify the goals and cost functions or to tune the loop dynamics (e.g., gains) to satisfy stability constraints. The population dynamics of a prey-predator system is a better analogy for the sensemaking dynamic than a classical servomechanism (e.g., thermostat). In the prey-predator system, the population levels emerge from the intrinsic coupling of prey and predators, rather than being imposed from without (e.g., as the temperature setting on a thermostat).

The critical point is that the sensemaking system must actively discover stable solutions through interaction with an ecology. This implies that the sensemaking system has to be capable of problem-solving in addition to control. In other words, the sensemaking system has to discover the appropriate values and the appropriate tuning of the perception-action dynamics through interaction with its ecology. In sensemaking systems, this typically involves a process of learning while doing that has been described by Peirce (1878) as *abduction* or *hypothesis* and by Lindblom (1959, 1979) as *muddling through* or *incrementalism*.

Why Representations Matter

A key implication of this perspective is that it is not sufficient to evaluate the interface relative to perceptual capabilities — its function is not simply for accessing raw information. In order to function as a quality comparator, the interface must allow the information to be evaluated relative to the functional values or goals and relative to the action possibilities. Note that in the simpler servo examples the comparator process is represented in terms of subtracting feedback from a goal signal to yield a value for commanding action. This representation is trivial relative to natural work systems — where the goals (e.g., a safe landing), the feedback (e.g., instruments and optical flow fields), and the actions (e.g., adjusting a control yoke, throttle, and pedals) will not be in commensurate terms. A primary function of interface representations will be to help operators/workers to solve the comparator problem — to translate multiple sources of feedback into actions that lead to satisfying consequences. *This is the sensemaking problem!*

The point is NOT that perceptual concerns are unimportant. The point is that *making sure that an interface's information is perceptible is not sufficient.* The interface information must also be *meaningful* relative to the values and action capabilities of the organization (Bennett & Flach, 2011). Thus, the goal of this chapter is to frame the interface design challenge in terms of creating representations that facilitate problem-solving.

PROBLEM-SOLVING AND EXPERTISE

Evaluating the impact of representations on thinking and problem-solving was a central focus of research for several Gestalt psychologists (Duncker, Seltz, and Wertheimer). Duncker's (1945) "traveling monk problem" provides a good illustration of this:

One morning a Buddhist monk sets out at sunrise to climb a path up the mountain to reach the temple at the summit. He arrives at the temple just before sunset. A few days later, he leaves the temple at sunrise to descend the mountain, traveling somewhat faster since it is downhill. Is there a spot along the path that the monk will occupy at precisely the same time of day on both trips?

Is the answer apparent to you? This is a difficult problem for most people. Is the difficulty inherent in the problem? Or, is the difficulty a function of the representation? Is there an alternative representation of the same problem that would allow most people to "see" the answer? Here is one alternative:

One morning two monks, one at the top of the mountain, the other at the bottom start their trips along the same narrow mountain path — one going down, the other going up. Will they meet?

Is the answer to this version apparent to you? Most people have little difficulty when the problem is framed in terms of two monks moving along the same path in different directions. Yet, the two narratives are describing essentially the same problem. The first version makes the essence of the problem (two monks moving in opposite directions on the same path) less salient than the second version. The first version tends to induce an analytical process of tracking space and times for two trips, in order to determine if there is a common place and time. In the second version, there is nothing to determine or figure out, the answer is intuitively obvious — no analysis is required (see Flach and Vorhoorst, 2016, for several other alternative representations for this problem).

Wertheimer's (1959) book on *Productive Thinking* is a compilation of many examples similar to Duncker's traveling monk problem. In these examples, he contrasts various representations in terms of whether they induce a deliberate, rote kind of analytical reasoning or whether they make the *deep structure* of a problem salient in a way that leads to sudden insights. These sudden insights have classically been referred to as *aha! experiences*. More recently, research on expertise has described these experiences as *expert intuition* or as *recognition-primed decision-making* (e.g., Klein, 1989, 1993, 2004, 2015). The common thread is that with experience, experts often discover representations that allow them to directly perceive solutions to situations that people with less experience tend to approach indirectly or analytically (Klein, 2004). Further, the indirect approaches often tax peoples' limited resources and often fail to lead to successful solutions, particularly in dynamic situations where the time to respond is limited.

The qualitative differences between rote thinking and *aha! insights* has led to many discussions of the relative benefits of analytic versus intuitive reasoning (Kahneman & Klein, 2009). It is our contention that this debate is miscast to reflect the surface differences of expert-novice problem-solving. This framing of the debate fails to emphasize the more fundamental difference, which is the quality of the representations that people are using to frame the problems. Experts tend to have representations that are more directly tuned to the problem situations that help to make solutions intuitively evident (i.e., that support productive thinking). Without such representations, novices have to fall back on more indirect forms of analysis that tax their limited computational resources. The indirect analytic approaches are *logical* but impractical for anything other than the toy problems often used in laboratory research on decision-making; and in fact, the laboratory studies clearly illustrate that, even on the toy problems, people have difficulty applying the indirect, more analytical styles of thinking (Kahneman, 2011).

We hypothesize that the key distinction between experts and novices is the quality of coupling with the problem ecology. The knowledge and skills of experts reflect internal representations (frames or schemata) that are directly tuned to the demands of the environment, whereas novices have not yet accommodated to the problem environment and have not yet discovered the deeper structures that might support productive thinking. Simon (1969, 1990) used the analogy of the ant on a beach to illustrate the importance of the coupling for understanding cognition and the analogy of a scissors to emphasize how the coupling of the representation with the problem determines whether thinking is productive (i.e., whether the scissors cut). Gigerenzer (e.g., Todd, Gigerenzer, and ABC Research Group, 2012) has extended Simon's intuitions to emphasize the positive aspects of heuristics in guiding expert decision-making — particularly in complex and uncertain environments.

In essence, the heuristics used by experts are the product of representations that make critical aspects of a problem salient (e.g., a particularly diagnostic dimension). The *aiming off* heuristic used by skilled orienteers is a good example of how a well-tuned representation can lead to robust solutions that bypass internal limitations to

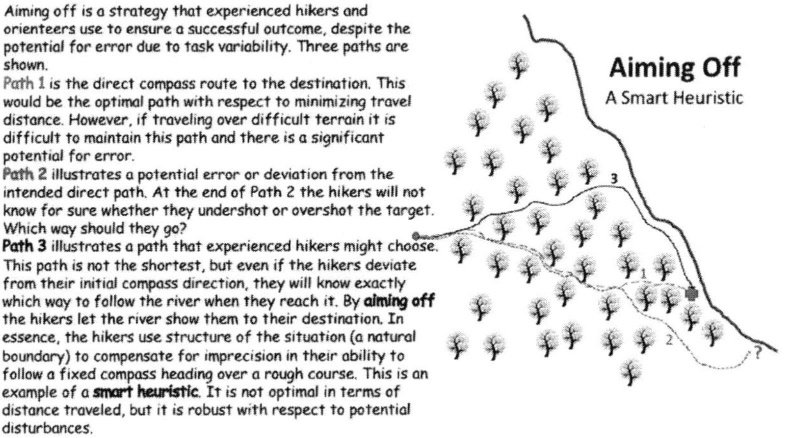

Why Representations Matter 73

Aiming Off
A Smart Heuristic

Aiming off is a strategy that experienced hikers and orienteers use to ensure a successful outcome, despite the potential for error due to task variability. Three paths are shown.
Path 1 is the direct compass route to the destination. This would be the optimal path with respect to minimizing travel distance. However, if traveling over difficult terrain it is difficult to maintain this path and there is a significant potential for error.
Path 2 illustrates a potential error or deviation from the intended direct path. At the end of Path 2 the hikers will not know for sure whether they undershot or overshot the target. Which way should they go?
Path 3 illustrates a path that experienced hikers might choose. This path is not the shortest, but even if the hikers deviate from their initial compass direction, they will know exactly which way to follow the river when they reach it. By **aiming off** the hikers let the river show them to their destination. In essence, the hikers use structure of the situation (a natural boundary) to compensate for imprecision in their ability to follow a fixed compass heading over a rough course. This is an example of a **smart heuristic**. It is not optimal in terms of distance traveled, but it is robust with respect to potential disturbances.

FIGURE 4.2 Aiming off is a smart heuristic used by skilled orienteers. It is used as an illustration of how experts use their deep understanding of problem constraints to reduce the computational demands of a problem.

ensure reliable solutions to problems that are computationally difficult (see Figure 4.2). In employing the aiming off heuristic, expert orienteers take advantage of structural properties of the problem (an extended boundary) to reduce both the computational demands and the consequences of potential computational errors. The experts replace the computational problem of maintaining a constant compass heading with the perceptual problem of following the river, which directly specifies the direction to the goal.

The implications for reframing the interface design problem in terms of problem-solving is that the focus of design becomes to discover the deep structure of a problem and then integrating that structure into representations that support productive, direct, intuitive, or recognition-primed ways of thinking. Relative to the expertise literature, the goal is to help operators to *see* or *frame* the problem through the eyes of experts. In other words, the goal is to externalize the internal representations that experts use, so that they are available to people with less experience. In essence, interface representations help to frame the problem for the worker. The hope is that by constructing interface representations that make the deep structure of problems salient, we will be helping people to adopt more productive ways of thinking about the problems — increasing the likelihood for direct insights and reducing the computational burden.

ECOLOGICAL INTERFACE DESIGN (EID)

The Gestalt research on problem-solving was an important influence inspiring Rasmussen and Vicente's (1989: Vicente & Rasmussen, 1992) *ecological interface design* approach to constructing representations. Rasmussen and Vicente framed the challenge of supporting *productive thinking* in terms of three modes of coupling between perception and action: skill-based, rule-based, and

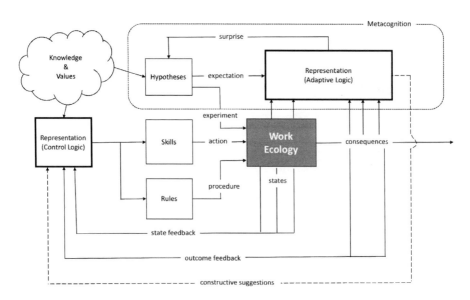

FIGURE 4.3 The sensemaking process is a multiloop, adaptive process that involves multiple modes of coupling with the work (or problem) ecology.

knowledge-based couplings. Figure 4.3 illustrates how these three modes might be integrated within a sensemaking system. Before considering the three modes of coupling, note that at the heart of this sensemaking system is the work ecology. It is the intuition that stable sensemaking requires coordination between a cognitive system and its *ecology* that motivated the label of *ecological* interface design. As noted earlier, Simon (1969) used the ant on the beach metaphor to emphasize the significance of the ecology (i.e., the beach) for understanding sensemaking processes. Also, note that in Figure 4.2, there are two boxes labeled representation. This does not necessarily imply that multiple representations are required, but it does suggest that there are two distinctive functions that representations serve with respect to sensemaking: 1) directly guiding *performatory action* in pursuit of goals and 2) supporting *exploratory action* in pursuit of deeper understanding and/or increasing quality of performance.

The first function of representations illustrated in Figure 4.3 is supporting performatory action. This function involves two modes of active coupling with the work ecology: skill-based control and rule-based control. *Skill-based control* is illustrated as the inner most loop in which actions are continuously adjusted based on both state and outcome feedback from the work ecology. This is the most direct coupling and is consistent with the intuitions of Gibson (1979) with respect to how animals control locomotion. Prototypical examples of this mode of control are piloting and driving. In both cases, continuous feedback about the motion of the vehicle is available and continuous control of the motion is possible. For example, pilots might monitor continuous feedback about air speed, altitude, and heading as they guide their aircraft toward a safe landing. In designing representations to support skill-based control, the key is to make the appropriate state variables salient to the pilots so that they can immediately "see" the consequences of control actions relative to their intended goal (i.e., a safe touchdown on the runway).

The *rule-based mode* of coupling is illustrated as the next loop beyond the inner skill-based loop. Rule-based coupling reflects situations where continuous direct state feedback is ambiguous or unavailable (e.g., due to time delays or lags). Rule-based modes of action are typical of process control, where there are long lags between action (e.g., initiating a heating process or docking a large ship) and the consequences of that action (e.g., reaching a target temperature or coasting in to dock). In these cases, the responses are typically procedures (i.e., sequences of actions) that are triggered by an intention or a situation. These procedures are typically carried out open-loop so that feedback is only evaluated intermittently — for example, setting your shower control to a particular position or cutting the engines of a large tanker, then waiting to see the results before making a second adjustment. To support rule-based control, the key is to make the appropriate triggers for initiating a sequence of actions salient. For example, for supertankers, an effective representation should directly specify when to cut the engines so that the tanker will coast to a minimal speed before reaching the dock.

These two inner loops play an important role for control, but the outermost loops are the more interesting dynamic when it comes to sensemaking or problem-solving in complex work ecologies. The outer loop involves exploratory action, learning, and adaptation. The function of the outer loop is to evaluate performance of the inner loops with respect to expectations and quality. Cognitive psychologists have referred to this function as *metacognition* to reflect the fact that this function is monitoring cognitive processes and tuning the dynamics of the inner loops to align with performance goals (e.g., Bransford, Brown & Cocking, 2000). In essence, the outer loop functions as a critic of the inner loop control in order to provide constructive feedback for changing the control logic when opportunities to increase the quality of performance are discovered. The dotted line on the most outer of the bottom loops indicates that the feedback along this path can actually alter or tune the logic governing the inner loops (i.e., skills and rules). For example, upon activating the brakes in a rental car, drivers may be surprised by the degree of braking that results. As a result of this surprise, they may alter when and how hard they press the brakes in order to slow down or stop (i.e., in essence altering the gain of their skill-based braking algorithm). It is the metacognitive loop that guides the changes to the inner loop dynamics so that it is better tuned to the dynamics of the rental car.

In the braking example, the quality of the logic guiding braking behavior is evaluated as the consequence of actively testing the brakes and being surprised by the resulting behavior of the car. However, the metacognitive processes can also be engaged as mental simulations in which an internal representation is manipulated without actually acting on the environment. For example, Klein (2004) observed that in high-risk environments, experts might simulate a solution using their mental model to test its plausibility before actively pursuing that solution. By doing this, experts can sometimes discover a problem with a solution before acting on it, and thus reject it and consider an alternative path. This process of simulating possible solutions before acting on or rejecting them is analogous to a process that de Groot

(1965) described as *progressive deepening*. He observed that expert chess players would mentally play out a sequence of moves and then either accept the initial move in the sequence or reject it to consider an alternative move and sequence. Thus, players only considered one move at a time and accepted or rejected that move based on the results from their mental simulations.

The goal of EID is to develop interface representations that support all three modes of processing. The interface representation needs to 1) provide direct continuous feedback about states and consequences (*signals*) in order to support skill-based interactions (e.g., for piloting), 2) make the *signs* for triggering procedures salient for initiating rule-based interactions) (e.g., for initiating flare at the end of an approach), and 3) allow the workers to monitor and evaluate the quality of the inner loops so they can intervene when the current skills and rules no longer apply (e.g., to call off an approach or to plan an alternative route when a landing is not proceeding as intended). Generally, EID attempts to do this with configural representations that make the relevant functional constraints for each mode of control available as nested geometric forms — with higher or global relations in the geometry reflecting higher order semantics useful for evaluating the quality of control (supporting metacognitive processes) and with more local relations and properties reflecting the states of the process for direct continuous feedback (skill-based control) or for triggering procedures (rule-based control).

THE LOGIC OF ABDUCTION OR MUDDLING

The framing of the representation design problem in terms of adaptive control raises a fundamental question about the norms for rationality. Classically, the norms for rationality have been derived from mathematical logic and economic theories — where the focus has been on the form of argument for analytically processing the output of perceptual processes in terms of validity or inference. However, in the context of adaptive control, we suggest that the norms for rationality need to be framed in the context of the direct empirical consequences of action and the impact on long-term stability. The central question is: Does the sensemaking system tend to converge to stable states (i.e., safe solutions) consistent with the organization's intentions (or values) or does the system diverge (either flailing as it chases noise or collapsing catastrophically)?

C. S. Peirce (1878) introduced *abduction* as an alternative to classical forms of logic. The key to abduction is that beliefs are empirically grounded in terms of the habits of action that they induce. Habits that lead to satisfying consequences validate (or reinforce) the beliefs that motivated them; and conversely, beliefs that lead to unsatisfying habits/consequences are invalidated or extinguished. This is also consistent with Piaget's (1973) dynamics of cognitive development in which schema are constantly being revised as a result of the empirical consequences of acting on them (accommodation). Thus, abduction is essentially a pragmatic process of trial-and-error learning.

A key to stability in closed-loop (trial-and-error) systems is the gain, which is essentially the aggressiveness of responses (trials) to either error in the inner loop or surprise in the outer loop. In order to maintain stability, a general rule is to not be too aggressive. That is, an incremental or conservative approach to action or change will typically be best. Note that, while changes of representations have been associated with sudden, aha! experiences, in the Gestalt literature, the point of representation design is to insure a good representation from the start not to change representations on the fly. So, the point is not to generate sudden insights but to provide a representation that frames the work context in a constructive way from the start to foster direct, intuitive decisions and actions (though, innovative improvements to conventional interfaces motivated by EID may lead to aha! reactions).

With regards to a conservative approach to change, an analogy has been made with friction in physical systems (Akerman, 1998). For example, Rochlin (1998) writes that "without the damping effect of friction, we would live in an impossibly kinetic world in which the consequences of every action would persist and multiply to the point of insanity" (p. 132). He continues the analogy to sensemaking systems: "In the realm of the social and political, morals, ethics, knowledge, history, and memory may all serve as the sources of 'social friction,' by which gross motions are damped, impetuous ones slowed, and historical ones absorbed" (p. 132). Thus, the point is that a well-tuned adaptive system should be somewhat conservative. It should rely heavily on the skills and rules (e.g., heuristics) that have worked in the past; and it should tend to resist large or dramatic changes in the underlying logic guiding action.

The benefit of friction is that it helps to "damp" or "filter out" impulsive changes that might seem like a good idea at the time, but that may not work as intended or that may have undesirable side effects. For example, the mental simulations described by Klein can be thought of as a kind of friction that damps an initial impulse. The cost is that there is also resistance to smart changes — so that even correct actions will be delayed. In essence, speed or efficiency is traded off for safety or stability. Stability in sensemaking requires a delicate balance between preventing rash actions while also avoiding the trap of paralysis by analysis (Schwartz, 2006). The appropriate balance will be determined by the dynamics of the work ecology in terms of the degree of uncertainty (e.g., noise) in the input and the speed at which windows of opportunity open and close (e.g., the signal frequencies).

Lindblom (1959, 1979) also recognized that there are many organizational obstacles to change, and thus he described the process of organizational decision-making as *incrementalism* or *muddling through*. In his second paper, he noted that not only was this a description of how organizations make decisions, but it also reflected a smart strategy for dealing with complex problems. Whereas, many others saw the muddling as a suboptimal, inefficient process that ought to be corrected or improved, Lindblom made the case that rather than correcting the muddling, we should be supporting it. He wrote that, "Most critics of incrementalism believe that doing better usually means turning away from incrementalism. Incrementalists believe that for complex problem-solving it usually means practicing incrementalism more skillfully and turning away from it only rarely" (Lindblom, 1979, p. 517). The point is that for solving complex problems, incremental muddling is a very smart process. This is consistent with the logic of abduction, the friction metaphor, and with the requirements for stability in adaptive control systems.

DESIGNING TO SUPPORT SKILLED MUDDLING

If you accept the adaptive control premise, then a goal for representation design will be to support skilled muddling. In framing the design problem this way, we would like to draw attention to three important contrasts with more conventional approaches to design:

1. Shaping mental models rather than matching mental models
2. Work analysis to identify constraints rather than task analysis to evaluate behaviors
3. Creativity and satisficing rather than conformity and optimality

Shaping Mental Models

One of the well-known mantras of the human factors (HF) profession has been: "Know thy user." This has typically meant that a fundamental role for human factors has been to make sure that system designers are aware of computational limitations (e.g., perceptual thresholds, working memory capacity, potential violations of classical logic due to reliance on heuristics) and expectations (e.g., population stereotypes, mental models) that bound human performance. It is important to note that these limitations have generally been validated with a wealth of scientific research. Thus, it is important that these limitations be considered by designers. It is important to design information systems so that relevant information is perceptually salient, so that working memory is not overtaxed, and so that expectations and population stereotypes are not violated.

The emphasis on the bounds of human rationality, however, tends to put human factors at the back of the innovation parade. While others are touting the opportunities of emerging technologies, HF is apologizing for the weaknesses of the humans. This feeds into a narrative in which automation becomes the "hero" and humans are pushed into the background as the weakest link — a source of error and an obstacle to innovation. From the perspective of the technologists, the world would be so much better if we could simply engineer the humans out of the system (e.g., get human drivers off the road in order to increase highway safety).

Of course, we know that this is a false narrative. Bounded rationality is not unique to humans; all technical systems are bounded (e.g., by the assumptions of their designers or, in the case of neural nets, by the bounds of their training/experience). It is important to understand that the bounds of rationality are a function of the complexity or requisite variety of nature (Flach & Behymer, 2016). It is the high dimensionality and interconnectedness of natural work domains (e.g., process control, aviation, or health care) that creates the bounds on any information processing system (human or robot/automaton) that is challenged to work in those domains. In natural work domains, there are always potential sources of information that will be beyond the limits of any computational system. There always will be situations that were not or could not have been anticipated (e.g., a stuck valve, loss of both engines of an airliner due to bird strikes, the first Ebola case in a U.S. hospital).

The implication for designing sociotechnical systems is that designers need to take advantage of whatever resources are available to cope with this requisite variety of nature. The creative problem-solving abilities of humans and human social systems should be considered to be one of the resources that designers should be leveraging. Thus, the muddling of humans (i.e., incrementalism) described by Lindblom is NOT considered to be a weakness but rather a strength of humans.

However, it is also important to note that most workers are not experts in their domains. And even the understanding of experts in a work domain can be constrained by conventional interfaces and technologies that limited the kinds of representations that were available (e.g., the steam gauge interfaces in nuclear power plants and aircraft prior to the development of high-speed computer graphics and configural displays). Thus, the challenge for designing representations is not to *match the existing mental models of users*, but rather it is to *shape the mental models in productive ways*. Today, advances in sensors, high-speed computers, and inexpensive graphic engines provide unique opportunities for innovating at the interface. There is an opportunity to allow people to access information that was not accessible before and to organize/integrate (i.e., chunk) that information into new configurations and patterns that map more directly into both the deep structure of the work ecology and into the three different modes of interaction illustrated in Figure 4.3.

Thus, we see representation design as an opportunity to innovate. The goal is to build representations that leverage the deeper insights that advances in sensor, computing, and display technologies offer to shape or bias workers to think about their work in more productive ways.

WORK ANALYSIS

A prerequisite for taking advantage of the opportunities for innovation that advanced sensor and display technologies open up is to map the work ecology to uncover the deep structures or meaningful properties. This is the point of *work analysis* (Vicente, 1999; Naikar, 2013). In contrast to more classical approaches to cognitive task analysis, which tend to focus on mental activities, the focus of work analysis is to uncover the functional constraints in the work ecology (in terms of Simon's ant analogy — the focus of work analysis is on a functional description of the beach not on any specific path an ant might take) (Flach & Bennett, 2018). In particular, the three categories of constraint identified in Figure 4.1 are particularly important in guiding the design of representations: satisfying, affording, and specifying.

Consistent with Sinek's (2009) recent insights into design, of these three categories of constraint, satisfying is preeminent when it comes to sensemaking. The constraints on satisfying — why the organization exists (i.e., the goals and value propositions) — set the broader context for determining the relevance of the constraints on action (affording) and the constraints on perception (specifying). Although goals and values are often subjective, they dominate the objective physical constraints (e.g., physical laws that constrain perception and action) when it comes to determining what is *meaningful*. For example, in designing aircraft interfaces, the laws of aerodynamics are particularly meaningful (e.g., Amelink, Mulder, van Paassen, & Flach, 2005); in designing nuclear process control interfaces, the laws of thermodynamics become particularly meaningful (e.g., Hajdukiewicz & Vicente, 2002); in designing interfaces for military command and control (e.g., Bennett, Posey & Shattuck, 2008), military science becomes more significant; and finally, in designing displays for health care (e.g., McEwen, Flach & Elder, 2014), models of health from the medical literature become particularly significant. Note that while our goals and values cannot change physical laws, they do determine which particular laws are most *meaningful*.

As noted earlier, a key aspect of skilled muddling is an incremental or conservative approach to action or change. Another way to think about this in control theoretic terms is that for stable control, it is necessary to anticipate the consequences of actions (to see the future). For example, in the aiming off heuristic, once the orienteer reaches the river, the river specifies where the bridge *will be*. In vehicle control, higher derivatives (e.g., rates of change) are critical for specifying the imminence of collision. A driver (or an automatic braking system) that gives heavier weight to velocity feedback will be much more conservative in initiating braking than a more aggressive driver who places less weight on velocity. Quickened display used in large tankers integrates higher derivative information with position to show the operators when to cut the engines, so that the ship will coast to a stop before reaching the dock (Jagacinski & Flach, 2003). Thus, one of the goals for work analysis is to identify the constraints (e.g., physical laws) that allow workers to anticipate the consequences of their actions. By integrating these constraints into geometric patterns in a representation, it is possible to help workers to "see" where things are going and to intervene if the current direction does not align with the goals to be achieved or values to maximize.

For example, Vicente (1999) incorporated principles of thermodynamics into his DURESS interface in terms of mass and energy balance that allow operators to see where the thermodynamic processes were going. Similarly, Amelink et al. (2005) integrated a total energy target to allow prospective control with respect to controlling landing. Bennett et al. (2008: Hall, Shattuck, & Bennett, 2012) included continuous feedback on the force ratio (i.e., combat power relative to the opposing force), so that army battalion commanders could evaluate how an engagement was trending (i.e., whether they were winning). Finally, McEwen et al. (2014) included longitudinal models of cardiovascular health to allow physicians to make treatment decisions in anticipation of potential risks.

In essence, anticipation means seeing current actions in light of future goals and risks. Thus, the better a sensemaking system is able to anticipate consequences, the more capable it will be in determining outcomes. This is also reflected in the mental simulations that Klein and de Groot described. Representations that are visual analogs of the dynamics of the work can function as simulations so that workers can use the interface to evaluate alternative approaches to the inner control loop (e.g., alternative heuristics or hypotheses) before acting on those alternatives. If the relations in a configural representation are visual analogs to the work dynamic, then these patterns can foster good hypotheses about what kinds of changes will lead to improved performance. Or, at least, the representation might help the worker to discover risks or unintended consequences of their choices early, so that they can recover before going too far down a dangerous path.

CREATIVITY AND SATISFICING

There is a tendency, rooted in Taylor's (1913) scientific management approach to work design and in a focus on behavior (rather than constraints), for designers to assume that there is "one best way" to solve a problem, to think, or to do work. Thus, a goal for

Why Representations Matter

design has been to ensure that workers conform to that particular path (e.g., that they comply with predetermined standard procedures or norms). In this context, human variability (deviations from standard procedures) has been seen as a problem to be eliminated. However, in work domains where sensemaking is important, there typically is no a priori single best way. In fact, it has been said that a good control system is one that always gets to the goal but never in the same exact way. For example, for pilots to fly consistently safe landing trajectories at different airports under varying weather conditions, they will have to act differently on every approach. The trajectory of each approach may be very similar, but this is because the pilots adapt their behaviors to compensate for the variations in conditions. With regards to conformity to standard procedures, Rasmussen has also observed that a strategy that unions use when bargaining for their workers is a "work to rules" strike (private communication). The result is that when everyone conforms perfectly to the rules, things tend to get very inefficient. The point being that most deviations from procedures (e.g., workarounds) are creative adaptations that generally improve performance of the organization. Eliminating the creative adaptations that people make in their daily work activities typically results in very brittle, clumsy, and inefficient processes.

The alternative to a focus on a single best way or to designing optimal control systems is to design sensemaking systems to be robust or resilient (e.g., Hollnagel, Woods, & Leveson, 2006). In reality, optimal control is only possible for very narrow ranges of situations. In contrast, a robust control system is designed with a much broader range of situations in mind. The robust controller may not be optimal for any particular situation, but it is designed to ensure stability over a broad range. In designing robust controllers, there is an implication that the breadth of situations can be specified in advance. Resilience, in contrast, suggests an ability to achieve stability even for situations that were not or could not have been anticipated in advance. Thus, resilience implies the capability of a sensemaking system to creatively adapt to surprise. Smart humans are a potential resource for achieving resilience in sensemaking systems. Thus, designing for resilience implies tapping into this resource and supporting human creativity. In EID terms, this means supporting the kinds of knowledge-based processing reflected in the metacognitive loop in Figure 4.3.

Thus, a goal for design in complex work ecologies should be to provide workers with resources for creative problem-solving. This requires representing the deep structures of the work ecologies that shape the possibilities for action and the consequences of those actions relative to an organization's functional goals. It means supporting an abductive reasoning process for testing hypotheses, both through creating patterns that allow workers to anticipate the consequences of actions and supporting direct manipulation capabilities to simulate choices prior to committing to them.

CONCLUSION

The major point of this chapter is that designing interface representations is not simply about ensuring that data are perceptible. It is also necessary to ensure that data are *meaningful* in relation to supporting a stable semiotic process. This involves specifying the affordances of a work ecology in relation to the functional goals of an organization. In doing this, much can be learned from analogies with resilient, adaptive control systems. Such systems involve both inner loops for directing action (skill- and rule-based modes of behavior) and outer loops for monitoring the quality of the inner loops in relation to both the changing demands of work ecologies and the values of the organization. These outer loops reflect a process of learning by doing (i.e., muddling) that tends toward increasingly stable (i.e., well-tuned) inner loop solutions to work situations. A major challenge for such sensemaking systems is to balance the need for quick effective action against the dangers of impulsive solutions with destabilizing consequences. This requires that representations specify potential consequences so that they can be anticipated in advance. Thus, an effective representation is one that couples the inner and outer loops of the adaptive sensemaking system into a resilient process capable of anticipating consequences in order to skillfully muddle through to satisfying solutions in spite of the complexity and noise (e.g., requisite variety) of natural work domains.

REFERENCES

Akerman, N. (Ed.) (1998). *The Necessity of Friction*. Boulder, CO: Westview Press.

Amelink, H. J. M., Mulder, M., van Paassen, M. M., and Flach, J. M. (2005). Theoretical foundations for total energy-based perspective flight-path displays for aircraft guidance. *International Journal of Aviation Psychology*, 15, 205–231.

Bennett, K. B., and Flach, J. M. 2011. *Display and Interface Design: Subtle Science, Exact Art*. London: Taylor & Francis Group.

Bennett, K. B., Posey, S. M., & Shattuck, L. G. (2008). Ecological interface design for military command and control. *Journal of Cognitive Engineering and Decision Making*, 2(4), 349–385.

Bransford, J. D., Brown A. L., and Cocking, R. R. (2000). *How People Learn: Brain, Mind, Experience, and School*. Washington, DC: National Academy Press.

De Groot, A. (1965). *Thought and Choice in Chess*. The Hague: Mouton.

Duncker, K. (1945). *On Problem Solving* (Psychological Monographs No. 270). Washington, DC: American Psychological Association.

Flach, J. M., & Behymer, K. J. (2016). From designing to enabling effective collaborations. *She Ji, the Journal of Design, Economics, and Innovation*, 2(2), 119–124.

Flach, J. M., & Bennett, K. B. (2018). Improving sensemaking through the design of representations. In P. Smith & R. Hoffman (Eds.), *Cognitive Systems Engineering: The Future for a Changing World* (pp. 165–180). Boca Raton, FL: CRC Press.

Flach, J. M., & Voorhorst, F. A. (2016). *What Matters?* Dayton, OH: Wright State University Library.

Gibson, J. J. (1979). *The Ecological Approach to Visual Perception*. Boston: Houghton Mifflin.

Hajdukiewicz, J. R., & Vicente, K. J. (2002). Designing for adaptation to novelty and change: The role of functional information and emergent features. *Human Factors*, 44, 592–610.

Hall, D. S., Shattuck, L. G., & Bennett, K. B. (2012). Evaluation of an ecological interface design for military command and control. *Journal of Cognitive Engineering and Decision Making*, 6(2), 165–193.

Hollnagel, E., Woods, D. D., & Leveson, N. (2006). *Resilience Engineering: Concepts and Precepts*. Boca Raton, FL: CRC Press.

Jagacinski, R. J., & Flach, J. M. (2003). *Control Theory for Humans: Quantitative Approaches to Modeling Performance*. Mahwah, NJ: Erlbaum.

Kahneman, D. (2011). *Thinking Fast and Slow*. New York: Farrar, Straus & Giroux.

Kahneman, D., & Klein, G. (2009). Conditions for intuitive expertise. *American Psychologist*, 64(6), 115–126.

Klein, G. A. (1989). Recognition-primed decisions. In W. B. Rouse (Ed.), *Advances in Man-Machine System Research*, Vol. 5 (pp. 47–92). Greenwich, CT: JAI Press.

Klein, G. A. (1993). A recognition-primed decision (RPD) model of rapid decision-making. In G. A. Klein, J. Orasanu, R. Calderwood, & C. E. Zsambok (Eds.), *Decision Making in Action: Models and Methods* (pp. 138–147). Norwood, NJ: Ablex.

Klein, G. (2004). *The Power of Intuition*. New York: Currency Book/Doubleday.

Klein, G. (2015) A naturalistic decision-making perspective on studying intuitive decision-making. *Journal of Applied Research in Memory and Cognition*, 4, 164–168.

Lindblom, C. E., 1959. The science of "muddling through," *Public Administration Review*, 19(2), 79–88.

Lindblom, C. E., 1979. Still muddling, not yet through. *Public Administration Review*, 39(6), 517–526.

McEwen, T., Flach, J. M., & Elder, N. (2014). Interfaces to medical information systems: Supporting evidence-based practice. *IEEE: Systems, Man, & Cybernetics Annual Meeting*, 341–346. San Diego, CA. (Oct 5–8).

Miller, G. A., Galanter, E., & Pribram, K. H. (1960). *Plans and the Structure of Behavior*. New York: Henry Holt and Company.

Naikar, N. 2013. *Work Domain Analysis*. Boca Raton, FL: CRC Press.

Peirce, C. S. (1878). Deduction, induction, and hypothesis. *Popular Science Monthly*, 13, 470–482.

Piaget, J. 1973. *The Child and Reality*. Translated by R. Arnold. New York: Grossman. Original addition: Problemes de psychologie genetique.

Rasmussen, J., and Vicente, K. J. (1989). Coping with human errors through system design: Implications for ecological interface design. *International Journal of Man–Machine Studies*, 31, 517–534.

Rochlin, G. I. (1998). Essential friction: Error-control in organizational behavior. In N. Akerman (Ed.), *The Necessity of Friction*. Boulder, CO: Westview.

Schwartz, B. (2006). More isn't always better. *Harvard Business Review*, 84(6), 22–22.

Simon, H. A. (1969). *The Sciences of the Artificial*. Cambridge, MA: MIT Press.

Simon, H. A. (1990). Invariants of human behavior. *Annual Review of Psychology*, 41, 1–19.

Sinek, S. (2009). *Start With Why: How Great Leaders Inspire Everyone to Take Action*. New York: Portfolio/Penguin.

Taylor, F. (1913). *The Principles of Scientific Management*. New York: Harper.

Todd, P. M., Gigerenzer, G., & ABC Research Group. (2012). *Ecological Rationality*. New York: Oxford University Press.

Vicente, K. J. (1999). *Cognitive Work Analysis*. Mahwah, NJ: Erlbaum.

Vicente, K. J., and Rasmussen, J. (1992). Ecological interface design: Theoretical foundations. *IEEE Transactions on Systems, Man, and Cybernetics*, SMC-22, 589–606.

Wertheimer, M. (1959). *Productive Thinking*. New York: Harper & Row.

5 Vigilance and Workload in Automated Systems: Patterns of Association, Dissociation, and Insensitivity

James L. Szalma & Victoria L. Claypoole

In the earlier edition of this book, Warm, Dember, & Hancock (1996) summarized the then extant literature on the perceived workload associated with vigilance. In the vigilance studies they reviewed, and those that have been published since, the most commonly used measure of perceived workload is the NASA Task Load Index (TLX; Grier, 2015; Wickens & Hollands, 2000), which provides a weighted average global workload score and six scale scores (ranging from 0–100) computed from six ratings of different sources of perceived workload (Hart & Staveland, 1988). Three of these scales reflect appraisals of task demands (mental demand, physical demand, and temporal demand), and three reflect appraisals of the person's response to the task (perceived performance, effort, and frustration). The TLX is generally considered to be among the best measures of perceived workload (Warm, Matthews, & Finomore, 2008).

Four major conclusions emerged from the review by Warm et al. (1996): (1) contrary to predictions of arousal theory, the workload of vigilance is high; (2) there is a consistent profile of the workload associated with vigilance, such that, in general, the largest contributors to perceived workload are mental demand and frustration; (3) the typical relationship between performance and workload in vigilance is one of *association*, that is, factors that improve performance also reduce perceived workload; and (4) the consistently high workload associated with vigilance supports a resource depletion account of the performance decrement.

The present chapter examines these conclusions in light of research conducted since Warm et al. (1996) published their seminal paper on the topic. In addition, the theoretical and practical implications of workload in vigilance are discussed. As most studies investigating the workload of vigilance have used the NASA-TLX, we restrict our attention here to studies that employed this measure of perceived workload.

THE WORKLOAD OF VIGILANCE IS HIGH

A consistent finding in vigilance research is that the decline in performance with time on watch (the vigilance decrement; Davies & Parasuraman, 1982; See, Howe, Warm, & Dember, 1995) is associated with higher levels of perceived workload. Warm et al. (1996) noted that global workload scores in vigilance tend to be high (e.g., 60–100 range on a 0–100 point scale). These effects have been observed in subsequent research (i.e., Grier, 2015; Warm, Parasuraman, Matthews, 2008), although there have been some cases in which global workload is at or below the midpoint of the scale (e.g., Arrabito et al., 2015; Finomore et al., 2013; Szalma et al., 2004).

THE PROFILE OF WORKLOAD IN VIGILANCE

In the earlier edition of this book, Warm et al. (1996) argued that there was a consistent profile of the workload associated with vigilance tasks, such that mental demand and frustration tend to be the greatest contributors to workload. In addition, they noted that when the event rate is high, temporal demand can also contribute substantially to the workload of vigilance. This profile has largely remained consistent over the last 20 years (i.e., Warm et al., 2008; 2017). However, it has recently been argued that this profile of workload may only be applicable to sensory-based vigilance tasks (Claypoole, 2018).

Cognitive vigilance tasks are defined as those in which symbolic or alphanumeric stimuli are used and the discrimination requires manipulation of these stimuli (e.g., lexical decisions, arithmetic computations) rather than a sensory discrimination (See, Howe, Warm, & Dember, 1995). Claypoole (2018) noted that when assessing the profile of workload in cognitive-based vigilance tasks, which are underrepresented in the extant literature, this profile tends to include effort as a primary contributor rather than frustration (although mental demand remains a substantial contributor to workload). This profile of workload have been similarly reported in other studies employing a cognitive vigilance task (Claypoole, Dever, Denues, & Szalma, 2018; Claypoole & Szalma, 2018).

Thus, it is possible that the profile of workload may be more dependent on task characteristics than previously suggested. In line with this assertion, recent research has demonstrated that when directly comparing the workload of cognitive and sensory-based vigilance tasks, two distinct profiles tend to emerge: cognitive vigils were associated with high mental demand and effort, while sensory vigils were associated with high mental demand and frustration (Claypoole, Neigel, Waldfogle, Fraulini, et al., 2018).

PERFORMANCE-WORKLOAD ASSOCIATIONS

In addition to the information provided by measuring perceived workload, further insight into the demands of a task can be found by examining the pattern of relationships between performance and perceived workload. *Associations* are defined as cases in which factors that impair performance also increase workload (Hancock, 1996; Yeh & Wickens, 1988). In such cases, performance and workload measures reflect mental capacity changes in the same direction, i.e., depletion would be expected to lead to impaired performance and increased workload.

However, patterns in which the measures do not agree can provide diagnostic information regarding the effects of a given factor or manipulation on the cognitive state of the person performing the task (Hancock, 1996; Hancock & Matthews, 2019). *Dissociation* can occur when a factor or experimental manipulation changes performance in a direction opposite that of perceived workload (Hancock, 1996; Yeh & Wickens, 1988). That is, one measure indicates a decline in capacity while the other indicates an increase in capacity. For instance, an experimental manipulation that impairs performance would be associated with lower workload. The former effect is indicative of resource depletion, while the latter effect reflects an increase in or recruitment of mental capacity. This may indicate disengagement from the task — a reduction in effort (Hancock, 1996; Hancock & Matthews, 2019). Dissociation can also occur when a factor improves performance but also increases workload. This indicates that performance is improved but at the cost of increased capacity demand (Hancock, 1996; c.f. Hockey, 1997).

Insensitivity occurs when a factor or experimental manipulation affects one measure but has no substantive effect on the other (Hancock, 1996). Although it is possible that a given factor may not show differences in performance or workload because of measurement issues (e.g., floor or ceiling effects; Hancock, 1996), these conditions too can be diagnostic. If performance is stable but workload increases as a function of an experimental manipulation, this suggests that stable performance is maintained at the cost of greater effort (a *performance insensitivity*). In some cases, this may reflect adaptation to task-based stress as predicted by stress theory (Hancock & Warm, 1989; Hockey, 1997). Cases in which performance is stable but workload decreases may be interpreted as an adaptive response in which skill acquisition or strategic shifts in effort allocation allow the person to maintain performance using fewer cognitive resources.

Workload insensitivity occurs when an experimental manipulation affects performance but has no substantive effect on workload. This pattern generally indicates that participants are insensitive to their own response to task demands. Performance increases or decreases as a function of the manipulation, but participants do not consider their performance (or they are unaware of the quality of their performance) when rating the workload of the task.

PERFORMANCE-WORKLOAD ASSOCIATIONS IN VIGILANCE

Research since the publication of Warm et al. (1996) has been somewhat less consistent with respect to performance-workload associations. There have been cases in which workload insensitivity was observed (e.g., Catanzaro & Scerbo, 1999; Claypoole & Szalma, 2018; Dillard et al., 2014; Epling, Russell, & Helton, 2016; Grier et al., 2003; Helton et al., 2005; Hitchcock et al., 1999; Szalma et al., 2004), fewer cases in which performance insensitivity was observed (e.g., Arrabito et al., 2015; Bunce & Sisa, 2002; Deaton & Parasuraman, 1993; Ross, Russell, & Helton, 2014; Sawin & Scerbo, 1995), and one case in which dissociation was observed (Claypoole & Szalma, 2018). However, in the majority of cases, performance-workload association is observed (e.g., Arrabito et al., 2015; Claypoole & Szalma, 2018; Finomore et al., 2013;

Gunn et al., 2005; Hollander et al., 2004; Kamzanova, Kustubayeva, & Matthews, 2014; Laurie-Rose, Curtindale, & Frey, 2017; Matthews, 1996; Temple et al., 2000; Teo & Szalma, 2011). These effects vary across different measures of performance (hit, false alarm, response time, sensitivity, response bias) as well as the task characteristics of the vigil.

VIGILANCE TAXONOMY

One of the major conclusions of Warm et al. (1996) was that the taxonomic factors of vigilance can moderate the workload experienced by the observer. Thus, each of the taxonomic categories has been examined for workload effects in at least one experiment. This work was in part intended to further validate the vigilance taxonomy, but it also served to establish that the workload of vigilance can be brought under psychophysical control by manipulation of the characteristics of the task. The research over the past 20 years has generally supported the contention that the workload of vigilance is moderated by factors that also influence performance. These cases are the aforementioned performance-workload associations. However, there have been exceptions (i.e., insensitivities and dissociations), and these may serve to identify potentially fruitful avenues for future research.

In addition to taxonomic factors, other stimulus factors have been examined in vigilance research, such as spatial uncertainty, signal regularity (i.e., temporal uncertainty), memory load, signal probability, and signal salience and feature presence versus absence. However, there have also been studies that manipulated non-stimulus factors, such as instruction manipulations, rest breaks, participant versus experimenter control of task parameters, social presence, and coactors. For purposes of this review, the associations, dissociations, and insensitivities will be examined for studies grouped into those that manipulated stimulus factors and those that manipulated non-stimulus factors. See Table 5.1 for a list of the specific manipulations.

TABLE 5.1
Experimental Manipulations Related to Stimulus and Non-Stimulus Factors

Stimulus Factors	Non-Stimulus Factors
Type of Discrimination (simultaneous/successive)	Coactors
Task Type (cognitive/sensory)	Instructions (detection vs. relaxation emphasis)
Event Rate	Rest Breaks/Task Switching
Sensory Modality	Social Presence
Source Complexity (Single vs. Multiple Displays)	Experimenter vs. Participant Control of Task
Feature presence/absence	KR/cueing
Feature integration/configuration	
Signal Probability	
Signal Salience	
Signal Regularity	
Temporal Uncertainty	
Spatial Uncertainty	
Working Memory Demand/Memory Load	
Single vs. Dual Task	
Response Instructions (Standard vs. SART)	
Demand Transitions	

STIMULUS FACTORS

CATEGORIES OF THE VIGILANCE TAXONOMY

Sensory Modality

Traditionally, sensory modality in vigilance is categorized as either auditory or visual stimuli, though visual tasks are most commonly used throughout the literature. Of the limited experiments directly comparing the perceived workload associated with auditory and visual vigilance, the results are mixed. For instance, Arrabito et al. (2015) reported that auditory tasks result in lower workload relative to visual vigilance tasks. These workload effects were accompanied by performance differences, such that performance on an auditory task was better than that of a visual task — an instance of performance-workload association. In contrast, Szalma et al. (2004) reported no significant difference between modalities in perceived workload (i.e., a workload insensitivity). This is particularly noteworthy as both experiments employed the same tasks. Thus, it is unclear exactly how sensory modality influences the workload associated with vigilance and the extent to which there are performance-workload associations or workload insensitivities.

Event Rate

Event rate, or the rate of presentation of nontarget stimulus events (Parasuraman, 1979), is typically inversely related to performance, such that as event rate increases, performance is impaired (Galinsky, Rosa, Warm, & Dember, 1993; Jerison & Pickett, 1964; Lanzetta, Dember, Warm, & Berch, 1987; Warm & Jerison, 1984). Event rate has also been argued to be directly related to perceived workload, such that as event rate increases, perceived workload also increases (e.g., Claypoole, Dever, et al., 2018). For relatively simple perceptual stimuli, Parasuraman and Davies (1977) have classified a "low" event rate as under 24 events per minute and a "high" event rate as over 24 events per minute, although most studies classifying event rate as "high" tend to utilize an event rate of over 30 events per minute (e.g., Galinsky, Dember, & Warm, 1989; Jerison & Pickett, 1964; Mouloua & Parasuraman, 1995; Parasuraman, 1979; Parasuraman & Giambra, 1991; Rose, Murphy, Schickedantz, and Tucci, 2001).

The effects of event rate on perceived workload are mixed, though research has indicated that higher event rates are generally associated with poorer performance and higher workload — a performance-workload association (e.g., Claypoole, Dever, et al., 2018; Tiwari, Singh, & Singh, 2009). However, it also been demonstrated that workload is not always affected by event rate (i.e., a workload insensitivity). For instance, one experiment demonstrated that while increasing event rate led to higher sensitivity, it did not influence perceived workload (Prytz & Scerbo, 2015). Similarly Yadav, Dubey & Singh (2016) reported that increasing event rate led to poorer detection performance, but again, workload was not affected.

Interestingly, one study indicated that event rate did not influence performance or workload (i.e., Bush, 2002). In this unpublished doctoral dissertation, low (i.e., 8 and 24) and high (i.e., 40 and 56) levels of event rate did not affect performance or perceived workload. However, it should be noted that correct detection rates were relatively high throughout the course of the vigil (i.e., higher than 85%) and workload was well below the midpoint (i.e., scores of 35 or less). Thus, in this case, other task characteristics likely influenced performance and workload more so than event rate.

Discrimination Type

Vigilance tasks are also categorized according to the type of signal and nonsignal discrimination required by the observer. Signal discrimination types include successive, or absolute, discriminations, and simultaneous, or comparative, discriminations (Parasuraman, 1979; Parasuraman & Davies, 1977; See, Howe, Warm, & Dember, 1995). It has been demonstrated that vigils requiring successive-discrimination result in poorer performance and more pronounced sensitivity decrements relative to simultaneous vigils, although this may vary as a function of event rate (See et al., 1995).This effect has been attributed to greater working memory load associated with successive relative to simultaneous tasks (Caggiano & Parasuraman, 2004; Warm & Dember, 1998); however, this effect has not always been observed (Grubb, Warm, Dember, & Berch, 1995; Miller, Warm, Dember, & Schumsky, 1998; Scerbo, Greenwald, & Sawin, 1993).

With respect to performance-workload relationships, some studies have indicated a pattern of performance-workload association for discrimination type, such that successive vigilance tasks typically result in better performance and lower perceived workload relative to simultaneous vigilance tasks (i.e., Grubb et al., 1995; Miller, 1998). However, Scerbo et al. (1993) reported an instance of insensitivity for performance and workload for discrimination type, such that neither performance nor workload differed significantly between the successive and simultaneous task. Thus, the effects of discrimination type on perceived workload is unclear as relatively few studies have singularly compared successive and simultaneous vigilances tasks without also investigating other factors of the taxonomy, such as event rate or source complexity.

Task Type

Although four dimensions comprised the original taxonomy (i.e., Davies & Parasuraman, 1977), a meta-analysis of the vigilance decrement found that the type of task, a sensory or cognitive discrimination, also moderated the decrement (i.e., See et al., 1995). Moreover, this new dimension of "task type" did so in the context of an interaction with discrimination type and event rate. Generally, cognitive-based vigilance tasks result in poorer detection performance relative to their sensory-based counterparts (i.e., Deaton & Parasuraman, 1993; Gunn et al., 2005; Teo & Szalma, 2011), though some cases have demonstrated a performance increment rather than a decrement (Dember, Warm, Bowers, & Lanzetta, 1984; See et al., 1995; Warm, Howe, et al., 1984).

With respect to the performance-workload relationship for task type, association tends to occur such that sensory-based detection performance is higher and perceived workload is lower relative to cognitive-based detection performance and workload (Deaton & Parasuraman, 1993; Gunn et al., 2005; Teo & Szalma, 2011). Moreover, cognitive vigils can be extremely effortful and demanding (Claypoole, Dever, et al., 2018; Mouloua & Parasuraman, 1995), especially relative to sensory-based vigilance tasks (Deaton & Parasuraman, 1993).

The profile of the six sources of workload in cognitive-based vigilance tasks may differ slightly than that previously observed in sensory-based vigils. For instance, while *mental demand* and *frustration* seem to be the key contributors to workload in sensory-based vigilance (Warm et al., 1996), *mental demand* and *effort* tend to be the greatest contributors in cognitive vigilance (Bunce & Sisa, 2002; Claypoole, 2018; Claypoole, Dever, et al., 2018; Claypoole & Szalma, 2018; Epling, Russell, & Helton, 2016; Laurie-Rose et al., 2017; Shaw, Finomore, Warm, & Matthews, 2012). Frustration does not seem to play a key role in the profile of workload for cognitive vigilance (but see Epling et al., 2016). Moreover, in one study that directly compared sensory and cognitive vigils, two distinct profiles of workload were observed: cognitive vigils were associated with high mental demand and effort, while sensory vigils were associated with high mental demand and frustration (Claypoole, Neigel, et al., 2018). However, there is a limited number of direct comparisons of the profile of workload associated with cognitive and sensory vigils.

Source Complexity

Source complexity refers to the number of displays that are monitored (e.g., one display versus four displays) during the vigil (Parasuraman, 1979; Parasuraman & Davies, 1977; See et al., 1995). Typically, increasing the number of displays results in diminished performance (Parasuraman & Davies, 1977; Parasuraman, Warm, & Dember, 1987) and can also impose additional workload (Devlin & Riggs, 2018; Donald, 2008; Grubb, Warm, Dember, & Berch, 1995; Teo & Szalma, 2011; Warm et al., 1996). For instance, Teo and Szalma (2011) reported that increasing the complexity from one display to two or four displays resulted in greater perceived workload and lower performance. Specifically, the four-display condition was associated with the lowest performance and highest perceived workload relative to the two-display and one-display conditions. Interestingly, while the two-display condition was associated with lower performance relative to the one-display condition, perceived workload for these conditions did not differ significantly from one another. This suggests that there may be a task load "threshold" before the workload effects of source complexity can manifest. That is, there may be levels of task load at which workload insensitivity occurs (performance differs but not workload), and that levels of task load higher than this are required for association to occur. This would be consistent with the dynamic adaptability model (Hancock & Warm, 1989), which argues that subjective comfort declines at less extreme levels of stress or task load than performance impairment.

It has been demonstrated that the effect of source complexity on perceived workload may vary as a function of discrimination type. Grubb et al. (1995) reported that the rate of gain in workload, as a function of the number of displays, was higher for a simultaneous, compared to successive, task. This workload effect paralleled a steeper decline in detections as a function of a simultaneous compared to a successive task. With respect to the individual scales, mental demand, temporal demand, and frustration were the strongest contributors, which is consistent with the previous profile of workload (Warm et al., 1996). Furthermore, mental and temporal demand were each higher for the four-display condition relative to the two- or one-display condition. Thus, increasing source complexity can increase global workload as well as the individual subscales that contribute most to that workload.

OTHER STIMULUS FACTORS

In addition to the taxonomic categories, extensive research has examined the detection performance and perceived workload of vigilance as a function of manipulations of the psychophysical characteristics of the task. These include signal saliency, signal probability, spatial and temporal uncertainty, and discriminations of target feature presence versus absence.

Signal Salience

Performance efficiency in vigilance varies with the discriminability, or the saliency, of the critical signals to be detected (Helton & Warm, 2008; See et al., 1995). Note that this has also been described as signal conspicuity in the previous literature (e.g., Warm & Jerison, 1984). It has been well established that as signal saliency decreases, performance is impaired (Helton, Shaw, Warm, Matthews, & Hancock, 2008; Helton & Warm, 2008; See et al., 1995; Shaw, Parasuraman, Sikdar, & Warm, 2009; Temple et al., 2008). The effect of signal saliency on vigilance performance is argued to support a resource theory approach to vigilance (i.e., Helton & Warm, 2008) such that as signal saliency decreases, resulting in a more demanding task, resources are depleted at a faster rate. Therefore, performance decrements are steeper in low signal saliency tasks (i.e., Helton & Warm, 2008; Helton et al., 2002; Temple et al., 2000; Warm & Jerison, 1984).

In regard to workload, an inverse relationship with signal saliency is typically observed (i.e., Temple et al., 2000). For instance, Temple et al. (2000) reported that lower signal salience was associated with fewer correct detections and higher perceived workload, with mental and temporal demand being the strongest contributors. The latter scale was likely high in this case because the event rate for the task employed was very high (57.1 events/minute) as high event rates have been associated with high perceived temporal demand (i.e., Warm et al., 1996). Moreover, previous research has used signal saliency as a manipulation of workload. For instance, Helton et al. (2008) lowered signal saliency to increase perceived workload in a demand transition task. In general, manipulations of signal saliency result in performance-workload associations.

Signal Probability

Signal probability refers to the frequency of signal events relative to the total number of events that occur. Traditionally, vigilance tasks involve responding to rare, or "low prevalence," targets over time (Sawyer & Hancock, 2018; Warm & Jerison, 1984). Previous research has indicated that increasing signal probability results in higher detection performance (Davies & Parasuraman, 1982; Methot & Huitema, 1998). However, event rate may offset this trend (i.e., Matthews, 1996; Warm & Jerison, 1984).

The previous, albeit very limited, research on the effects of signal probability on perceived workload have indicated a performance-workload association. For instance, Matthews (1996) reported that in a high event rate vigil, high signal probability was associated with poorer detection performance and higher perceived workload relative to low signal probability. These results indicate that when high signal probability is combined with high event rate, workload is exacerbated, which supports a resource-depletion account (Matthews, 1996). However, it is unclear whether workload is influenced by signal probability when event rate is low. Thus, future work should examine how the effects of signal probability may be moderated by other taxonomic factors.

Spatial Uncertainty

The difficulty of vigilance tasks can also be affected by the spatial uncertainty in where in a display signals may appear (Warm et al., 1996; Warm & Jerison, 1984). Stimuli can be displayed in either a fixed location or a fixed sequence of locations (i.e., spatial certainty), or they can appear in one of several possible locations within a display (i.e., spatial uncertainty). Typically, detection performance is poorer when the spatial location of the target is uncertain relative to when the spatial location is fixed or predictable (e.g., Helton, Weil, Middlemiss, & Sawers, 2010; Laurie-Rose et al., 2017; Mouloua & Parasuraman, 1995; Sullivan, 1991; Warm, Dember, Murphy, & Dittmar, 1992; Warm et al., 1996; Warm et al., 2008; cf. Head & Helton, 2013).

Moreover, previous research has demonstrated a performance-workload association for spatial uncertainty, such that uncertainty results in poorer performance and higher workload relative to spatial certainty (Laurie-Rose et al., 2017; Sullivan, 1991). For instance, Laurie-Rose et al. (2017) assessed the effects of spatial uncertainty on vigilance performance and perceived workload in children. Their results indicated that, similar to the effects observed in adults, children reported higher workload and performed worse on the task when targets were positioned in a spatially uncertain location relative to a fixed location.

Signal Irregularity (Temporal Uncertainty)

Several investigations have established that uncertainty regarding the timing of signal presentation results in poorer performance relative to a condition in which signals appear at regular or predictable intervals (e.g., Funke et al., 2017; Hollander et al., 2002; Shaw, Finomore, Warm, & Matthews, 2012). However, with one exception, these studies did not report a measure of perceived workload. In the one exception, Helton et al. (2005) reported that performance was better for regular signals relative to the performance of those who experienced signals that appeared on an irregular schedule. However, there were no significant effects on perceived workload, indicating a workload insensitivity.

Presence versus Absences of Stimulus Features

Signal detection in sustained attention paradigms is also sensitive to the search asymmetry phenomenon (Treisman & Gormican, 1988) such that feature-absent detections are more difficult and require more processing time than feature-present detections (Warm & Parasuraman, 2007). Previous research has indicated that detecting the absence of a feature is more capacity demanding than detecting the presence of a feature (Helton & Russell, 2011; Schoenfeld & Scerbo, 1997, 1999). Moreover, it results in slower response times (Schoenfeld & Scerbo, 1997, 1999) and poorer detection performance (Finomore et al., 2013; Helton & Russell, 2011;

Hollander et al., 2004; cf. Catanzaro & Scerbo, 1999). This effect is moderated by event rate such that as event rate increases, performance decrements are steeper for feature-absent tasks relative to feature-present tasks (Hollander et al., 2004; Warm & Parasuraman, 2007).

The lower performance observed in feature-absence tasks is accompanied by higher workload relative to feature-presence tasks, indicating a performance-workload association (Finomore et al., 2013; Hollander et al., 2004; Schoenfeld & Scerbo, 1997, 1999). For instance, Hollander and colleagues (2004) examined the effects of feature presence or absence on a 40-minute vigilance task that varied event rate (i.e., 6, 12, or 24). Their results demonstrated that detection performance varied inversely with event rate for the feature-absent condition but, interestingly, not the feature-present condition. Moreover, workload was higher in the feature-absent conditions relative to the feature present conditions, regardless of event rate. Thus, the results of the previous research indicate that feature-absent tasks are more resource demanding and result in higher workload relative to the traditional feature-present tasks typically observed in vigilance.

Non-Stimulus Factors

In addition to the taxonomic categories and psychophysical characteristics of stimuli and tasks, there are non-stimulus factors that influence the performance and workload of vigilance. These include training the knowledge of results (KR) or cueing, providing rest breaks, manipulating motivation by instructions, control of task settings (specifically, event rate), social presence, and the presence of a coactor. Of the studies that manipulated non-stimulus factors, only three reported performance-workload associations. Hitchcock et al. (1999) and Kamzanova et al. (2014) each reported that cueing participants regarding the appearance of a critical signal improved performance and reduced the perceived workload associated with the task.

In the third study, Claypoole and Szalma (2018) found that performing a vigilance task in the presence of a coactor improved performance (reduced false alarms) and also reduced perceived workload. However, in another study using the same task, Claypoole and Szalma (2017) reported that performing the task in the presence of an evaluative observer (in the role of a supervisor) was associated with higher performance (perceptual sensitivity) but also higher workload. This pattern represents a dissociation. Taken together, the two studies indicate that the form of social presence may be influenced by the social role of the person who is present while a participant completes a vigil.

Performance insensitivities were obtained in the context of manipulations of instructions that emphasized detections versus relaxation (Sawin & Scerbo, 1995), of manipulation of rest breaks (Arrabito et al., 2015; Ross et al., 2014), social presence (Claypoole & Szalma, 2018), and providing success versus failure feedback (Singh, Tiwari, & Singh, 2010). In three of these cases, perceived workload differences were observed in the expected direction (i.e., relaxation instructions resulted in lower workload than detection instructions; rest breaks reduced workload). However, Singh et al. (2010) reported that success feedback was associated with higher workload than failure feedback, and Claypoole and Szalma (2018) reported that the presence of a coactor reduced the workload associated with vigilance.

However, these results are not always consistent. Funke et al. (2016) reported that participants in a coactor condition performed poorer than those who completed the task alone, but the perceived workload for these conditions did not substantially differ from one another (workload insensitivity). In contrast to Arrabito et al. (2015) and Ross et al. (2014), Finkbeiner et al. (2016) reported that there were no differences in perceived workload between a group that was provided a rest break and one that did not. Further, Helton and Russell (2012) reported that providing rest breaks did not affect performance or workload.

Three other manipulations known to affect motivation also have yielded workload insensitivities in some studies. Thus, KR has been associated with better performance without affecting perceived workload (Hitchcock et al., 1999; Szalma, 1997). Allowing participants to not only respond to signals but to take subsequent action to deal with them (Parsons, 2007), and allowing participants to determine the pace of stimulus presentation (Scerbo et al., 1993), each improves performance but has no effect on perceived workload.

Providing instructions to respond to nonsignals rather than signals (i.e., the Sustained Attention to Response Task, "SART") is associated with poorer performance relative to the standard instructions (i.e., to respond to signals and withhold response to nonsignals), but response type does not affect perceived workload (Dillard et al., 2014; Grier et al., 2003). However, in two studies (Dillard et al., 2014; Grier et al., 2003) response type did not substantively affect performance or workload.

Patterns of Association, Dissociation, and Insensitivity

Performance-workload association is the most common finding in vigilance research. Of the 56 performance-workload relationships identified in published research, 20 (36%) were associations. However, the dominance of association has been observed in studies that employed psychophysical manipulations, i.e., the characteristics of the display or stimuli. Where associations were not observed, workload insensitivity (17 of 56; 30%) was more common than performance insensitivity (10 of 56; 18%). Dissociation only occurred in two cases. In both of these studies, the tasks used were simultaneous sensory tasks. One of these studies manipulated stimulus characteristics, and the other study manipulated a non-stimulus factor. Apparently, dissociation is the exception rather than the rule in vigilance.

Association is less common in studies that manipulate non-stimulus factors. Of the 56 relationships examined here, the majority (35) were in studies that investigated the effects of stimulus manipulations. The remaining relationships (21) were from studies that manipulated non-stimulus factors, such as instructions, rest breaks, training, or social presence. As shown in Figure 5.1, the pattern of relationships differs according to the nature of the experimental manipulations. For studies involving stimulus manipulations, 49% exhibited performance-workload association, with the majority of the other cases being workload insensitivity (26%). By contrast, manipulation of non-stimulus factors tend not to yield association but rather insensitivities. Only 14% of relationships in studies that manipulated non-stimulus factors were associations. The majority were either workload insensitivity (38%) or performance insensitivity (29%). Hence, the effects of non-stimulus factors on resource capacity seem to be less clear than those associated with stimulus factors.

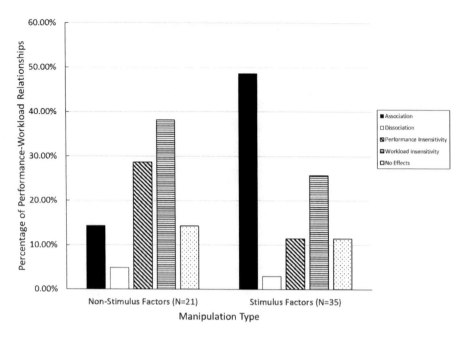

FIGURE 5.1 Percentage of different performance-workload relationships as a function of manipulation type.

The dominance of associations in studies that manipulate stimulus characteristics may be due to these manipulations having direct effects on resource capacity, which is reflected in both performance and workload measures. Non-stimulus manipulations may affect more than just capacity and thus show effects on performance and workload that are distinct from one another. Given the larger proportion of performance insensitivities for non-stimulus manipulations (29%) relative to stimulus manipulations (11%), it may be that the workload differences reflect compensatory effort or dimensions of cognitive state not directly related to resource capacity or effort.

Workload insensitivity is more difficult to interpret given that one cannot generally rule out measurement artifacts (e.g., floor or ceiling effects; Hancock, 1996). However, the fact that a larger proportion of studies showing insensitivities were workload insensitivity for both stimulus and non-stimulus manipulations suggests that there are some causes of performance change that are not detectable from measures of perceived workload. Inconsistencies in associations versus insensitivities across studies that employed similar manipulations indicate that these effects are likely moderated by other characteristics of the task or of the context in which the task is performed.

Note that of the 56 performance-workload relationships examined, only two dissociations occurred, and both of these involved sensory tasks requiring simultaneous discriminations. One of these was in the context of a stimulus manipulation, and the other was associated with a non-stimulus manipulation. Thus, the divergence in the measures of performance and perceived workload (as measured by the TLX) are relatively rare. However, slightly more than one-third of these relationships (20/56) were associations. Hence, the majority of performance-workload relationships display some form of insensitivity (27/56) or no effects of the manipulation on performance or workload (7/56). Together, these are observed in 61% of the relationships that have been reported in the literature. It is likely that although performance and workload measures both capture resource capacity or effort allocation, each also reflects other structures or processes not detectable via the other measure.

CONCLUSIONS

Over 20 years ago, Warm et al. (1996) summarized the evidence that the perceived workload of vigilance could be brought under psychophysical control, such that manipulations that impaired performance also increased workload. The pattern of associations provided evidence in support of the capacity depletion account of the vigilance decrement and evidence against the mindlessness and mind-wandering theories. If these latter theories are correct, the perceived workload of vigilance would be low, particularly for mental demand and effort. A large number of studies have clearly established that this is not the case. Further, the workload effects of vigilance are moderated by the characteristics of the task in a manner consistent with a capacity depletion explanation of the decrement, although there are inconsistencies in which insensitivities were observed rather than associations. These inconsistencies should be resolved by further empirical investigation as there are some task manipulations that have been examined by only a few studies (e.g., sensory modality, source complexity).

Most of the effects observed in vigilance research have been either associations or insensitivities, but the type of manipulation may affect which of these patterns are observed. Specifically, the dominant effect of stimulus factors seems to be associations, while the dominant effect of non-stimulus factors seems to be insensitivity. The former pattern indicates that performance and perceived workload are each manifestations of cognitive capacity available for allocation to the task. However, the dominance of insensitivity for non-stimulus factors suggests that for these manipulations, workload and performance may not reflect the same underlying processes. Perceived workload seems to be affected more by the features of the display rather than by factors related more to the context of the task. This distinction may be useful for differentiating vigilance effects on both performance and workload in future research.

The patterns of effects discussed in this chapter also indicate that there are likely interactive effects of stimulus or non-stimulus factors on performance-workload relationships. That is, the effects of a given manipulation may be moderated by variations in other stimulus or non-stimulus factors. The current literature is not sufficient in size to thoroughly evaluate these moderating effects, so future research efforts should be directed toward generating sufficient empirical evidence for such analyses.

REFERENCES

Arrabito, G. R., Ho, G., Aghaei, B., Burns, C., & Hou, M. (2015). Sustained attention in auditory and visual monitoring tasks: Evaluation of the administration of a rest break or exogenous vibrotactile signals. *Human Factors*, 57(8), 1403–1416.

Bunce, D., & Sisa, L. (2002). Age differences in perceived workload across a short vigil. *Ergonomics*, 45(13), 949–960.

Bush, J. M. (2002). The effect of event rate on sustained attention and stress states in a simultaneous vigilance task paradigm (Doctoral dissertation, Texas Tech University).

Caggiano, D. M., & Parasuraman, R. (2004). The role of memory representation in the vigilance decrement. *Psychonomic Bulletin & Review*, *11*(5), 932–937.

Catanzaro, J. M., & Scerbo, M. W. (1999). Searching for the presence and absence of features in vigilance: The decrement persists. In *Proceedings of the Human Factors and Ergonomics Society Annual Meeting* (Vol. 43, No. 23, pp. 1285–1288). Los Angeles, CA: SAGE Publications.

Claypoole V. L. (2018). Understanding human performance and social presence: Toward a taxonomic framework and analysis of social facilitation and vigilance (Doctoral dissertation, University of Central Florida).

Claypoole, V. L., & Szalma, J. L. (2017). Examining social facilitation in vigilance: A hit and a miss. *Ergonomics*, *60*(11), 1485–1499.

Claypoole, V. L., & Szalma, J. L. (2018). Independent co-actors may improve performance and lower workload: Viewing vigilance under social facilitation. *Human Factors*, *60*(6), 822–832.

Claypoole, V. L., Dever, D. A., Denues, K. L., & Szalma, J. L. (2018). The effects of event rate on cognitive-based vigilance tasks. *Human Factors*, *61*(3), 440–450. https://doi.org/10.1177/0018720818790840

Claypoole, V. L., Neigel, A. R., Fraulini, N. W., Hancock, G. M., & Szalma, J. L. (2018). Can vigilance tasks be administered online? A replication and discussion. *Journal of Experimental Psychology: Human Perception and Performance*, *44*(9), 1348–1355.

Davies, D. R., & Parasuraman, R. (1977). Cortical evoked potentials and vigilance: A decision theory analysis. In Mackie, R.R. (Eds.) *Vigilance* (pp. 285–306). Boston, MA: Springer.

Davies, D. R., & Parasuraman, R. (1982). *The Psychology of Vigilance*. London, UK: Academic.

Deaton, J. E., & Parasuraman, R. (1993). Sensory and cognitive vigilance: Effects of age on performance and subjective workload. *Human Performance*, *6*(1), 71–97.

Dember, W. N., Warm, J. S., Bowers, J. C., & Lanzetta, T. (1984). Intrinsic motivation and the vigilance decrement. *Trends in Ergonomics/Human Factors*, *1*, 21–26.

Devlin, S. P., & Riggs, S. L. (2018). The effect of video game experience and the ability to handle workload and workload transitions. In *Proceedings of the Human Factors and Ergonomics Society Annual Meeting* (Vol. 62, No. 1, 736–740). Los Angeles, CA: SAGE Publications.

Dillard, M. B., Warm, J. S., Funke, G. J., Funke, M. E., Finomore Jr, V. S., Matthews, G., ... & Parasuraman, R. (2014). The sustained attention to response task (SART) does not promote mindlessness during vigilance performance. *Human factors*, *56*(8), 1364–1379.

Donald, F. M. (2008). The classification of vigilance tasks in the real world. *Ergonomics*, *51*(11), 1643–1655.

Epling, S. L., Russell, P. N., & Helton, W. S. (2016). A new semantic vigilance task: Vigilance decrement, workload, and sensitivity to dual-task costs. *Experimental Brain Research*, *234*(1), 133–139.

Finkbeiner, K. M., Russell, P. N., & Helton, W. S. (2016). Rest improves performance, nature improves happiness: Assessment of break periods on the abbreviated vigilance task. *Consciousness and cognition*, *42*, 277–285.

Finomore, V. S., Jr., Shaw, T. H., Warm, J. S., Matthews, G., & Boles, D. B. (2013). Viewing the workload of vigilance through the lenses of the NASA-TLX and the MRQ. *Human Factors*, *55*(6), 1044–1063.

Funke, G. J., Warm, J. S., Baldwin, C. L., Garcia, A., Funke, M. E., Dillard, M. B., ... & Greenlee, E. T. (2016). The independence and interdependence of coacting observers in regard to performance efficiency, workload, and stress in a vigilance task. *Human factors*, *58*(6), 915–926.

Funke, M. E., Warm, J. S., Matthews, G., Funke, G. J., Chiu, P. Y., Shaw, T. H., & Greenlee, E. T. (2017). The neuroergonomics of vigilance: Effects of spatial uncertainty on cerebral blood flow velocity and oculomotor fatigue. *Human factors*, *59*(1), 62–75.

Galinsky, T. L., Dember, W. N., & Warm, J. S. (1989). Effects of event rate on subjective workload in vigilance performance. In *Meeting of the Southern Society for Philosophy and Psychology*, New Orleans, LA.

Galinsky, T. L., Rosa, R. R., Warm, J. S., & Dember, W. N. (1993). Psychophysical determinants of stress in sustained attention. *Human Factors*, *35*(4), 603–614.

Grier, R. A. (2015). How high is high? A meta-analysis of NASA-TLX global workload scores. In *Proceedings of the Human Factors and Ergonomics Society Annual Meeting* (Vol. 59, No. 1, pp. 1727–1731). Los Angeles, CA: SAGE Publications.

Grier, R. A., Warm, J. S., Dember, W. N., Matthews, G., Galinsky, T. L., Szalma, J. L., & Parasuraman, R. (2003). The vigilance decrement reflects limitations in effortful attention, not mindlessness. *Human factors*, *45*(3), 349–359.

Grubb, P. L., Warm, J. S., Dember, W. N., and Berch, D. B. (1995). Effects of multiple signal discrimination on vigilance performance and perceived workload. *Proceedings of the Human Factors and Ergonomics Society*, *39*, 1360–1364.

Gunn, D. V., Warm, J. S., Nelson, W. T., Bolia, R. S., Schumsky, D. A., & Corcoran, K. J. (2005). Target acquisition with UAVs: Vigilance displays and advanced cuing interfaces. *Human Factors*, *47*(3), 488–497.

Hancock, P. A. (1996). Effects of control order, augmented feedback, input device and practice on tracking performance and perceived workload. *Ergonomics*, *39*, 1146–1162.

Hancock, P. A., & Warm, J. S. (1989). A dynamic model of stress and sustained attention. *Human factors*, *31*(5), 519–537.

Hancock, P.A., & Matthews, G. (2019). Workload and performance: Associations, insensitivities, and dissociations. *Human Factors*, *61*(3), 374–392.

Hart, S. G., & Staveland, L. E. (1988). Development of NASA-TLX (Task Load Index): Results of empirical and theoretical research. In: P.A. Hancock and N. Meshkati (Eds.). *Human Mental Workload* (pp. 139–183). Amsterdam, The Netherlands: Elsevier.

Head, J., & Helton, W. S. (2013). Perceptual decoupling or motor decoupling? *Consciousness and Cognition*, *22*(3), 913–919.

Helton W. S., & Russell, P. N. (2011). Working memory load and the vigilance decrement. *Experimental Brain Research*, *212*, 429–437.

Helton, W. S., & Russell, P. N. (2012). Brief mental breaks and content-free cues may not keep you focused. *Experimental brain research*, *219*(1), 37–46.

Helton, W. S., & Warm, J. S. (2008). Signal salience and the mindlessness theory of vigilance. *Acta Psychologica*, *129*(1), 18–25.

Helton, W. S., Hollander, T. D., Warm, J. S., Matthews, G., Dember, W. N., Wallaart, M., ... & Hancock, P. A. (2005). Signal regularity and the mindlessness model of vigilance. *British Journal of Psychology*, *96*(2), 249–261.

Helton, W. S., Shaw, T., Warm, J. S., Matthews, G., & Hancock, P. (2008). Effects of warned and unwarned demand transitions on vigilance performance and stress. *Anxiety, Stress, & Coping*, *21*(2), 173–184.

Helton, W. S., Warm, J. S., Matthews, G., Corcoran, K. J., & Dember, W. N. (2002). Further tests of an abbreviated vigilance task: Effects of signal salience and jet aircraft noise on performance and stress. In *Proceedings of the Human Factors and Ergonomics Society Annual Meeting* (Vol. 46, No. 17, pp. 1546–1550). Los Angeles, CA: SAGE Publications.

Helton, W. S., Weil, L., Middlemiss, A., & Sawers, A. (2010). Global interference and spatial uncertainty in the Sustained Attention to Response Task (SART). *Consciousness and Cognition*, *19*(1), 77–85.

Hitchcock, E. M., Dember, W. N., Warm, J. S., Moroney, B. W., & See, J. E. (1999). Effects of cueing and knowledge of results on workload and boredom in sustained attention. *Human factors, 41*(3), 365–372.

Hockey, G. R. J. (1997). Compensatory control in the regulation of human performance under stress and high workload: A cognitive-energetical framework. *Biological Psychology, 45*(1-3), 73–93.

Hollander, T. D., Warm, J. S., Matthews, G., Shockley, K., Dember, W. N., Weiler, E. M., Tripp, L. D., & Scerbo, M. W. (2004) Feature presence/absence modifies the event rate effect and cerebral hemovelocity in vigilance performance. *Proceedings of the Human Factors and Ergonomics Society, 48,* 1943–1947.

Jerison, H. J., & Pickett, R. M. (1964). Vigilance: The importance of the elicited observing rate. *Science, 143*(3609), 970–971.

Kamzanova, A. T., Kustubayeva, A. M., & Matthews, G. (2014). Use of EEG workload indices for diagnostic monitoring of vigilance decrement. *Human Factors, 56*(6), 1136–1149.

Lanzetta, T. M., Dember, W. N., Warm, J. S., & Berch, D. B. (1987). Effects of task type and stimulus heterogeneity on the event rate function in sustained attention. *Human Factors, 29*(6), 625–633.

Laurie-Rose, C., Curtindale, L. M., & Frey, M. (2017). Measuring sustained attention and perceived workload: A test with children. *Human Factors, 59*(1), 76–90.

Matthews, G. (1996). Signal probability effects on high-workload vigilance tasks. *Psychonomic Bulletin & Review, 3*(3), 339–343.

Methot, L. L., & Huitema, B. E. (1998). Effects of signal probability on individual differences in vigilance. *Human Factors, 40*(1), 102–110.

Miller, L.C., Warm, J. S., Dember, W. N., & Schumsky, D. A. (1998). Sustained attention and feature-integrative displays. *Proceedings of Human Factors and Ergonomics Society, 42,* 1585–1589.

Mouloua, M., & Parasuraman, R. (1995). Aging and cognitive vigilance: Effects of spatial uncertainty and event rate. *Experimental Aging Research, 21*(1), 17–32.

Parasuraman, R. (1979). Memory load and event rate control sensitivity decrements in sustained attention. *Science, 205,* 924–927.

Parasuraman, R., & Davies, D. (1977). A taxonomic analysis of vigilance performance. In: R. Mackie (Ed.), *Vigilance: Theory, Operational Performance and Physiological Correlates.* New York: Plenum Press (pp. 559–574).

Parasuraman, R., & Giambra, L. (1991). Skill development in vigilance: Effects of event rate and age. *Psychology and Aging, 6*(2), 155.

Parasuraman, R., Warm, J. S., & Dember, W. N. (1987). Vigilance: Taxonomy and utility. In Huston R. L. (Eds.) *Ergonomics and Human Factors, Recent Research in Psychology* (pp. 11–32). New York: Springer.

Parsons, K. S. (2007). Detection-action sequence in vigilance: Effects on workload and stress. Doctoral dissertation, University of Cincinnati.

Prytz, E. G., & Scerbo, M. W. (2015). Changes in stress and subjective workload over time following a workload transition. *Theoretical Issues in Ergonomics Science, 16*(6), 586–605.

Rose, C. L., Murphy, L. B., Schickedantz, B., & Tucci, J. (2001). The effects of event rate and signal probability on children's vigilance. *Journal of Clinical and Experimental Neuropsychology, 23*(2), 215–224.

Ross, H. A., Russell, P. N., & Helton, W. S. (2014). Effects of breaks and goal switches on the vigilance decrement. *Experimental Brain Research, 232*(6), 1729–1737.

Sawin, D. A., & Scerbo, M. W. (1995). The effects of instruction type and boredom proneness in vigilance: Implications for boredom and workload. *Human Factors, 37,* 752–765.

Sawyer, B. D., & Hancock, P. A. (2018). Hacking the human: The prevalence paradox in cybersecurity. *Human Factors, 60*(5), 597–609.

Scerbo, M. W., Greenwald, C. Q., & Sawin, D. A. (1993). The effects of subject-controlled pacing and task type on sustained attention and subjective workload. *The Journal of General Psychology, 120*(3), 293–307.

Schoenfeld, V. S., & Scerbo, M. W. (1997). Search differences for the presence and absence of features in sustained attention. In *Proceedings of the Human Factors and Ergonomics Society Annual Meeting* (Vol. 41, No. 2, pp. 1288–1292). Los Angeles, CA: SAGE Publications.

Schoenfeld, V. S., & Scerbo, M. W. (1999). The effects of search differences for the presence and absence of features on vigilance performance and mental workload. In Scerbo, E. and Mouloua, M. (Eds.) *Automation Technology and Human Performance: Current Research and Trends* (pp. 177–182). Mahwah, NJ: Lawrence Erlbaum Associates.

See, J. E., Howe, S. R., Warm, J. S., & Dember, W. N. (1995). Meta-analysis of the sensitivity decrement in vigilance. *Psychological Bulletin, 117*(2), 230–249.

Shaw, T. H., Funke, M. E., Dillard, M., Funke, G. J., Warm, J. S., & Parasuraman, R. (2013). Event-related cerebral hemodynamics reveal target-specific resource allocation for both "go" and "no-go" response-based vigilance tasks. *Brain and Cognition, 82*(3), 265–273.

Shaw, T. H., Parasuraman, R., Sikdar, S., & Warm, J. (2009). Knowledge of results and signal salience modify vigilance performance and cerebral hemovelocity. *Proceedings of the Human Factors and Ergonomics Society, 53*, 1062–1065.

Singh, A. L., Tiwari, T., & Singh, I. L. (2010). Performance feedback, mental workload and monitoring efficiency. *J Indian Acad Appl Psychol, 36*(1), 151–158.

Sullivan, T. E. (1991). Effects of spatial uncertainty on perceived workload and mood in the vigilance performance of high and low resourceful individuals (Doctoral dissertation, University of Cincinnati).

Szalma, J. L. (1997). *Intraclass and interclass transfer in training for vigilance.* Unpublished Master's Thesis, University of Cincinnati.

Szalma, J. L., Warm, J. S., Matthews, G., Dember, W. N., Weiler, E. M., Meier, A., & Eggemeier, F. T. (2004). Effects of sensory modality and task duration on performance, workload, and stress in sustained attention. *Human Factors, 46*(2), 219–233.

Temple, J. G., Warm, J. S., Dember, W. N., Jones, K. S., LaGrange, C. M., & Matthews, G. (2000). The effects of signal salience and caffeine on performance, workload, and stress in an abbreviated vigilance task. *Human Factors, 42*(2), 183–194.

Teo, G., & Szalma, J. L. (2011). The effects of task type and source complexity on vigilance performance, workload, and stress. In *Proceedings of the Human Factors and Ergonomics Society Annual Meeting* (Vol. 55, No. 1, pp. 1180–1184). Los Angeles, CA: Sage Publications.

Tiwari, T., Singh, A. L., & Singh, I. L. (2009). Task demand and workload: Effects on vigilance performance and stress. *Journal of the Indian Academy of Applied Psychology, 35*(2), 265–275.

Treisman, A., & Gormican, S. (1988). Feature analysis in early vision: Evidence from search asymmetries. *Psychological Review, 95*(1), 15.

Warm, J. S. Parasuraman, R., & Matthews, G. (2008). Vigilance requires hard mental work and is stressful. *Human Factors, 50*(3), 433–441.

Warm, J. S., & Dember, W. N. (1998). Tests of vigilance taxonomy. In R. R. Hoffman, M. F. Sherrick, & J. S. Warm (Eds.), *Viewing Psychology as a Whole: The Integrative Science of William N. Dember* (pp. 87–112). Washington, DC: American Psychological Association.

Warm, J. S., & Jerison, H. J. (1984). The psychophysics of vigilance. In J. S. Warm (Ed.), *Sustained Attention in Human Performance* (pp. 15–59). Chichester, UK: Wiley.

Warm, J. S., & Parasuraman, R. (2007). Cerebral hemodynamics and vigilance. In R. Parasuraman and M. Rizzo Eds.). *Neuroergonomics. The Brain at Work* (pp. 146–158). New York, NY: Oxford University Press.

Warm, J. S., Dember, W. N., & Hancock, P. A. (1996). Vigilance and workload in automated systems. *Automation and Human Performance: Theory and Applications*, 183–200.

Warm, J. S., Dember, W. N., Murphy, A. Z., & Dittmar, M. L. (1992). Sensing and decision-making components of the signal-regularity effect in vigilance performance. *Bulletin of the Psychonomic Society*, *30*(4), 297–300.

Warm, J. S., Howe, S. R., Fishbein, H. D., Dember, W. N., & Sprague, R. L. (1984). Cognitive demand and the vigilance decrement. In A. Mital (Ed.). *Trends in Ergonomics/Human Factors*. Amsterdam, The Netherlands: Elsevier (North-Holland).

Warm, J. S., Matthews, G., & Finomore, V. S. (2008). Vigilance, Workload, and Stress. In P.A. Hancock and J.L. Szalma (Eds.), *Performance under Stress*, (pp. 115–141). Aldershot, Hampshire: Ashgate Publishing.

Wickens, C. D., & Hollands, J. G. (2000). Signal detection, information theory, and absolute judgment. *Engineering Psychology and Human Performance*, *2*, 24–73.

Yadav, A. K., Dubey, S., & Singh, I. L. (2016). Event rate and vigilance: A psychophysiological investigation of mental workload. *Journal of the Indian Academy of Applied Psychology*, *42*(2), 328–336.

Yeh, Y. Y., & Wickens, C. D. (1988). Dissociation of performance and subjective measures of workload. *Human Factors*, *30*(1), 111–120.

6 Theoretical Perspectives on Adaptive Automation

Mark W. Scerbo

Editor's Note: This chapter appeared in the original edition of this book and offered a description of automation from a user performance perspective. The research literature was reviewed, and current topics and emerging trends were discussed. The chapter appears here much as it did in its original format and offers an historical perspective. An updated discussion of these topics by the same author appears in the Parasuraman and Rizzo book, *Neuroergonomics: The Brain at Work*, (2007). Oxford: Oxford University Press.

INTRODUCTION

Adaptive automation is a form of automation that is flexible or dynamic in nature. In adaptive systems, decisions regarding the initiation, cessation, and type of automation are shared between the human operator and machine intelligence. Unlike more traditional forms of automation, adaptive automation can adjust its method of operation based on changing situational demands. Although still in its infancy, researchers and developers alike have begun to espouse the virtues and promise of adaptive automation (Morrison, Gluckman, & Deaton, 1991; Rouse, 1988).

Numerous theories, models, and platforms for delivering adaptive automation have already been proposed. The literature also contains the concerned voices of those who see the potential danger in adaptive automation evolving from a technological impetus instead of a human-centered design philosophy (Billings & Woods, 1994; Wiener, 1989). This is fortunate, for as we shall see, the ultimate success of adaptive automation may rest more with what we know about ourselves than with what we know about technology. This chapter describes the theoretical basis of adaptive automation and surveys research and development efforts aimed at validation of this approach to automation design.

AUTOMATION

Automation can be thought of as the process of allocating activities to a machine or system to perform (Parsons, 1985). These can be entire activities or portions thereof. As Davis and Wacker (1987) have suggested, an activity is automated when it can be performed without human assistance under normal operating conditions. A truly automatic device is typically one that can both detect changes in the environment and affect the environment. A thermostat and the automatic pilot in a cockpit are both examples of automatic devices. Each can detect deviations from some reference point and can take action to return the system to that reference point. What distinguishes the automatic pilot from the thermostat is the number and variety of inputs, the complexity of the processing, and the number and variety of responses the device can make.

Wickens (1992) has described three general classes of automation that serve different purposes. First, automation can perform functions that are beyond the ability of humans. For example, the faceted appearance of Lockheed's F-117 stealth fighter makes the aircraft almost impossible to fly. Automated systems, however, "interpret" the control stick movements of the pilot in order to control the unusual aerodynamics of the aircraft. Second, automation can perform tasks that humans do poorly. An example would be the autoexposure system in a camera. This feature eliminates the need for photographers to make estimates about the amount of light in a scene as well as having to calculate the proper shutter speed and lens opening. Third, automation can assist humans by performing undesirable activities. For instance, an automatic transmission eliminates the clutch pedal and shifts the gears for those of us who find that aspect of driving burdensome.

BENEFITS OF AUTOMATION

There are numerous benefits to automation. As machines, devices, or systems become capable of performing more and more activities, there are fewer for the human to do. This is ideal in situations where the operator is overloaded with activities. In fact, several researchers have reported that some of the more successful applications of automation are in cases where it relieves the human from having to deal with nuisance or housekeeping activities (Gaba, 1994; Rouse, 1991).

Automation can also increase the flexibility of operations or permit control of more complex systems (Woods, 1994). On the other hand, automation can attenuate the variability associated with human performance and thereby significantly reduce errors.

Within the aviation industry, cockpit automation has made it possible to reduce flight times, increase fuel efficiency, navigate more effectively, and extend or improve the pilot's perceptual and cognitive capabilities (Wiener, 1988). In medicine, the availability of automatic ventricular fibrillation in the field has made it possible to save lives (Thompson, 1994).

COSTS OF AUTOMATION

The benefits derived from automation come at a price. Several researchers have indicated that automation brings with it a different set of problems (Billings, 1991; Wiener & Curry, 1980; Woods, 1994). As noted previously, when tasks become highly automated there may be fewer activities for the operator to perform and his or her role shifts from that of an active participant to one of a passive monitor. The unfortunate consequence of this changing role is that humans are not well suited to monitor sources of information for extended periods of time (Parasuraman, 1986; Warm, 1984). Parasuraman and his colleagues have demonstrated that the ability to detect automation failures deteriorates under automatic as opposed to manual operating conditions (Parasuraman, Molloy, & Singh, 1993; Parasuraman, Mouloua, & Molloy, 1994).

Another problem concerns workload. At first glance, it seems that automation should help reduce mental workload. However, evidence shows that this is does not necessarily happen. Instead, Woods (1994) argued that automation merely changes how work is accomplished. In fact, Wiener (1989) claims that in some instances, the introduction of automation may even increase workload. He cautions that too often automated systems may operate well under periods of low workload and become a burden during high-workload periods.

Still, another problem often associated with automation concerns the maintenance of skills. Wickens (1992) argued that manual skills may deteriorate in the presence of long periods of automation. Along similar lines, others have argued that automation removes the operator from the "loop" leading to decreases in situation awareness (Sarter & Woods, 1992). Consequently, overreliance on automation may make the operator less aware of what the system is doing at any given moment, leaving the operator ill-equipped to deal with a failure of automation.

In complex systems, Woods (1994) claimed that automation can lead to incongruent goals among subsystems. Also, when subsystems are highly coupled, it may be more difficult to isolate the locus of a problem if the system as a whole should fail.

Finally, there appears to be a good deal of skepticism about automation among its users. Confidence, both in one's self and the automation, impact on its usage (Lee & Moray, 1992, Muir, 1987). Extensive experience with the automation is needed for operators to assess its reliability. In his work with pilots, Rouse (1991) learned that indices of reliability had to be much higher than 95% for automatic systems to be considered useful.

Thus, it appears that some degree of confidence in automation is necessary for users to embrace it. However, too much confidence brings with it another set of problems. Once a sense of trust has developed and an automatic system has become an accepted method of operation, there lies the potential danger that individuals will become overly reliant on the automation. This may lead operators to be less willing to evaluate or even monitor the automated activities, a situation that has been described as automation induced "complacency" (Parasuraman, Molloy, & Singh, 1993).

Consequently, there are both benefits and costs associated with automation. Woods (1994) suggested that it may not be appropriate to view automation in terms of costs and benefits. Rather, he argued that automation *transforms* the nature of work. Offloading a task to an automated system to perform does not leave a gap in the operator's responsibilities. Instead, it changes his or her responsibilities, often requiring a redistribution of resources.

LEVELS OF AUTOMATION

DUAL-MODE

The simplest form of automation operates in two modes, manual and automatic. For example, a thermostat automatically maintains a desired temperature. If it is broken in the dead of winter, you must manually turn on the heating system each time you begin to feel cold and turn it off when you are warm.

Surprisingly, little is known about the effects of automation on human performance. In fact, in 1985, an entire issue of *Human Factors* was devoted to the topic of automation but offered little comparative data regarding performance on manual and automatic tasks.

Researchers have examined how aspects of automation impact performance by themselves and in concert with other activities. Several studies have addressed monitoring efficiency under manual and automatic modes. For example, Parasuraman, Molloy, and Singh (1993) had subjects perform a systems monitoring task in which they were to watch a set of gauges for periodic deviations from prescribed levels and reset them. Subjects were quite capable of performing the task by itself or in conjunction with two other tasks (a compensatory tracking task and a resource management task). In another condition, however, the systems monitoring task was automated, i.e., the system corrected the deviations when they occurred. The subjects were responsible for monitoring the automation in addition to performing the tracking and resource management tasks. Periodically, the automated task failed to take corrective action. Parasuraman and his colleagues found that under these conditions, monitoring performance declined dramatically after only 20 minutes. These findings indicate that operators may be unable or unwilling to continue monitoring automation while performing other tasks. Further, monitoring behavior appears to become increasingly inefficient with longer and longer periods of automation.

In another study, Parasuraman, Mouloua, Molloy, and Hilburn (1992) again asked subjects to monitor automation. However, in this experiment, the monitoring task was interrupted with a brief period requiring manual control. The investigators found that overall monitoring performance improved under these conditions and remained fairly stable for the rest of the study. Parasuraman et al. (1992) argued that a temporary return to manual operations may act as a countermeasure against poor monitoring behavior induced by automation.

In addition, it appears that there may be benefits to shorter periods of automation. Parasuraman, Hilburn, Molloy, and Singh (1991) found no costs associated with 10-minute cycles of manual and automated control. On the contrary, when subjects had to work at three tasks simultaneously, they found that performance on any of the nonautomated tasks benefitted from automation on the other two tasks. Similar results were reported by Gluckman, Carmody, Morrison, Hitchcock, and Warm (1993) who found that tracking performance was superior in the context of an automated fuel management task.

To date, little is known about how the frequency and duration of automation cycles affect performance. In their study, Scallen, Duley, and Hancock (1994) found superior performance on a tracking task that cycled between manual and automated modes every 15 seconds as compared to every 60 seconds. By contrast, Hilburn, Molloy, Wong, and Parasuraman (1993) observed a progressive deterioration in tracking performance with increases in the frequency of shifts between automated and manual modes. It is important to note that the pattern of shifts between modes in the Hilburn et al. study was derived from operator performance. Therefore, it is unlikely that the timing parameters surrounding mode shifts in this study were as consistent as those in the Scallen et al. (1994) study, which followed a precise temporal schedule. Thus, the benefits accruing shorter cycles reported by Scallen et al. may be attributable, in part, to a more predictable schedule of mode shifts.

Other investigators have been concerned about potential carry-over effects emerging after operating under periods of automation (Morris & Rouse, 1986). This refers to the notion of an "automation deficit," that is, a degradation in manual performance after a period of automated control of a task. Glenn et al. (1994) explored this issue by asking subjects to perform a compensatory tracking task, a tactical assessment task, and a communication task simultaneously over three 10-minute periods. In some conditions, one of the tasks was automated during the second period. The researchers compared performance in period 3 with that of period 1 but found no evidence of improvements or deficits attributable to automation in the second period. Parasuraman, Bahri, Molloy, and Singh (1991) also failed to find any positive or negative residual effects after 10-minute periods of automation.

These findings contradict those of Ballas, Heitmeyer, and Perez (1991a) who reported an automation-related deficit in response time on a tactical assessment task. This effect, however, was observed on the first response after return to manual operation. Thus, it is possible that any automation-related carry-over effects may be short-lived. Furthermore, Ballas, Heitmeyer, and Perez (1991b) later indicated that residual effects may be tied to interface design.

MULTIPLE MODES

In more complex systems, there may be several different levels and/or modes of operation. The level of automation can vary along a continuum from none at all (i.e., manual operations) to fully automatic. Billings (1991) distinguished among seven levels of automation. Differences along the continuum reflect different levels of autonomy. At the low end of the spectrum, activities and functions are under control of the operator. At the other end are systems capable of executing complete functions and monitoring their own actions. Somewhere between these two extremes are systems that might perform some portions of an activity or make recommendations to the operator about a course of action.

Rouse and Rouse (1983) distinguish three unique modes of automation based upon how the technology might assist the operator. For example, entire tasks can be allocated to the system to perform. Or, a task could be partitioned so that the system and operator would each be responsible for controlling some portion of the task. In the third mode, a task might be transformed to make it easier for the operator (e.g., changing the format used to present information to the operator). The allocation and partition modes would be high in autonomy because the system has control over all or some part of the task. Transformations, however, are low in autonomy because the operator is still responsible for performing the task.

In complex systems, it is not uncommon to find multiple modes of automation. The flight management system (FMS) found in the most advanced commercial aircraft is responsible for eight unique kinds of functions including navigation, guidance, system monitoring, and management of its own systems (Billings, 1991). The Honeywell MD-11 FMS can be configured for at least 12 separate modes of operation (Billings, 1991).

Complex systems with multiple modes of automation like the FMS present operators with a real challenge. First, they are difficult to learn. Pilots must learn the boundaries of each mode of operation and how to fly under each. Second, they may increase workload because the pilot must be cognizant of the operating procedures associated with each mode but must also be aware of which mode is active at all times. Indeed, Sarter and Woods (1994a) claim that in the presence of multiple modes of automation, flying becomes a task of orchestrating a "suite of capabilities" for different sets of circumstances.

Endsley and Kiris (1994) examined the impact of different levels of automation on performance. The investigators had subjects use an expert system to help them make decisions about automobile navigation. Groups of subjects were asked to use one of four systems that differed in level of autonomy. Systems at the low end either recommended a course of action or sought concurrence on a suggested action. On the high end, systems made decisions and acted upon them. Subjects worked through a set of four problems; after which the system failed, and they were required to solve two more problems on their own. Performance on the last two problems was compared to that of a "manual" group that made all of their decisions without use of an expert system. The investigators found that the decision times of subjects working with more autonomous systems were longer than those in the manual condition. Their findings suggest that higher levels of autonomy remove the operator from the task at hand and can lead to poorer performance during automation failures.

ADAPTIVE AUTOMATION

There continues to be a growing interest in the merits of automation that is dynamic or adaptive in nature (Hancock & Chignell, 1987; Morrison, Gluckman, & Deaton, 1991; Rouse, 1976). In adaptive automation, the level or mode of automation or the number of systems that are automated can be modified in real time. Further, *both* the human and the machine share control over changes in the state of automation. Parasuraman, Bahri, Deaton, Morrison, and Barnes (1992) have argued that adaptive automation represents an optimal coupling of the level of automation to the level of operator workload. This, of course, presupposes that levels of operator workload can be specified to permit suitable adjustments in automation.

Adaptive Mechanisms

Perhaps the most critical challenge facing developers of adaptive automation concerns how changes among modes or levels of automation will be accomplished. Morrison and Gluckman (1994) discussed several potential candidates for triggering changes among levels of automation. One candidate is the operator's own performance. The operator's interactions with the system interface would be monitored and evaluated against some standard to determine when to change levels of automation. At present, there is much debate about how to derive the standards against which performance is to be assessed. Presumably, one could create a database of human performance information that would then be accessed to evaluate operator performance online. However, such an approach would be of limited utility because the system would be entirely reactive. Further, it is unlikely that a comprehensive database could ever be assembled.

A more promising alternative lies with operator performance modeling. This approach enables the system to generate standards derived from models of human performance. More important, performance models provide the capability for a more proactive system. For example, Geddes (1985) and his colleagues (Rouse, Geddes, & Curry, 1987) offered a model to invoke automation based upon information about the current state of the system, external events, and expected operator actions. The operator's intentions are predicted from patterns of his or her activity. In a well-structured environment (i.e., one with a rigid script of goals and plans), the system is capable of changing operation modes to meet future demands.

Hancock and Chignell (1987, 1988) have also proposed an adaptive aiding mechanism. In their model, tasks or subtasks are allocated to the human or the system based on both current and future levels of operator workload. Unique to this approach is the idea that current levels of workload are determined in part by deviations from an ideal state. The prioritization and scheduling of activities needed to achieve desired states as well as the activities themselves all contribute to current estimates of workload. The discrepancy between current and expected states coupled with current levels of performance and the predicted levels of performance needed to reach the desired state drive the adaptive allocation of tasks. (Other models have also been proposed. For a more thorough review of operator performance models see Parasuraman et al., 1992; Rouse 1988.)

Morrison and Gluckman (1994) have also proposed using biopsychometrics to invoke adaptive automation. Under this method, physiological signals that reflect autonomic or brain activity and presumably changes in workload might serve as the trigger for shifting among the automation modes. The advantage to biopsychometrics is that such measures can be obtained continuously with little or no impact on operator performance. Several physiological measures have been shown to reflect operator workload and are potential candidates for triggering adaptive automation, including heart rate variability, eye movements, and event-related potential (Byrne & Parasuraman, 1996).

The third possibility for invoking adaptive automation outlined by Morrison and Gluckman (1994) is to monitor the activities of the mission itself. For example, Barnes and Grossman (1985) advocate the development of systems that would monitor ongoing activities for the occurrence of critical events. Detection of such events would, in turn, activate the automation. These critical events might be emergency situations or predetermined periods of high workload. Alternatively, the priority and sequencing of functions or the allocation of activities between the pilot and the cockpit might be scheduled to take place at prespecified points in the mission. This method of invocation may be the most immediately accessible of the three described by Morrison and Gluckman. Unfortunately, as Parasuraman et al. (1992) have indicated, activity monitoring systems are not very sophisticated and are only loosely coupled to operator workload or performance.

System Responsiveness

As noted earlier, adaptive automation is still in its infancy, and as a result there are many issues that will need to be addressed before such systems become truly viable. One of these issues is system responsiveness. Early human factors work with computers revealed that system response time was an important determinant of both usability and user acceptance (Bailey, 1982). The issue is likely to be more critical with adaptive automation. Successful adaptive automation will require the proper amount of automation at the proper times. Under high workload conditions, this may necessitate instantaneous system response. Proponents of operator performance models argue that system response times such as these are unlikely to be achieved without the ability to predict future workload demands (Greenstein & Revesman, 1986; Rouse, Geddes, & Curry, 1986).

Timing

Although predictive modeling is necessary for ideal system response times, it is not sufficient. As noted previously, developers of adaptive automation will also need to be concerned about cycles of automation and the frequency of changes among states. Furthermore, the timing parameters surrounding fluctuations in automation have yet to be worked out. There are two strategies that might be employed. One would be to match as closely as possible the changes in task demands with changes in automation. This procedure would maximize the chances of having the appropriate mode of automation in operation at any moment. However, this strategy might be inappropriate under conditions of rapidly fluctuating workload demands. Clearly, as the time spent under any mode of automation grows shorter, a point will be reached where the operator can no longer be effective. Furthermore, rapidly shifting modes of automation would impose an additional burden on the individual to maintain awareness of a constantly changing set of unique operational parameters. The alternative strategy would be to keep the number of mode changes to a minimum. This strategy would produce greater discrepancies between task demands and the appropriate automation mode at any moment but would reduce the need to keep track of a rapidly fluctuating set of operating procedures. Ultimately, a compromise between these strategies must be reached. The optimal choice, however, is likely to depend on the nature of the task as well as the individual.

AUTHORITY AND INVOCATION

Operator Authority

Adaptive automation makes it possible for a system to have autonomous control over changes among modes of automation. This has led some researchers to ask whether a system *should* have that authority.

There are many who argue that the operator should always have authority over the system (Billings & Woods, 1994; Malin & Schreckenghost, 1992). Pilots, too, have argued that because they have responsibility for the aircraft, themselves, and any passengers, they should have the authority to initiate changes in automation (Billings, 1991).

Although research in this area is scarce, there is performance-based evidence to support operator-initiated invocation of automation. For instance, Harris, Hancock, Arthur, and Caird (1991) had subjects perform a resource management task, systems monitoring task, and compensatory tracking task simultaneously. Under one condition, the subjects performed all tasks manually, whereas in a second condition, the tracking task was automated. In a third condition, the subjects were encouraged to invoke automated tracking to help maintain optimal performance on the other two tasks. The results showed that subjects were more efficient at the resource management task when they had control over invoking automation. Hilburn, Molloy, Wong, and Parasuraman (1993) also reported an advantage for operator-invoked automation on a similar suite of tasks.

System Authority

There are also arguments to support system-initiated control over automation invocation. For instance, the operator may need to change automation modes at the precise time he or she is too busy to make the change (Sarter & Woods, 1994b; Wiener, 1989). Harris, Goernert, Hancock, and Arthur (1994) gathered evidence that validates this concern. These investigators had subjects perform a systems monitoring task, resource management task, and tracking task simultaneously. The tracking task was under operator control in one condition and under system control in another. In the operator-controlled condition, subjects were asked to use the automation whenever they encountered difficulties with the other two tasks. Throughout the course of the experiment there were predetermined changes in task difficulty. Half of these task changes were preceded with a warning. Harris and his colleagues found that there was little difference in use of automation among conditions. However, there was an important interaction between automation and warning conditions in the resource management task. Specifically, subjects were less able to manage resources in the operator-initiated condition when workload increases came without warning. These data support the idea that operators may be unprepared to adequately manage automation in the context of abrupt, unexpected changes in workload.

Still, another concern is whether the operator is the best judge of when and if automation is needed (Morris & Rouse, 1986). Harris, Hancock, and Arthur (1993) observed that as subjects became more fatigued under multitask conditions, they showed a tendency to use less automation. Moreover, the decline in automation use was accompanied by a decline in performance on a nonautomated task. These researchers argued that operator-initiated automation is unlikely to be of any value if it is not used precisely when it is needed.

Finally, there are situations where it would be very beneficial for the system to have authority over automation invocation. If an operator's life or the lives of others were in danger or the continued operation of a system were to result in serious damage, clearly one would want the system to intervene and circumvent the threat or minimize the potential damage. For example, it is not uncommon for many of today's fighter aircraft to sustain higher levels of G forces than the pilot can withstand physically (Buick, 1989). Despite the presence of anti-G protective equipment, high enough G forces can still render the pilot of an armed and fast-moving aircraft unconscious for periods of up to 12 seconds (Whinnery, 1989). Thus, it is not surprising that situations such as these have been held up as ideal candidates for applications of adaptive automation (Whinnery, 1988).

RESEARCH AND DEVELOPMENT IN ADAPTIVE AUTOMATION

The concept of adaptive automation originated with work in artificial intelligence in the 1970s. Specifically, efforts were aimed at developing adaptive aids to facilitate decisions about allocating tasks between a human and computer (Rouse, 1976; 1977). Later, efforts were directed at applying the adaptive aiding ideas to aviation systems (Hammer & Rouse, 1982).

In the mid-1980s, the Defense Advanced Research Projects Agency (DARPA), Lockheed Aeronautical Systems Company, McDonnell Aircraft Company, and what was then Wright Research and Development Center combined efforts to utilize state-of-the-art intelligent systems to assist pilots of advanced fighter aircraft. This effort was called the Pilot's Associate program, and its primary objective was to provide the pilot with "the right information, in the right format, in the right location, at the right time" (Judge, 1991, p. 86).

The Pilot's Associate (PA) was a consortium of cooperative knowledge-based systems capable of monitoring and assessing events and generating plans to respond to problems (Hammer & Small, 1995). Situation assessment systems monitored events in the external environment as well as the status of internal systems. This information could be evaluated and presented to the pilot. Planning systems informed the pilot of potential actions the system might take. These actions ranged from short-term tactics (e.g., adjusting the throttle) to changes in the overall mission (e.g., modifying the route). In addition, the PA included an intelligent interface capable of selecting the appropriate information to present to the pilot at the appropriate time. The intelligent pilot-vehicle interface was a complex communications system capable of conversing with the pilot in different modes and at different levels of detail (Rouse, Geddes, & Hammer, 1990).

The PA program was a successful showcase of current intelligent system capabilities. Much knowledge of human information processing was needed to demonstrate the adaptive automation concepts. The initial stages of design drew heavily upon the current models of human information processing and performance. Interestingly, many had to be abandoned by the second phase because they were too restrictive in a more contextually rich information environment (see Hammer & Small, 1995).

The original objective of the PA program was to demonstrate how artificial intelligence might be used to assist fighter pilots. It was never, however, conceived of as a plan for how adaptive automation *ought* to be implemented in a fighter cockpit. Later, the Navy initiated its Adaptive Function Allocation for Intelligent Cockpits (AFAIC) program (Morrison, Gluckman, & Deaton, 1991). The primary objective of this effort is to understand the human performance issues surrounding adaptive automation and to develop guidelines for its implementation. Morrison, Cohen, and Gluckman (1993) have outlined a three-fold classification scheme for understanding how various strategies for implementing adaptive automation affect performance (see Figure 6.1). On one dimension of the taxonomy are the factors that either invoke or change modes of automation: critical external events (mission), changes in human performance, and changes in physiology. The second dimension is based on Rouse and Rouse's (1983) ideas about how functions might be changed. Entire tasks or portions thereof can either be allocated to the system, the pilot, or transformed. The

Perspectives on Adaptive Automation

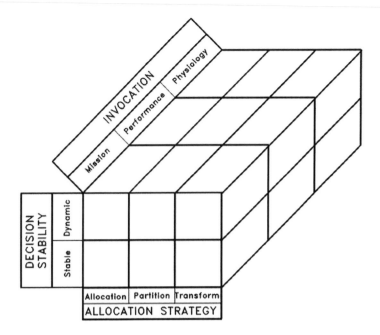

FIGURE 6.1 Taxonomy for the implementation of adaptive automation. (Adapted from Morrison, Cohen, & Gluckman, 1993.)

third dimension of the taxonomy reflects the complexity of the decisions and actions required of automated tasks. Stable decisions are those with consistent stimuli and actions, whereas dynamic decisions have more varied inputs and a wider range of actions. Efforts to validate the taxonomy are still underway, although the AFAIC program has generated some preliminary guidelines for implementing adaptive automation (see Morrison, 1993; Morrison & Gluckman, 1994).

IMPLEMENTING ADAPTIVE AUTOMATION

Although adaptive automation is still very much a concept that exists in laboratories, that has not kept researchers and designers from speculating about what form it ought to take and the human performance issues associated with it. Consider some of the unique issues that are likely to arise in systems equipped with adaptive automation. The operator will have to learn the capabilities and limitations of the system, and the system, in turn, will have to learn those of the operator. The operator may have to log extensive hours of training with the system to achieve optimal levels of performance. The operator and the system are likely to have unique responsibilities but will collaborate on many others. Moreover, the system should be capable of assuming some of the operator's responsibilities during periods of high workload or anticipating the operator's actions and adjusting its own behavior accordingly. Perhaps, most importantly, the operator and system will have to be able to exchange information freely and effortlessly.

Activities like collaboration, backing up one another, and communication suggest the notion of teamwork. Indeed, several researchers have appealed to team concepts when referring to advanced automation. Hammer and Small (1995), for instance, viewed the Pilot's Associate as "an electronic crew member." Woods (1994) has described advanced automation as a human-machine cooperative system with cognitive work distributed across agents and discusses the need for "automated agents" to coordinate activities among all agents.

Others have been more forthright in their conceptualizations. Malin et al. (1991) and Malin and Schreckenghost (1992) considered the design issues necessary to make an intelligent system a team player in space operations. Space teams have common goals. The individuals on those teams must coordinate, collaborate, and communicate effectively. Team members perform tasks both individually and cooperatively. In addition, members have some ability to back up one another's specialties.

Malin and her colleagues argue that an intelligent system may be considered a team member if certain criteria are met (Malin & Schreckenghost, 1992). An intelligent system can be included as a team member if it: is reliable, can communicate effectively, can coordinate activities, and can be guided by a coach. The system must be able to do what is necessary without causing harm or interference. Furthermore, the system must also be capable of being repaired: that is, the information within the system, its activities, and its ability to reason must be modifiable. Second, the system must communicate effectively by providing only as much information as is necessary when it is needed. Third, the system must coordinate activities to optimize performance among all team members and to minimize the potential for interference. Team members must be able to monitor the system and the system must be able to monitor the activities of the team members. This requires the need to exchange information about goals, plans, intentions, and beliefs. Finally, teams are managed or coached. The humans not only have their own responsibilities but must also manage the collective activities of all members, including the system. Thus, the human must be fully aware of the capabilities and limitations of the system, what it is doing at any given time, and how the system impacts the activities of the other team members.

The criteria outlined by Malin and her colleagues are the foundation for a set of guidelines aimed at developing intelligent systems that might behave more like a human in a team context. Scerbo (1994), on the other hand, argued that an understanding of team dynamics should be the force guiding the development of adaptive automation.

A team can be considered as two or more individuals who interact dynamically and interdependently to achieve a common goal (Salas, Dickinson, Converse, & Tannenbaum, 1992). Team members are dependent upon one another. Furthermore, there is a need for individual members to coordinate and synchronize their activities and to communicate with one another.

It is probably no accident that those characteristics that define human teams (e.g., coordination, communication, etc.) found their way into the "team player" criteria proposed by Malin et al. (1991). Accordingly, Scerbo (1994) believed that an examination of the team literature might provide important insights about interactions with adaptive automation.

To fully appreciate team behavior, one has to understand those factors that affect team performance. Nieva, Fleishman, and Rieck (1978) have discussed four such factors. First, there are the member resources. These are the skills, abilities, and the knowledge that individual members contribute to the team. Second, there are task characteristics that dictate how the individual and interdependent activities will impact the performance of the team as a whole. Third, there are also team characteristics. These refer to those aspects of the team itself that impact performance. Fourth, the member resources and the task and team characteristics are each affected by external conditions. These might be environmental or organizational in nature.

These factors have been embodied in a model of team behavior that emphasizes both individual and collective contributions to performance (Fleishman & Zaccaro, 1992). According to this model, there are seven categories of team functions: orientation, resource distribution, timing, response coordination, systems monitoring, motivation, and procedure maintenance. Orientation functions describe those activities associated with the acquisition and distribution of information. This may include details about team goals or the overall mission, potential problems, constraining influences, and the prioritization of activities. Information can either be collected from external sources and distributed to team members or gathered from team members and disseminated externally. Resource distribution functions refer to those activities necessary to match member resources with appropriate tasks and to distribute work equitably. Timing functions determine the pace of work at the individual and team levels. Response coordination functions refer to the sequencing and timing of member responses. The systems monitoring functions address how progress is measured at the individual and team levels. Work may be adjusted or tasks reallocated after performance is evaluated. Motivational functions refer to the establishment of team objectives and performance norms as well as the processes necessary to motivate members to meet those objectives. These functions also include measures needed to balance individual competition against team objectives and resolve performance-relevant disputes. The last category, procedure maintenance, refers to monitoring individual and team activities for compliance with standard procedures or organizational policies.

Fleishman and Zaccaro's (1992) taxonomy was derived from many years of studying humans in team situations. Scerbo (1994), however, argued that there are analogs for these functions in the adaptive automation arena. Consider orientation functions. With respect to aviation, the pilot and the aircraft would need to know the capabilities and limitations of one another. The tasks needed to complete a mission would have to be allocated between the pilot and the aircraft. Resource distribution functions, for example, in an aircraft with adaptive automation refer to task allocation and workload distribution. Systems monitoring functions for teams are similar to performance assessment functions in adaptive automation. Procedure maintenance in teams and adaptive automation would also be similar. The one area where the taxonomy does not have a suitable analog in adaptive automation concerns motivational functions. These functions do not readily translate into performance issues for adaptive automation. However, the need for pilots to get to know and trust the automation as they would another human team member should not be overlooked.

Scerbo (1994) used this taxonomy as a framework to identify those team functions addressed by the two of the most ambitious research efforts in adaptive automation at the time, the PA and the AFAIC programs. He found that in both programs, considerable effort had been directed at developing methods to allocate functions between the pilot and the cockpit and to distribute pilot workload more uniformly (i.e., resource distribution functions). Systems monitoring functions have also been studied; however, efforts to assess pilot performance have received more emphasis than that of the cockpit or the combined performance of the two. The PA program also addressed many orientation functions in order to build an intelligent planning system. The PA program also considered numerous response coordination functions in the creation of their intelligent interface system.

Scerbo (1994) also noted that several functions had not been addressed. Perhaps the most important oversight was the need to monitor the collective performance of the pilot and the aircraft. Overall mission effectiveness would ultimately be determined by how well the pilot and aircraft operated as a team. Neither program had directly addressed the timing functions, although it was possible that the PA considered pacing and task sequencing in their planning system. The timing functions, however, become more critical when activities must be performed in parallel. In fact, Amalberti and Deblon (1992) claim that under emergency situations, flying becomes an exercise in scheduling and prioritizing activities. It is unclear whether either program addressed procedure maintenance functions; however, monitoring the pilot's and cockpit's compliance with standard protocols and policies could probably be handled with little difficulty. Also, neither program considered motivational functions. However, research in this area might make a compelling argument for adaptive automation. Helmreich and Foushee (1993) indicate that the social dynamics of the flight crew have contributed to its share of aviation mishaps. In situations like these, adaptive automation may have an advantage over human teams. Issues such as individual competition, individual acceptance of team norms, compliance with authority, or disregard for protocol would be meaningless for adaptive automation.

That there are many issues surrounding adaptive automation that are still unexplored should not be taken as short-sightedness on the part of either the PA or AFAIC programs. To their credit, they have addressed many important team functions without formally acknowledging a pilot-cockpit team. Instead, the gaps identified by Scerbo (1994) should be viewed as support for approaching adaptive automation from a broader, team perspective.

FUTURE DIRECTIONS

It is clear that automation has become increasingly sophisticated and the issues surrounding it more complex. Some researchers and developers describe future incarnations of automation as "electronic team members" (Emerson, Reising, Taylor, & Reinecke, 1989; Malin et al., 1991; Scerbo, 1994). It is likely that as both the systems and the interactions among people and systems become more complex, drawing upon the human factors and cognitive engineering literatures will no longer be sufficient to address the myriad issues that will arise. In fact, Hammer and Small (1995) came to a similar conclusion in their review of PA. They discovered that traditional human factors knowledge did not adequately address many of the human performance issues that arose during the course of the project. Hammer and Small concluded that "an examination of how humans decide to share tasks and information may be a more fruitful area in which to develop a theory of human-associate interaction" (p. 29).

In the following section, I raise several issues that are apt to increase in importance if we are to realize the potential of adaptive automation. I continue to use the team analogy because it provides a broader perspective of how work is accomplished. The following topics should not be considered exhaustive but merely representative of the kind of issues that will need to be considered for successful applications of adaptive automation.

MENTAL MODELS

A mental model is an individual's cognitive representation of how a system operates. Mental models enable an individual to describe, explain, and make predictions about system operations (Rouse & Morris, 1986; Wickens, 1992).

Several researchers have argued that it is advantageous for operators to form accurate mental models of complex systems (Carroll & Olson, 1988; Wickens & Flach, 1988). Doing so enables them to anticipate future system states, formulate plans, and troubleshoot effectively (Wickens, 1992). In fact, Wickens (1992) and Woods (1994) claimed that adding automation increases the complexity of systems. Thus, the need for an accurate mental model may be all the more important for automated systems.

There has also been a growing interest among some researchers studying team performance in the potential value of mental models in their work (Cannon-Bowers, Salas, & Baker, 1991; Rouse, Cannon-Bowers, & Salas, 1992). An accurate mental model may be particularly important to teams that must act upon information that is integrated from separate sources and different members. Cannon-Bowers et al. (1991) argue that team effectiveness is apt to be determined by the degree to which all members work from a common mental model.

It should be noted that the utility of mental models has not gone unquestioned. Norman (1983), for instance, has argued that mental models require time to evolve, they may be incomplete, and even imprecise. Also, mental models about particular systems may be structured from knowledge of other systems, from prototypes, or even faulty information. In his observations of people using calculators, Norman found numerous instances of "superstitious" behavior that he argued was the result of an improper mental model.

These problems are likely to become more serious in the context of adaptive systems. This is because adaptive automation adds another layer of complexity to mental model formation. In addition to knowledge about how a system operates under various conditions, the mental model must now include self-evaluative information. In other words, a complete understanding of an adaptive system will require knowledge of one's own behavior, how it affects the system, and how the system reacts to it.

There is little in the human factors literature that addresses this type of mental model. Again, it may be necessary to consider other avenues of research outside the traditional core of our discipline.

At this level of complexity, the relationship between the user and the system becomes truly interdependent. Interactions with adaptive systems begin to resemble interpersonal interactions. Therefore, an understanding of how individuals come to know one another's, strengths, weaknesses, and behavior patterns may provide some clues to mental model construction of adaptive systems. In this regard, theories about the development of relationships, communication patterns and self-disclosure, and the development of team skills may prove valuable.

Although research along these lines may hold some promise for understanding how mental models of adaptive systems might be formulated, one should not underestimate the challenge that lies ahead. Current knowledge of interpersonal behavior is incomplete, at best. Moreover, our ability to understand the intentions and actions of others is generally poor. Therefore, one should not be surprised to learn that operators will find it quite difficult to form a representation of an adaptive system, if it can be done at all.

TRAINING AND PRACTICE

As noted earlier, learning a complex system with multiple modes of automation is a lengthy process. As it is, pilots often indicate that they would welcome additional training (Orlady, 1993). Introducing adaptive automation will only increase training requirements. Organizations utilizing adaptive systems should understand the need for and make available additional time to train their users. It has been said that incomplete knowledge of operations leads to questions about what the system is doing and why the system is behaving the way it is (Wiener, 1989). Systems appear to take on a life of their own and act independently (Sarter & Woods, 1994a).

Again, if one adopts a team perspective, then learning a system with adaptive automation is not unlike learning to work with a new team member. Research shows that training can improve team performance particularly if it is directed at communication or coordinated activities (Salas et al., 1992). Practice sessions allow members to come to know the boundaries of their teammate's abilities. This can benefit overall team performance in two ways. First, it allows team members to capitalize on one another's strengths and address each other's weaknesses. The second advantage of team practice is that as players come to know the skills and abilities of one another, the quality of their communication changes. For example, Foushee, Lauber, Baetge, and Acomb (1986) observed that flight crews that had spent longer periods of time together communicated more, verbalized more of their intentions, and were more likely to acknowledge one another's communications. Moreover, this pattern of communication was also associated with better performance when compared to other crews who were unfamiliar with one another.

Team members must also learn to coordinate their activities. This is one of the primary differences between individual and team performance. Within an individual, the brain "knows" what the right and left hand are doing, and in some instances, may require few, if any attentional resources to coordinate their activities (see Schneider & Shiffrin, 1977). In a team context, the quality of any executive or managerial process to coordinate individual activities is grossly inferior to what the human brain can accomplish. Consequently, an enormous amount of effort and resources are needed to monitor, plan, and schedule the activities of individual members to achieve overall objectives. This requires that goals and objectives be stated clearly, that plans and intentions be communicated, and that team members have the ability to observe one another's performance. Extensive practice enables team members to build models of one another's behavior and provide information and assistance at the times they are needed (Kanki & Palmer, 1993).

COMMUNICATION

The success of adaptive systems will in large part be determined by the interface. I use the term *interface* to include all methods of information exchange. Most researchers agree that effective communication among team members is critical for overall success (Kanki & Palmer, 1993; Streufert & Nogami, 1992). Individual team members each possess knowledge and information that the other members do not. Thus, individuals must share information with one another in order to make decisions and carry out actions. Team members must also coordinate their activities, thereby requiring each to communicate their intentions.

To ensure the successful exchange of information, humans use any and all available means to communicate with one another. Consider the wide range of options individuals have available to them. They can use spoken and written language. They can draw diagrams and pictures. They may also use hand gestures, facial expressions, and eye contact. In fact, they may even resort to physical contact. All of these are important forms of communication. Individuals constantly make decisions about when to use which form or forms of communication to best express themselves in different situations. It follows that an adaptive system designed with only one method of information exchange (e.g., an alphanumeric interface) is apt to severely limit interaction with that system. It would be analogous to asking team members to restrict their repertoire of communication to passing notes back and forth.

Designers of adaptive systems, must at the outset, attempt to include as many information formats as possible (e.g., text, graphics, voice, and video). There are at least two advantages to this strategy. First, multiple modes allow information to flow more freely because users can communicate more naturally. Second, workload can be reduced by eliminating the requirement to translate all information into only one or two formats.

FINAL THOUGHTS

The purpose of this chapter has been to survey developments and research on adaptive automation. Particular attention was paid to the hopes and concerns surrounding adaptive automation and its potential impact on human performance. Before closing, however, I would like to revisit an idea raised earlier. Woods (1994) argued that automation must be viewed from the broader context in which it is applied. Given that a great deal of automation is introduced on the job, it might be instructive to consider a more global view of work in order to gain a better understanding of the impact of automation.

There is a natural flow to the progression of work in organizations. The introduction of automation or new tools or procedures disrupts this flow. Some activities may be performed faster. Some activities may no longer be needed. New activities may be added. Productivity may even fall off temporarily until employees can fully incorporate the new methods or technology into their routines; but, the flow of work continues.

What this perspective shows is that automation is neither inherently good nor bad. It does, however, change the nature of work, and in doing so, solves some problems while creating others (Wiener & Curry, 1980). Both the developers and consumers of automation need to be aware of this. Designers must not incorporate automation into systems without considering how it will impact other activities in the flow of work. This may require extensive testing to ensure that the desired benefits of automation exceed the unintentional consequences of automation. Likewise, organizations must prepare their employees for the restructuring of work brought about by automation and provide proper training for them to adapt to the new processes.

Adaptive systems represent one step in the evolution of automation, and it is a big step. This type of technology will have a profound impact on how work is performed. Thus, it is not surprising to find the literature filled with cries of caution. Many have argued that automation, in general, necessitates a human-centered approach to design (Billings, 1991; Rouse, 1991; Wickens, 1992; Wiener, 1988). Adaptive automation, on the other hand, may require a *social-centered* approach. As noted earlier, a team perspective is likely to provide important insights into human interaction with adaptive automation. However, social and organizational issues may also have some bearing on the successful operation and acceptance of adaptive automation. This may be particularly true when multiple users must interact with the same adaptive system.

At present, adaptive automation is still in its conceptual stages. Although prototypes do exist, it will take many years for the technology to mature. Fortunately, this gives designers, cognitive engineers, and psychologists a chance to begin studying the many issues that surround adaptive automation before implementation of the technology is widespread. We have a real opportunity at this point in time to guide the development of adaptive automation from an understanding of human requirements instead of from restrictions imposed by current technological platforms.

REFERENCES

Amalberti, R., & Deblon, F. (1992). Cognitive modelling of fighter aircraft process control: A step towards an intelligent on-board assistance system. *International Journal of Man-Machine Studies, 36,* 639–671.

Bailey, R. W. (1982). *Human Performance Engineering: A Guide for System Designers.* Englewood Cliffs, NJ: Prentice-Hall, Inc.

Ballas, J. A., Heitmeyer, C. L., & Perez, M. A. (1991a). Interface styles for adaptive automation. In R. S. Jensen (Ed.), *Proceedings of Sixth International Symposium on Aviation Psychology* (pp. 96–101). Columbus, OH: The Department of Aviation, The Aviation Psychology Laboratory, The Ohio State University.

Ballas, J. A., Heitmeyer, C. L., & Perez, M. A. (1991b). Interface styles for the intelligent cockpit: Factors influencing automation deficit. *Proceedings of AIAA Computing in Aerospace 8* (pp. 657–667).

Barnes, M., & Grossman, J. (1985). *The intelligent assistant concept for electronic warfare systems* (NWC TP 5585). China Lake, CA: Naval Weapons Center.

Billings, C. E. (1991). *Human-centered aircraft automation philosophy: A concept and guidelines* (Technical memorandum no. 103885). Moffett Field, CA: NASA.

Billings, C. E., & Woods, D. D. (1994). Concerns about adaptive automation in aviation systems. In M. Mouloua & R. Parasuraman (Eds.) *Human Performance in Automated Systems: Current Research and Trends* (pp. 264–269). Hillsdale, NJ: Lawrence Erlbaum Associates.

Buick, F. (1989). *+Gz protection in the future — review of scientific literature.* (Technical report DCIEM 89-RR-47). Downsview, Ontario, Canada: Defense and Civil Institute of Environmental Medicine.

Byrne, E. A., & Parasuraman, R. (1996). Psychophysiology and adaptive automation. *Biological Psychology, 42,* 249–268.

Cannon-Bowers, J. A., Salas, E., & Baker, C. V. (1991). Do you see what I see? Instructional strategies for tactical decision making teams. In *Proceedings of the 13th Annual Interservice/Industry Training Systems Conference* (pp. 214–220). Washington, DC: National Security Industrial Associates.

Carroll, J. M., & Olson, J. R. (1988). Mental models in human-computer interaction. In M. Helander (Ed.), *Handbook of Human-Computer Interaction* (pp. 45–65). Amsterdam, The Netherlands: North Holland.

Davis, L. E., & Wacker, G. J. (1987). Job design. In G. Salvendy (Ed.), *Handbook of Human Factors* (pp. 431–452). New York: Wiley.

Emerson, T., Reising, J. M., Taylor, R. M., & Reinecke, M. (1989). *The human-electronic crew: Can they work together?* (WRDC-TR-89-7008). Wright-Patterson Air Force Base, OH: Wright Research and Development Center.

Endsley, M. R., & Kiris, E. O. (1994). The out-of-the-loop performance problem: Impact of level of automation and situation awareness. In M. Mouloua & R. Parasuraman (Eds.), *Human Performance in Automated Systems: Current Research and Trends* (pp. 50–56). Hillsdale, NJ: Lawrence Erlbaum Associates.

Fleishman, E. A., & Zaccaro, S. J. (1992). Toward a taxonomy of team performance functions. In R. W. Swezey & E. Salas (Eds.), *Teams: Their Training and Performance* (pp. 31–56). Norwood, NJ: Ablex.

Foushee, H. C., Lauber, J. K., Baetge, M. M., & Acomb, D. B. (1986). *Crew factors in flight operations III: The operational significance of exposure to short-haul air transport operations* (NASA Technical Memorandum 88322). Moffet Field, CA: NASA-Ames Research Center.

Gaba, D. M. (1994). Automation in anesthesiology. In M. Mouloua & R. Parasuraman (Eds.), *Human Performance in Automated Systems: Current Research and Trends* (pp. 57–63). Hillsdale, NJ: Lawrence Erlbaum Associates.

Geddes, N. D. (1985). Intent inferencing using scripts and plans. In *Proceedings of the First Annual Aerospace Applications of Artificial Intelligence Conference* (pp. 160–172). Wright-Patterson Air Force Base, OH: U.S. Air Force.

Glenn, F., Barba, C., Wherry, R. J., Morrison, J., Hitchcock, E., & Gluckman, J. P. (1994). Adaptive automation effects on flight management task performance. In M. Mouloua & R. Parasuraman (Eds.), *Human Performance in Automated Systems: Current Research and Trends* (pp. 33–39). Hillsdale, NJ: Lawrence Erlbaum Associates.

Gluckman, J. P., Carmody, M. A., Morrison, J. G., Hitchcock, E. M., & Warm, J. S. (1993). Effects of allocation and partitioning strategies of adaptive automation on task performance and perceived workload in aviation relevant tasks. In R. S. Jensen (Ed.), *Proceedings of the Seventh International Symposium on Aviation Psychology* (pp. 150–155). Columbus, OH: The Department of Aviation, The Aviation Psychology Laboratory, The Ohio State University.

Greenstein, J. S., & Revesman, M. E. (1986). Development and validation of a mathematical model of human decision making for human-computer communication. *IEEE Transactions on Systems, Man, and Cybernetics, 16,* 148–154.

Hammer, J. M., & Rouse, W. B. (1982). Design of an intelligent computer-aided cockpit. In *Proceedings of the 1982 IEEE Conference on Systems, Man, and Cybernetics* (pp. 449–452). New York: IEEE.

Hammer, J. M., & Small, R. L. (1995). An intelligent interface in an associate system. In W.B. Rouse (Ed.), *Human/Technology Interaction in Complex Systems*, Vol. 7, (pp. 1–44). Greenwich, CT: JAI Press.

Hancock, P. A., & Chignell, M. H. (1987). Adaptive control in human-machine systems. In P.A. Hancock (Ed.), *Human Factors Psychology* (pp. 305–345). Amsterdam, The Netherlands: Elsevier.

Hancock, P. A., & Chignell, M. H. (1988). Mental workload dynamics in adaptive interface design. *IEEE Transactions on Systems, Man, and Cybernetics, 18,* 647–658.

Harris, W. C., Goernert, P. N., Hancock, P. A., & Arthur, E. J. (1994). The comparative effectiveness of adaptive automation and operator initiated automation during anticipated and unanticipated taskload increases. In M. Mouloua & R. Parasuraman (Eds.), *Human Performance in Automated Systems: Current Research and Trends* (pp. 40–44). Hillsdale, NJ: Lawrence Erlbaum Associates.

Harris, W. C., Hancock, P. A., & Arthur, E. J. (1993). The effect of taskload projection on automation use, performance, and workload. In R. S. Jensen (Ed.), *Proceedings of the Seventh International Symposium on Aviation Psychology* (pp. 178–184). Columbus, OH: The Department of Aviation, The Aviation Psychology Laboratory, The Ohio State University.

Harris, W. C., Hancock, P. A., Arthur, E. J. & Caird, J. K. (1991, Sept.). *Automation influences on performance, workload, and fatigue.* Paper presented at the 35th Annual Meeting of the Human Factors Society, San Francisco, CA.

Helmreich, R. L., & Foushee, H. C. (1993). Why crew resource management? Empirical and theoretical bases of human factors training in aviation. In E. L. Wiener, B. G. Kanki, & R. L. Helmreich (Eds.), *Cockpit Resource Management* (pp. 3–45). San Diego, CA: Academic Press.

Hilburn, B., Molloy, R., Wong, D., & Parasuraman, R. (1993). Operator versus computer control of adaptive automation. In R. S. Jensen (Ed.), *Proceedings of the Seventh International Symposium on Aviation Psychology* (pp. 161–166). Columbus, OH: The Department of Aviation, The Aviation Psychology Laboratory, The Ohio State University.

Judge, C. L. (1991). Lessons learned about information management within the Pilot's Associate program. In R. S. Jensen (Ed.), *Proceedings of the Sixth International Symposium on Aviation Psychology* (pp. 85–89). Columbus, OH: The Department of Aviation, The Aviation Psychology Laboratory, The Ohio State University.

Kanki, B. G., & Palmer, M. T. (1993). Communication and crew resource management. In E. L. Wiener, B. G. Kanki, & R. L. Helmreich (Eds.), *Cockpit Resource Management* (pp. 99–136). San Diego, CA: Academic Press.

Lee, J. D., & Moray, N. (1992). Trust, control strategies and allocation of function in human-machine systems. *Ergonomics, 35,* 1243–1270.

Malin, J. T., & Schreckenghost, D. L. (1992). *Making intelligent systems team players: Overview for designers.* (NASA Technical Memorandum 10475). Houston, TX: Johnson Space Center.

Malin, J., Schreckenghost, D., Woods, D., Potter, S., Johannesen, L., Holloway, M., & Forbus, K. (1991). *Making intelligent systems team players: Case studies and design issues. Vol. 1: Human-computer interaction design; Vol. 2: Fault management system cases.* (NASA Technical Report 104738). Houston, TX: Johnson Space Center.

Morris, N. M., & Rouse, W. B. (1986). *Adaptive aiding for human-computer control: Experimental studies of dynamic task allocation.* (Tech. Report AAMRL-TR-86-005). Wright-Patterson Air Force Base, OH: Armstrong Aerospace Medical Research Laboratory.

Morrison, J. G. (1993). *The Adaptive Function Allocation for Intelligent Cockpits (AFAIC) program: Interim research and guidelines for the application of adaptive automation.* (Technical Report NAWCADWAR-93031-60). Warminster, PA: Naval Air Warfare Center-Aircraft Division.

Morrison, J. G., Cohen, D., & Gluckman, J. P. (1993). Prospective principles and guidelines for the design of adaptively automated crewstations. In J. G. Morrison (Ed.), *The adaptive function allocation for intelligent cockpits (AFAIC) program: Interim research and guidelines for the application of adaptive automation.* (Technical Report No. NAWCADWAR-93031-60). Warminster, PA: Naval Air Warfare Center, Aircraft Division.

Morrison, J. G., & Gluckman, J. P. (1994). Definitions and prospective guidelines for the application of adaptive automation. In M. Mouloua & R. Parasuraman (Eds.), *Human Performance in Automated Systems: Current Research and Trends* (pp. 256–263). Hillsdale, NJ: Lawrence Erlbaum Associates.

Morrison, J. G., Gluckman, J. P., & Deaton, J. E. (1991). *Program plan for the Adaptive Function Allocation for Intelligent Cockpits (AFAIC) program.* (Final Report No. NADC-91028-60). Warminster, PA: Naval Air Development Center.

Muir, B. M. (1987). Trust between humans and machines, and the design of decision aids. *International Journal of Man-Machine Studies, 27,* 527–539.

Nieva, V. F., Fleishman, E. A., & Rieck, A. M. (1978). *Team dimensions: Their identity, their measurement, and their relationships.* Washington, DC: ARRO.

Norman, D. A. (1983). Some observations on mental models. In D. Gentner & A. Stevens (Eds.), *Mental Models* (pp. 7–14). Hillsdale, NJ: Lawrence Erlbaum Associates.

Orlady, H. W. (1993). Airline pilot training today and tomorrow. In E. L. Wiener, B. G. Kanki, & R. L. Helmreich (Eds.), *Cockpit Resource Management* (pp. 447–477). San Diego, CA: Academic Press.

Parasuraman, R. (1986). Vigilance, monitoring, and search. In K. Boff, L. Kaufman, & J. Thomas (Eds.), *Handbook of Perception and Performance* (Vol. II, pp. 43.1–43.9). New York: Wiley.

Parasuraman, R., Bahri, T., Deaton, J. E., Morrison, J. G., & Barnes, M. (1992). *Theory and design of adaptive automation in aviation systems* (Progress Report No. NAWCADWAR-92033-60). Warminster, PA: Naval Air Warfare Center, Aircraft Division.

Parasuraman, R., Bahri, T., Molloy, R., and Singh, I. (1991). Effects of shifts in the level of automation on operator performance. In R.S. Jensen (Ed.), *Proceedings of the Sixth International Symposium on Aviation Psychology* (pp. 102–107). Columbus, OH: The Department of Aviation, The Aviation Psychology Laboratory, The Ohio State University.

Parasuraman, R., Hillburn, B., Molloy, R., & Singh, I. (1991). *Adaptive automation and human performance: III. Effects of practice on the benefits and costs of automation shifts.* (Technical Report CSL-N91-2). Washington, DC: The Catholic University of America, Cognitive Science Laboratory.

Parasuraman, R., Molloy, R., & Singh, I. L. (1993). Performance consequences of automation-induced "complacency." *International Journal of Aviation Psychology, 3,* 1–23.

Parasuraman, R., Mouloua, M., & Molloy, R. (1994). Monitoring automation failures in human-machine systems. In M. Mouloua & R. Parasuraman (Eds.), *Human Performance in Automated Systems: Current Research and Trends* (pp. 45–49). Hillsdale, NJ: Lawrence Erlbaum Associates.

Parasuraman, R., Mouloua, M., & Molloy, R., & Hilburn, B. (1992). *Training and adaptive automation II: Adaptive manual training.* (Technical Report CSL-N92-2). Washington, DC: Cognitive Science Laboratory, Catholic University of America.

Parsons, H. M. (1985). Automation and the individual: Comprehensive and comparative views. *Human Factors, 27,* 99–111.

Rouse, W. B. (1976). Adaptive allocation of decision making responsibility between supervisor and computer. In T. B. Sheridan & G. Johannsen (Eds.), *Monitoring Behavior and Supervisory Control* (pp. 295–306). New York: Plenum Press.

Rouse, W. B. (1977). Human-computer interaction in multi-task situations. *IEEE Transactions on Systems, Man, and Cybernetics, SMC-7,* 384–392.

Rouse, W. B. (1988). Adaptive aiding for human/computer control. *Human Factors, 30,* 431–443.

Rouse, W. B. (1991). *Design for Success: A Human-Centered Approach to Designing Successful Products and Systems.* New York: Wiley.

Rouse, W. B., Cannon-Bowers, J. A., & Salas, E. (1992). The role of mental models in team performance in complex systems. *IEEE Transactions on Systems, Man, and Cybernetics, 22,* 1296–1308.

Rouse, W. B., Geddes, N. D., & Curry, R. E. (1986). An architecture for intelligent interfaces: Outline of an approach to supporting operators of complex systems. *Human Computer Interaction, 3,* 87–122.

Rouse, W. B., Geddes, N. D., & Curry, R. E. (1987). An architecture for intelligent interfaces: Outline of an approach to supporting operators of complex systems. *Human-Computer Interaction, 3,* 87–122.

Rouse, W. B., Geddes, N. D., & Hammer, J. M. (1990). Computer-aided fighter pilots. *IEEE Spectrum, 27* (3), 38–41.

Rouse, W. B., & Morris, N. M. (1986). On looking into the black box: Prospects and limits in the search for mental models. *Psychological Bulletin, 100,* 349–363.

Rouse, W. B., & Rouse, S. H. (1983). *A framework for research on adaptive decision aids.* (Technical Report AFAMRL-TR-83-082). Wright-Patterson Air Force Base, OH: Air Force Aerospace Medical Research Laboratory.

Salas, E., Dickinson, T. L., Converse, S. A., & Tannenbaum, S. I. (1992). Toward an understanding of team performance and training. In R. W. Swezey & E. Salas (Eds.), *Teams: Their Training and Performance* (pp. 3–29). Norwood, NJ: Ablex.

Sarter, N. B., & Woods, D. D. (1992). Pilot interaction with cockpit automation: Operational experiences with the Flight Management System. *International Journal of Aviation Psychology, 2,* 303–322.

Sarter, N. B., & Woods, D. D. (1994a). Decomposing automation: Autonomy, authority, observability and perceived animacy. In M. Mouloua & R. Parasuraman (Eds.), *Human Performance in Automated Systems: Current Research and Trends* (pp. 22–27). Hillsdale, NJ: Lawrence Erlbaum Associates.

Sarter, N. B., & Woods, D. D. (1994b). Pilot interaction with cockpit automation II: An experimental study of pilots' mental model and awareness of the Flight Management System (FMS). *International Journal of Aviation Psychology, 4,* 1–28.

Scallen, S. F., Duley, J. A., & Hancock, P. A. (1994). Pilot performance and preference for cycles of automation in adaptive function allocation. In M. Mouloua & R. Parasuraman (Eds.), *Human Performance in Automated Systems: Current Research and Trends* (pp. 154–160). Hillsdale, NJ: Lawrence Erlbaum Associates.

Scerbo, M. W. (1994). Implementing adaptive automation in aviation: The pilot-cockpit team. In M. Mouloua & R. Parasuraman (Eds.), *Human Performance in Automated Systems: Current Research and Trends* (pp. 249–255). Hillsdale, NJ: Lawrence Erlbaum Associates.

Schneider, W., & Shiffrin, R. M. (1977). Controlled and automatic human information processing I: Detection, search, and attention. *Psychological Review, 84,* 1–66.

Streufert, S., & G. Nogami. (1992). Cognitive complexity and team decision making. In R.W. Swezey & E. Salas (Eds.), *Teams: Their Training and Performance* (pp. 127–151). Norwood, NJ: Ablex.

Thompson, J. M. (1994). Medical decision making and automation. In M. Mouloua & R. Parasuraman (Eds.), *Human Performance in Automated Systems: Current Research and Trends* (pp. 68–72). Hillsdale, NJ: Lawrence Erlbaum Associates.

Warm, J. S. (1984). *Sustained attention in human performance*. Chichester, UK: Wiley.

Whinnery, J. E. (1988). *Considerations on aircraft autorecovery based on +Gz-induced loss of consciousness*. (Technical report NADC-88091-60). Warminster, PA: Naval Air Warfare Center.

Whinnery, J. E. (1989). Observations on the neurophysiologic theory of acceleration (+Gz) induced loss of consciousness. *Aviation, Space, and Environmental Medicine, 6,* 589–593.

Wickens, C. D. (1992). *Engineering psychology and human performance*, 2nd ed. New York: Harper Collins.

Wickens, C. D., & Flach, J. M. (1988). Information processing. In E. L. Wiener & D. C. Nagel (Eds.), *Human Factors in Aviation* (pp. 111–155). San Diego, CA: Academic Press.

Wiener, E. L. (1988). Cockpit automation. In E. L. Wiener & D. C. Nagel (Eds.), *Human Factors in Aviation* (pp. 433–461). San Diego, CA: Academic Press.

Wiener, E. L. (1989). *Human factors of advanced technology ("glass cockpit") transport aircraft*. (Technical report 117528). Moffett Field, CA: NASA Ames Research Center.

Wiener, E. L., & Curry, R. E. (1980). Flight-deck automation: Promises and problems. *Ergonomics, 23,* 995–1011.

Woods, D. D. (1994). Automation: Apparent simplicity, real complexity. In M. Mouloua & R. Parasuraman (Eds.), *Human Performance in Automated Systems: Current Research and Trends* (pp. 1–7). Hillsdale, NJ: Lawrence Erlbaum Associates.

7 Fatigue, Automation, and Autonomy: Challenges for Operator Attention, Effort, and Trust

Gerald Matthews, Ryan Wohleber, Jinchao Lin, & April Rose Panganiban

Human work increasingly depends on automated systems across a range of industrial, transportation, and military settings (Parasuraman & Wickens, 2008). Advancements in technology will increase further the scope of automation, including applications for intelligent, autonomous systems. The spread of automation in turn raises a variety of human factors issues, including the threat to operator performance from fatigue. Broadly, we anticipate a reciprocal relationship between operating automated systems and fatigue. That is, automation may be a source of fatigue, and fatigue may influence the operator's use of automation. This chapter reviews some of the principal human factors issues arising from the interplay of fatigue and automation, including the looming challenge of human teaming with autonomous systems.

Vigilance research provides a starting point for understanding the interplay between automation and fatigue (Warm, Parasuraman, & Matthews, 2008). Automation tends to shift the operator's role from active system control agency to passive monitoring of the functioning of the automation (Banks, Stanton, & Harvey, 2014; Parasuraman & Wickens, 2008). For example, the "driver" of an automated vehicle may have nothing more to do than to watch for situations requiring human intervention. As recent, well-publicized autonomous vehicle crashes show, humans are regrettably prone to loss of task engagement, distraction, and overreliance on the automation to maintain safety (Endsley, 2018). The primary function of the operator becomes one of maintaining vigilance for critical events. People typically become rapidly fatigued during vigilance tasks, and signal detection deteriorates over time (the vigilance decrement: Warm et al., 2008). Similar concerns have been expressed about loss of situation awareness in applied settings including aviation (Billings, 1996) and future autonomous vehicles (Banks & Stanton, 2016; Endsley, 2018). An additional human factors issue, beyond the scope of conventional vigilance studies, is the role of trust in technology and whether fatigue interferes with trust optimization.

Figure 7.1 outlines a conceptual model that serves as a guide for the structure of this review. On the left-hand-side are shown factors that influence the onset and

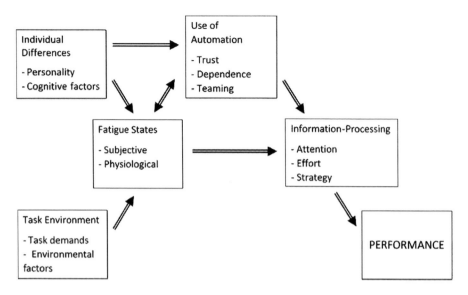

FIGURE 7.1 Fatigue processes influence performance in automated system operation.

progression of fatigue as the operator performs the task. These include task factors such as those that provoke boredom and loss of task interest. Automation is included among these factors. They also include occupational health factors such as chronic stress, long work hours, and lack of sleep, and individual difference factors such as personality traits that may moderate the impact of external factors on fatigue. Fatigue states are considered to impact information processing and hence performance through multiple pathways. First, fatigue may influence basic attention allocation efficiency, leading to loss of performance when the task is sufficiently demanding. Second, fatigue may affect task strategy, especially the operator's willingness to exert effort to maintain performance standards. Third, fatigue may affect trust and operators' utilization of automated systems (where they have the capability scope to do so). As systems become more autonomous, influences on teaming assume increasing importance. The operator interacts with the task environment dynamically so that there are various feedback processes, too many to show in a single illustration. Of special importance is feedback from use of automation to changes in fatigue state.

The focus of this review is on evidence from controlled experimental studies, mostly performed in laboratory environments. Four topics will be covered in more detail:

- *Influences on fatigue states during performance.* The focus of this chapter is on task-induced fatigue states rather than on sleep loss or circadian effects. In fact, there are different forms of fatigue that may be experienced during performance, such as active and passive fatigue (Desmond & Hancock, 2001). We will describe a multidimensional perspective concerning fatigue states (Matthews et al., 2002), delineate some of the features of automated systems that may provoke fatigue, and discuss the role of individual difference factors.

- *Effects of fatigue on performance: Attention and effort.* Vigilance tasks appear to be sensitive to both shortfalls in attentional resources (Warm, Dember, & Hancock, 1996) and to effort and motivation (Dewar, Fraulini, Claypoole, & Szalma, 2016). We will examine the extent to which mechanisms for vigilance decrement also contribute to performance impairment in automated task environments.
- *Fatigue and trust.* An important but somewhat neglected research issue is whether fatigue has systematic effects on trust and use utilization of automation. We will examine the limited evidence available, focusing on whether fatigue tends to produce over- or under-dependence on automation.
- *Intelligent autonomous systems.* Advances in machine intelligence support autonomous systems capable of decision-making without explicit operator direction. We will preview some of the novel fatigue issues that may arise and how they can be countered.

We conclude by highlighting some of the similarities and differences in fatigue effects on conventional cognitive task performance and on task performance with automated and autonomous systems. The important topic of fatigue countermeasures is largely beyond the present scope of this chapter, but we briefly indicate how improved a better understanding of the different facets of fatigue and their concomitant psychological processes can help to combat the ever-evolving dangers of fatigue.

INFLUENCES ON FATIGUE STATES DURING PERFORMANCE

TASK FATIGUE AS A PATTERNED RESPONSE

Subjective fatigue symptoms experienced during sustained attention tasks include not only tiredness but also various forms of stress and mind-wandering. The symptoms experienced are characterized within a multivariate model of subjective states experienced during performance (Matthews et al., 2002). Psychometric evidence identifies multiple dimensions of affective, motivational, and cognitive states that support a higher-order, three-factor structure that includes comprising task engagement, distress, and worry. Low task engagement corresponds to the prototypical fatigue state, defined by low energetic arousal, low task motivation, and lack of concentration. Fatiguing tasks may also induce distress, defined by negative emotions and perceptions of lack of control, and worry, i.e., predominantly negative, self-referencing, intrusive thoughts. The Dundee Stress State Questionnaire (DSSQ: Matthews et al., 2002) assesses patterns of fatigue and stress response across multiple state dimensions.

Signal detection tasks reliably reduce task engagement. Short, high-demand tasks of 12-min duration produce moderate losses in engagement of around 0.5 standard deviations (SDs) in magnitude (e.g., Shaw et al., 2010) whereas longer tasks of 60-min duration produce large effect sizes exceeding 1 SD (e.g., Matthews, Warm, Shaw, & Finomore, 2014). Similar losses in task engagement that have been observed are found for simulated vehicle driving under monotonous conditions (Matthews & Desmond, 2002; Matthews, Neubauer, Saxby, Wohleber, and

Lin, 2019). However, changes in distress and worry are more variable. Vigilance tasks typically elevate distress, but the magnitude of change varies across studies (Matthews, Szalma, Panganiban, Neubauer, and Warm, 2013). Performing vigilance tasks tends to depress worry as attention is reoriented from personal concerns prior to the task to task stimuli, but in some cases, change in worry is small or even positive. A study of truck drivers found an increase in cognitive interference, a marker for worry, during drives of approximately 12 hours (Desmond & Matthews, 2009). Subjective changes in task engagement are paralleled by changes in psychophysiological indices of low cortical arousal including cardiac response and low-frequency alpha power (Borghini, Astolfi, Vecchiato, Mattia, & Babiloni, 2014; Kamzanova, Kustubayeva, & Matthews, 2014), as well as in hemodynamic measures of frontal brain metabolism (Warm, Tripp, Matthews, & Helton, 2012).

Explaining Fatigue Impacts of Sustaining Attention

Why does sustaining attention lead to task disengagement? In fact, there are three, complementary, types of explanation (corresponding to different levels of explanation within cognitive science; Matthews, Lin, & Wohleber, 2017). First, repetitive stimuli that provide little positive reinforcement affect neural systems that influence energy and fatigue, contributing to boredom (Cummings, Gao, & Thornburg, 2016; Scerbo, 2001). Brain-imaging studies link boredom and mind-wandering to the default mode network (DMN) that includes the posterior cingulate cortex, precuneus, and areas of the prefrontal cortex, as well as parietal and temporal areas (e.g., Danckert & Merrifield, 2017). A second type of explanation is provided by the resource theory of vigilance decrement (Warm, Dember, et al., 1996, Warm, Parasuraman, et al., 2008a). Prolonged signal detection under high cognitive load leads to depletion of resources and loss of vigilance. Subjective task engagement appears to be a reliable marker for resource availability (Matthews et al., 2010, 2013). However, resource depletion may not explain performance deficits in low-workload task environments.

Third, from the perspective of the transactional model of stress and emotion (Lazarus, 1999; Matthews, 2001), fatigue reflects the appraisal and coping processes that mediate the operator's interaction with the task environment. Loss of task engagement during performance is associated with low challenge appraisal, low use of task-focused coping, and high use of avoidance (Matthews et al., 2013). This pattern of cognitive processing corresponds to a core relational theme (Lazarus, 1999) of effort-minimization; the operator perceives little benefit in striving for excellence in performance. Occupational studies suggest that the impact of boredom is also dependent on the operator's coping strategies (Cummings et al., 2016). The transactional model explains why the patterning of subjective state response varies across different task environments. Fatiguing tasks that also impose high cognitive load beyond the operator's immediate control, such as high event rate vigilance tasks, tend to elevate appraisals of lack of control and distress (Matthews et al., 2002). Similarly, those that afford scope for self-reflection, typically lower workload tasks, support maintenance of worry (Matthews et al., 2013).

The transactional model has implications for fatigue response during automated system operation. Fatigue results when commitment of effort is perceived as unrewarding (when the task is boring) or ineffective (when task constraints limit the payoff for effort). In many cases, automated systems will provoke perceptions of this kind. The operator's surrender of agency to the machine is likely to increase both boredom, as work becomes more limited and repetitive, and reluctance to exert effort, since striving to maintain vigilance tends to be subjectively aversive (Warm, Matthews, & Finomore, 2008). Automation may similarly disrupt team processes through external imposition of task pacing (Bowers, Oser, Salas, & Cannon-Bowers, 1996).

Fatigue Responses in Highly Automated Systems

Studies in our lab have explored the onset of fatigue in automated system operation. Saxby, Matthews, Warm, Hitchcock, and Neubauer (2013; Study 1) investigated subjective state response to simulated driving. They tested Desmond and Hancock's (2001) theory of active and passive fatigue, two qualitatively different forms of fatigue associated with overload and underload, respectively. Active fatigue was induced by adding wind gusts requiring frequent control responses to the drive, and passive fatigue via full vehicle automation, with a requirement to monitor for automation failures. Length of drive was manipulated independently (10, 30, or 50 min). Passive fatigue elicited a rapid, large-magnitude task engagement response, confirming the potential for automation to produce substantial fatigue. Distress increased moderately but not more than in a control drive (normal driving). By contrast, active fatigue produced smaller though significant decreases in task engagement, and large-magnitude increases in distress, presumably reflecting the high workload of steering the vehicle. Worry declined a little in all conditions but was lower after 10 min than at the longer task durations.

Saxby et al. (2013; Study 1) also assessed appraisal and coping. Compared to active fatigue, passive fatigue was associated with lower scores on appraisals of threat, challenge, and uncontrollability, and lower task-focused coping. Changes in task engagement over time corresponded most closely to changes in challenge. These findings also highlight how system automation induces maladaptive cognitions of the task environment. The detrimental effects of vehicle automation on task engagement proved consistently highly reliable in subsequent studies (Matthews et al., 2019). Studies of individual differences also confirmed that task-induced changes in task engagement and other state factors are reliably associated with the operator's appraisal and coping (Matthews et al., 2002, 2013). Personality factors that influence fatigue-proneness also influence the transactional stress process (Shaw et al., 2010).

Supporting the operator's sense of control in automated task environments may be beneficial (Banks & Stanton, 2016; Harris, Hancock, Arthur, & Caird, 1995). In the vigilance context, Parsons, Warm, Nelson, Riley, and Matthews (2007) compared a detection-action scenario, in which target detection allowed the participant to use the mouse to destroy the threat to their unmanned aerial system (UAS) with a conventional detection-only condition. Scope for action elevated task engagement and reduced distress, and also eliminated vigilance decrement. Similarly, in a simulated process control study, Sauer, Kao, and Wastell (2012) found that adaptable automation (operator controls levels of automation [LOA]) produced less subjective fatigue, less anxiety, and lower workload than adaptive automation (machine controls LOA, depending on performance). By contrast, in an autonomous-driving simulation, Neubauer, Langheim, Matthews, and Saxby (2012) provided drivers with the option to switch on fully automatic control of the vehicle for 5-min intervals. Drivers low in task engagement prior to the drive were more likely to initiate automated driving, implying that fatigue encouraged avoidance of effort. Interestingly, voluntary use of automation did not mitigate task disengagement and actually increased distress, implying that the strategy was actually counterproductive. Thus, greater control does not necessarily mitigate fatigue, especially in lower-workload environments; indeed, fatigued operators may find it difficult to make effective use of control opportunities.

Fatigue and UAS Operation

Future systems will require control of multiple UASs, possible only through use of automation to support functions such as vehicle routing, maintaining vehicle "health status," and targeting (Calhoun, Goodrich, Dougherty, & Adams, 2016; Mouloua, Gilson, & Hancock, 2003). Our research has utilized the Adaptive Levels of Automation (ALOA) simulation (Calhoun, Ruff, Draper, & Wright, 2011). The participant controls four UASs, and performs up to eight subtasks, with automation support. Our first study configured ALOA for a surveillance mission; a high-priority task activity was to open images of objects on the ground photographed by the vehicles as they followed their routes. Participants were required to discriminate friendly and hostile tanks on the basis of their appearance (sensory task) and to count the number of symbols on a map display (cognitive task). The sensory task was designed to resemble visually degraded vigilance tasks (e.g., Shaw et al., 2010). For both surveillance subtasks, an automated targeting system recommended a response, which the participant could override.

Lin et al. (2015) compared high and low cognitive demand conditions by manipulating the event rate of multiple subtasks but keeping the event rate of the surveillance subtasks constant. As expected, high demand elevated subjective workload and distress. By contrast with standard vigilance, task engagement was maintained over a 1-hour performance interval, despite the vigilance requirement of the task. The multitasking element of the simulation provided sufficient challenge and scope for task-focused coping to override the development of fatigue.

A second study configured ALOA to induce fatigue (Wohleber et al., 2019). Task duration was extended to 2 hours, and the event rates of the various subtasks were decreased substantially. Reliability of the automation supporting the surveillance tasks was manipulated (86.7% versus 60%). This task configuration produced subjective passive fatigue. A large-magnitude decline ($d = 1.31$) in task engagement was observed from pre- to post-task, with no change in distress. Worry increased when automation reliability was low. Mean overall workload on the NASA-TLX (Task Load Index; Hart & Staveland, 1988) was 39.7 in the low reliability condition and 32.2 when reliability was high. Thus, the simulation appeared to induce passive fatigue. As in the automated driving simulation studies (Matthews et al., 2019; Saxby et al., 2013), participants experienced low workload and large declines in task engagement, without substantial distress. Unreliable automation appeared to elicit intrusive thoughts (worry) but did not elevate distress. The event rate in this study was low enough that participants could usually process stimuli from the various subtasks serially. By contrast, in the higher-workload study (Lin et al., 2015), task pacing required executive control of subtask scheduling, which may have elevated challenge, and prevented the passive fatigue observed in the second study.

The buildup of fatigue over time was monitored using eye tracking in a subsample of 39 individuals (Calhoun, Funke, et al., 2016). Metrics including Percentage of Eye Closure (PERCLOS) suggested that it took time — at least an hour — for fatigue to develop on ALOA. Thus, the onset of fatigue was slower than that seen in both driving (Saxby et al., 2013) and typical high-demand vigilance tasks (e.g., Szalma et al., 2004). The ALOA task configuration required relatively infrequent interaction with the simulation but may have provided a sufficient requirement for control activities to mitigate against passive fatigue until the second half of the task. As discussed in the next section, performance data also support this interpretation.

EFFECTS OF FATIGUE ON PERFORMANCE: ATTENTION AND EFFORT

The Role of Workload: Resource Limitation and Effort Regulation

Some fatigue effects in automated systems may reflect resource-allocation and effort-regulation mechanisms observed during established in-laboratory studies of performance. To the extent that automation makes complex tasks more similar to vigilance, mechanisms controlling signal detection will generalize to the automation setting. Specifically, as for vigilance tasks (Warm et al., 1996; Warm, Parasuraman, et al., 2008), performance may reflect attentional resource availability. Poorly designed automation might also raise cognitive workload, contributing to active fatigue and attentional overload.

If automation lowers cognitive workload, resource availability may no longer limit performance, and regulation of task-directed effort may become more important (Hockey, 2012; Saxby et al., 2013). When control activities are infrequent, workload is low, and task duration is long, passive fatigue is likely (Desmond and Hancock, 2001). In Hockey's (2012) theory, operators adapt to fatigue by lowering their performance goals, allowing a commensurate reduction in task effort. However, automation effects on this process may be moderated by motivational factors. Vehicle automation often reduces workload, especially with higher levels of automation, but effects on situation awareness are variable (De Winter, Happee, Martens, & Stanton, 2014). Automation may enhance situation awareness if drivers are motivated or instructed to attend to the external environment, but engagement in distracting tasks lowers situation awareness.

VIGILANCE-LIKE EFFECTS IN AUTOMATED DRIVING

Recently, Greenlee, DeLucia, and Newton (2018) demonstrated a temporal loss of performance similar to classic vigilance decrement in an automated driving simulation. During a 40-min, fully automated drive, participants were required to detect stopped vehicles that protruded into the driver's lane. Detection rate declined by about 30% during the drive. The monitoring task also imposed a fairly substantial workload. Mean mental demand on the NASA-TLX (Hart & Staveland, 1988) was 84.0. These findings fit comfortably into a resource theory interpretation, i.e., workload was sufficient to deplete resources over time. Similar to standard vigilance (Warm, Matthews et al., 2008), the authors also confirmed decreased subjective task engagement and increased distress, along with a decrease in worry. Contrary results were obtained by Heikoop, de Winter, van Arem, and Stanton (2017) who simulated driving in an automated platoon. During the drive, participants had to detect red cars in multiple lanes. Evidence for fatigue was obtained from loss of task engagement on the Dundee Stress State Questionnaire (DSSQ), and concurrent eye-tracking and electrocardiogram (ECG) measures. However, there was no temporal change in detection. In this case, accuracy of detection was quite high (greater than 90%) and workload levels were modest: 39 for mental demands. Thus, their resource depletion may have been insufficient to produce performance decrement.

Contrasting data were also obtained from two studies that combined automated driving with an auditory oddball task, i.e., responding with a button-press to the less frequent of two tones. Körber, Cingel, Zimmermann, and Bengler (2015) found no effect of time on task on response times, whereas Solís-Marcos, Galvao-Carmona, and Kircher (2017) reported a U-shaped temporal trend with response times decreasing over the first part of the drive (20 min), then increasing thereafter. These authors reported NASA-TLX mental demand ratings of approximately 35, much lower than in the Greenlee et al. (2018) study, suggesting fatigue was of the passive type. Körber et al. (2015) also reported higher cognitive interference (worry) on the DSSQ. The lack of vigilance decrement might reflect either resource-insensitivity of the oddball task (error rates were very low) or lack of resource depletion due to insufficient mental demand. Both studies also assessed fatigue psychophysiologically. Körber et al. (2015) reported increasing blink frequency and duration, increasing PERCLOS (trend only), and decreasing pupil diameter — changes indicative of fatigue. Solís-Marcos et al. (2017) measured evoked potential (EP) response to the oddball stimuli. P3 response amplitude was lower in automatic than manual driving, and P3 amplitude declined over the course of the drive. Solís-Marcos et al. (2017) point out that temporal change could reflect changes in either resource allocation or in effort regulation, so that it is difficult to disentangle the roles of the two mechanisms.

Data from studies of driver fatigue in conventional vehicles are also open to contrasting interpretations. Monotonous driving tasks often induce changes indicative of low arousal also seen in vigilance studies, such as increasing alpha power in the electroencephalogram (EEG) and heart rate variability (Borghini et al., 2014; Matthews et al., 2019). Reinerman, Warm, Matthews, and Langheim (2008) showed that a 36-min monotonous simulated drive induced temporal declines in cerebral bloodflow velocity (CBFV), a response taken as a diagnostic of declining resource allocation in vigilance studies (Warm et al., 2012).

Matthews and Desmond (2002) induced fatigue with a demanding period of dual-task driving and then examined performance on reversion to normal single-task driving. Task demands were manipulated by contrasting straight and curved road sections. Resource theory predicts greater fatigue-induced impairment on the more demanding curves, but in fact, fatigue increased heading error only on straight paths. These findings suggest an effort-regulation mechanism; easier driving conditions encourage reduced lowering of performance standards when fatigued. Consistent with this interpretation, fatigue lowered steering activity on straight sections but not curves. Notably, workload was high. The NASA-TLX mental demand rating ranged from 58.5–64.3 in control conditions and 70.1–77.2 in fatigue conditions. Thus, even when resource depletion is likely, effort regulation sometimes provides a better explanation for performance effects.

In the automation context, Jamson, Merat, Carsten, and Lai (2013) reported a compatible finding. They showed that, consistent with other driving research (De Winter et al., 2014), drivers in the automated vehicle are inclined to divert attention to in-vehicle entertainment. However, drivers were more attentive to the roadway when traffic was heavy rather than light. Eye movement data (PERCLOS) also suggested greater fatigue during light compared to heavy traffic.

Fatigue and the Transition from Automated to Manual Operation

Studies of vigilance are most relevant to Level 2 vehicle automation (SAE International, 2016), at which some functions are automated but drivers are required to monitor automation continuously. By contrast, at Level 3, drivers do not have to monitor either the driving environment or the automated system performance until control reverts to a lower level. Thus, a key issue is whether fatigue influences manual take-over following automated driving (Matthews et al., 2019). The general human factors challenges of take-over such as lack of situation awareness, distraction from off-task activities, and reorienting gaze to the roadway have been well-documented (e.g., Gold, Happee, & Bengler, 2018). Fatigue potentially degrades influences efficiency of take-over, but thus far studies have not shown a consistent effect of fatigue on takeover time (Weinbeer, Bill, Baur, & Bengler, 2018), unless the driver is already sleep-deprived (Vogelpohl, Kühn, Hummel, Gehlert, & Vollrath, 2018).

However, speed of physical reengagement with the controls may not pick up on impairments in situation awareness (Vogelpohl et al., 2018). Saxby et al. (2013; Study 2) had drivers operate an automated vehicle and then revert to manual control. After 2.5 min, a van pulled out in front of the driver, requiring a rapid steering or braking response. Drivers in the automation (passive fatigue) condition were slower to respond than those in control or active fatigue conditions and more likely to crash. Subsequent studies, reviewed by Matthews et al. (2019), confirmed that automation-induced slowing of response is a reliable effect. The theoretical issue is whether the slowing effect is mediated by depleted attentional resources or effort regulation. Given that active fatigue — a high workload intervention — tended to speed rather than slow responses, Saxby et al. (2013) favored the effort-regulation hypothesis. That is, in passive fatigue, drivers are disinclined to allocate effort to the task; in the present instance, passive-fatigued drivers may have failed to search actively for potential hazards.

Consistent with the effort-regulation explanation, Feldhütter, Gold, Schneider, and Bengler (2017) found that a longer period of automation led to slower response times following a manual takeover request. Eye movement data showed that time watching the driving scene decreased during the automated period, consistent with decreasing motivation, although Feldhütter et al. (2017) suggested that distraction from a visual secondary task may have also played a role. Vogelpohl et al. (2018) also found a gaze effect; automation slowed the latency of drivers' first glance at the speed display following a takeover request. Such data suggest reduced effort in scanning for task-relevant information following a period of automation, but more research is needed to tease apart performance effects mediated by resource and effort-regulation mechanisms.

VIGILANCE IN A MULTI-UAS SIMULATION

In the Wohleber et al. (2019) multi-UAS study described previously, participants showed a vigilance decrement in surveillance. Supporting a resource-depletion interpretation, the decrement was seen only with the more difficult task, requiring a sensory discrimination with degraded images. Decrement was also larger when automation reliability was low. Lin et al. (2015) confirmed the sensitivity of the surveillance task to task demand manipulations, consistent with resource theory.

Other findings pointed more toward an effort-regulation explanation. In typical vigilance tasks, much of the performance decline takes place over the first 15-30 min. In the multi-UAS study, though, performance was fairly stable during over the first 45 min or so before declining sharply. It also showed a pronounced "end effect" with performance returning to its initial level in the final 15-min block. Participants could see on a map display that their vehicles were coming to the end of their routes. By increasing effort in this final block, participants could eliminate large-magnitude vigilance decrement effects apparent in preceding blocks. By contrast, while end effects and motivational impacts are found in conventional vigilance (e.g., Dewar et al., 2016), they rarely prevent the decrement entirely. Wohleber et al. (2019) concluded that both mechanisms play a role in decrement. Resources become depleted, but because overall workload was quite low, the depletion process is relatively slow and resource loss is modest. The impact of resource depletion depended critically on effort regulation. Over time, participants lost motivation to maintain performance standards, up until the final block at which a final burst of effort restores normal performance. Further investigation of the interplay between resource loss and compensatory effort might be of value in other task domains also.

FATIGUE AND TRUST

TRUST CALIBRATION FOR AUTOMATED SYSTEMS

The performance consequences of fatigue discussed thus far fit into a larger body of research on attention and effort. In this section, we address the further and novel issue of the relationship between fatigue and operator utilization of automation, especially calibrating trust. Both over-trust and under-trust may impair performance, via over- and under-dependence on automation, respectively (Lee & See, 2004). However, trust may be miscalibrated in different ways. Trust calibration research comes mostly from two automation paradigms, alerting systems and decision aids (Parasuraman & Manzey, 2010). In the former context, an autonomous vehicle may signal a need for the human to take over control based on its analysis of traffic conditions. In such scenarios, trust is defined in terms of compliance and reliance (Dixon & Wickens, 2006; Wickens, Clegg, Vieane, & Sebok, 2015). Compliance refers to the response to the alert. It is optimal when the operator accurately evaluates the validity of the alert and responds appropriately, e.g., taking over manual control if and only if it is necessary to do so. Reliance refers to operator behavior in the absence of an alert; for example, the autonomous vehicle driver must remain vigilant to manually assume control when the automated system has failed to detect a problem.

By contrast, decision aids provide recommendations for higher-level cognitive tasks, such as navigation in the vehicle case (Parasuraman & Manzey, 2010). The compliance/reliance perspective, although developed to understand responses to alerts, also applies to other implementations of automation, including decision aids (Boubin, Rusnock, & Bindewald, 2017). For example, a driver using a GPS may get into trouble both through overcompliance (following an incorrect route recommendation) and overreliance (failing to check for a turn on the assumption that the GPS will provide an instruction).

FATIGUE AND AUTOMATION-DEPENDENCE

Operator state characteristics such as motivation, stress, sleep loss, and boredom may impact trust and automation usage (Hoff & Bashir, 2015), but there is little empirical evidence on fatigue effects. Both resource and effort-regulation perspectives suggest possible mechanisms. From a resource theory perspective, the key issue is how cognitive workload influences trust. In Parasuraman and Manzey's model (2010), high task load, together with additional situational and person factors, leads to suboptimal resource allocation expressed in loss of situation awareness, increased complacency (overreliance), and automation bias (overdependence on decision aids). How might fatigue play into such a process? The answer may be different for active and passive fatigue. As a high-workload state, active fatigue may exacerbate such over-trust issues, especially if the task has vigilance elements that lead to resource depletion. However, in passive fatigue, attention is not directly vulnerable to overload.

Fatigue is associated with use of strategies for minimizing task-directed effort (Hockey, 2012; Sauer, Wastell, Robert, Hockey, & Earle, 2003). If automation does in fact remove some of the workload burden, fatigued operators may become especially dependent on automation and prone to delegate excessive responsibility for alerts and/or decision-making to it, minimizing cognitive demands. It is a little unclear how such effects might interact with task demands. On the one hand, high workload might increase motivation to off-load task functions onto the automation. On the other hand, vehicle driving studies (Jamson et al., 2013; Matthews & Desmond, 2002; Saxby et al., 2013) suggest that fatigue is most likely to lead to effort reduction in lower task load conditions, implying that automation use may be most vulnerable to disruption of trust calibration in passive rather than active fatigue states.

Thus, leading theories appear to broadly suggest that fatigue encourages overdependence on automation, although they appear to offer different perspectives on the respective roles of active and passive fatigue. Unfortunately, the limited available research provides somewhat conflicting results. In a study of decision aid usage, Reichenbach, Onnasch, and Manzey (2011) found that, contrary to an effort-minimization hypothesis, sleep-deprived operators were more attentive toward possible automation failures and showed less automation bias. However, sleep loss also impaired multitasking, suggesting loss of resources. In this case, awareness of a resource shortfall may have led to attempts to compensate for fatigue strategically through more effective use of automation.

The multi-UAS simulation study by Wohleber et al. (2019) also investigated automation dependence, defined as the percentage of trials on which the participant agreed with the recommendation of the automation on the surveillance tasks. Results were both clear-cut and surprising (see Figure 7.2). For the more demanding task (discriminating tank images), automation dependence declined steadily across the 2-hour run. As expected, dependence was lower in the lower-reliability condition. In addition, dependence declined more markedly when reliability was low. Figure 7.2 shows the optimal dependence for each condition, matched to the reliability values (86.7% versus 60%). Participants tend to be overdependent initially and underdependent by the end of the run. Certainly, increasing fatigue did not produce complacency.

Wohleber et al. (2019) suggested that managing the automation is itself perceived by participants as an additional subtask. Banks et al. (2014) argued that

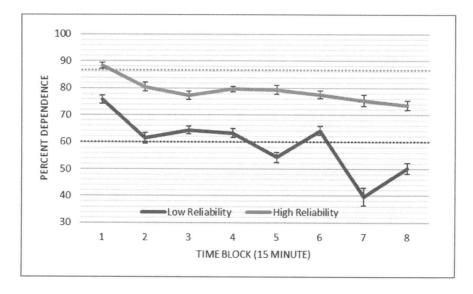

FIGURE 7.2 Changes in automation dependence over time on a multi-UAS simulation task, at two levels of automation reliability. Broken horizontal lines indicate optimal dependence at each reliability level.

use of automation may actually increase task complexity because of the additional need to monitor subsystems of automation on top of the primary task. Consistent with the effort-regulation hypothesis, fatigued participants may have been increasingly inclined to shed this task and rely on their own judgment, however fallible. However, the influence of task demands on the effect suggests a role for resources also. Resource depletion may have motivated task-shedding only when the surveillance task itself was sufficiently difficult. In any case, underdependence on automation under fatigue would be a significant operational issue.

FATIGUE AND TEAMING WITH INTELLIGENT AUTONOMY

In both commercial and government sectors, much of research and development is aimed toward evolving assistive technologies from basic alerting systems to advanced decision-making partners. Like humans, such partners process information at multiple stages including encoding and analysis, decision-making, and action selection and implementation (Parasuraman, Sheridan, & Wickens, 2000). Multiple advances in artificial intelligence (AI) such as advanced statistical classifiers, machine learning, algorithms for reasoning with uncertain information, and effector technologies are progressively increasing the range of functions at which machines can match or exceed humans, at least in certain domains. Thus, the future of automation increasingly lies with Intelligent Autonomous Systems (IASs) that function as teammates for the human operator (Weiss, 2011). IASs may include robots, vehicles, software agents, and other systems. Additionally, autonomy can be further categorized based on human-interaction style as either an adaptive system, adjustable autonomy, or mixed-initiative system based on the amount of control they assume from the operator (Barnes, Chen, & Hill, 2017).

Autonomous system technologies raise numerous human factors issues beyond our present scope. In this section, we focus on specific challenges for the fatigued operator, referring to previous work on human-human teaming. We caution that much of this work focuses on sleep deprivation rather than task fatigue, which has been neglected in the teaming literature. Given the novelty of IASs, this section is necessarily somewhat speculative, but we aim to define some focal issues for future research.

Matthews et al. (2016) defined three types of challenge that may be unique to IASs: cognitive, social, and self-regulative. First, cognitive complexity is higher than conventional automation; for example, sources of system malfunction may be as diverse as a sensor failure, faulty programming of high-level routines for goal prioritization, or hacking by an adversary. Second, to the extent that the system is perceived as a teammate, its interactions take on a social aspect. Human and machine must communicate to develop shared situational awareness, including awareness of the functional status of each other. Third, working with increasingly intelligent machines capable of evaluating the human operator raises novel self-regulative challenges, such as being negatively evaluated by the system. Fatigue may potentially compromise management of each of these challenges.

We focus here primarily on the social and teaming challenges of working with IAS, but we first briefly mention the other two types of challenge identified by

Matthews et al. (2016). Fatigue may impair performance of complex cognitive tasks requiring executive control (van der Linden, Frese, & Sonnentag, 2003), potentially leading to errors in activities such as fault diagnosis or mission planning. However, fatigued operators may sustain performance effectively when the task is interesting and supports intrinsic motivation (Matthews, Davies, Westerman, & Stammers, 2000). Thus, operators who enjoy the challenge of working with IASs are likely to be resistant to fatigue, especially passive fatigue. Active fatigue remains potentially problematic, e.g., if the operator must struggle with a system that persistently behaves unpredictably. In addition, both cognitive and social features of the IAS may pose self-regulative challenges. In a worst-case scenario, the human might consider appraising the machine to be as beyond their personal control and resent the machine for attributing mission failure to shortcomings in the human's performance, a stressful breakdown in teaming. Stress and fatigue may be mutually reinforcing, especially in active fatigue (Saxby et al., 2013).

Some clues to the influence of fatigue on the social aspects of teaming with an IAS are found from work on human-human teams. Teaming is comprised of the coordinating and organizing activities (i.e., teamwork) between teammates to manage the goals and specific tasks (i.e., taskwork) of the team (Cannon-Bowers & Bowers, 2011; Marks, Mathieu, & Zaccaro, 2001). Effective teaming is the result of a well-calibrated exchange across both of these team processes. Performance may suffer when poor coordination of taskwork compounds cognitive demands across the team, particularly if an individual team member's tasks are dependent on her/his partner's actions. In the context of adaptive automation, appropriately timed alerts may be seen as "etiquette" possessed by the automated system (Parasuraman & Miller, 2004). Designing etiquette may build transactive memory, i.e., an understanding of the knowledge possessed by the teammate and a collective knowledge of team function that can contribute positively to group performance (Austin, 2003) in both taskwork and teamwork.

A critical element of teamwork is maintaining awareness of the knowledge, aims, and functional capabilities of one's teammates, and providing support and backup where needed, either reactively (on request) or proactively (self-initiated). Cohesive teaming between humans and IASs requires appropriate design. For the IAS to provide reactive support, there must be effective communication channels for the human to signal what is needed. To provide proactive support additionally requires the IAS to detect and mitigate human performance vulnerabilities, e.g., by taking on additional tasks. An IAS that delivers information based on accurate analysis of demands on the human operator (Dorneich, Ververs, Mathan, Whitlow, & Hayes, 2012) may build trust through being perceived as a good teammate while functionally assisting in the maintenance of overall taskwork and team performance. Conversely, in *2001: A Space Odyssey*, the malfunctioning HAL 9000 computer advises the astronaut "This mission is too important for me to allow you to jeopardize it," leading to a catastrophic breakdown in trust.

Fatigue may disrupt teaming with an IAS in various ways, depending on the implementation of the autonomy. In human-human teaming, fatigue is a potential disruptive factor to be addressed in crew resource management (e.g., Flin & Maran, 2015), but there is little research on how it may influence specific teamwork processes.

One study showed that sleep deprivation impaired teamwork during crisis management in a simulated operating room (Neuschwander et al., 2017). Mechanisms for impairments in teaming are unclear, but studies of stress and team decision-making implicate distinct stages of defining the problem, sharing information, weighting and integrating inputs, and arriving at a final decision (Burke, Priest, Salas, Sims, & Mayer, 2008). Because research is limited concerning the effects of fatigue on each of these components and human-IAS teaming in general, we will focus on outlining promising lines of future research on communication, motivation, and team management that would help to provide insight on this topic. Communication may impose fatigue-sensitive cognition, especially if the human must adhere to protocols necessary for the artificial system to understand human speech or text. A system such as Alexa or Siri, which requires consistent communication to feed its filtering and search algorithms, may reduce the efficiency of an operator by consuming resources required for the communication task. Van der Linden et al. (2003) found increased rigidity, defined as performing an unsuccessful action repeatedly, in mentally fatigued individuals performing a computer based scheduling task. The task used a think aloud procedure, similar to communicating plans and intentions with a partner. Similar problem-solving deficits may occur under fatigue with an IAS requiring verbal updates.

Fatigue also encourages social loafing, i.e., reductions in individual effort, especially when the task is dull and the individual's responsibility is not salient (Hoeksema-van Orden, Gaillard, & Buunk, 1998). Human operators may similarly offload responsibility onto the IAS when fatigued, especially under circumstances likely to provoke overdependence such as high system reliability and high workload (Parasuraman & Manzey, 2010). In a team study using a command and control simulation, Chaiken and colleagues (2011) found that by the end of the 36-hour-long series of missions, performance in teams suddenly dropped below solo operators. Detrimental effects of fatigue may have included loss of the motivation to coordinate actions to maximize overall team efforts, further increasing task load across the team. This outcome may have been avoided if teams had initially reorganized into separate, serial versus parallel, and similar roles (Chaiken et al., 2011).

Motivational impacts may be more subtle than simple effort reduction. Barnes and van Dyne (2009) discriminated several fatigue-sensitive workload management strategies that may be implemented in teams depending on the individual's decision latitude and self-efficacy. For example, shifting workload to others may be most prevalent when decision latitude is high and self-efficacy is low. By extension, and somewhat counterintuitively, the fatigued IAS operator may be most likely to offload responsibility onto the machine when the level of autonomy is adaptable and the machine is perceived as unresponsive or uncooperative (i.e., reducing self-efficacy). Other strategies for workload reduction identified by Barnes and van Dyne (2009) include utilizing familiar responses, following the lead of others, and pursuing risky options requiring little processing. All are potentially maladaptive in the IAS teaming context, at least in some contexts.

Barnes and Hollenbeck (2009) identified horizontal and vertical differentiation as two factors that moderate the impact of fatigue on team management in the organizational context. Horizontal differentiation refers to the degree of specialization among team members. When team members are highly differentiated, each having their own unique skill, fatigue impacts are more likely because other team members cannot substitute for the person impaired by fatigue. On the principle that humans and artificial systems have different cognitive strengths and weaknesses, human-IAS teams are likely to be differentiated and hence fatigue-vulnerable. Specifically, the IAS may not be able to compensate for human failings, even if it recognizes that the person is impaired. Vertical differentiation is the extent to which team members differ in power and authority (formally or informally); the military chain of command imposes differentiation of this kind. According to Barnes and Hollenbeck (2009), fatigue may be most damaging to decision-making with vertical differentiation and an impaired leader; subordinates cannot readily overrule poor decisions. Humans typically exercise authority over machines, corresponding to this case; the fatigued human may not be able to direct an IAS effectively. However, increasing machine intelligence may reduce vertical differentiation, at least in contexts in which the machine is the better decision-maker, potentially mitigating such issues. Indeed, it may be needful to develop protocols for the IAS "assuming command" if the human is severely compromised.

Set against these instances in which fatigue may disrupt teaming, intelligent systems may also be explicitly designed to mitigate fatigue. An autonomous partner, designed with intelligence to carry out operator tasks especially vulnerable to fatigue, could adapt to operator needs, building trust and the operator's sense of control over team management. Häusser et al. (2016), in a human teaming study, found that sleep-deprivation increased advice-taking from an advisor, suggesting that fatigued operators might also be inclined to trust a competent IAS. Mixed-initiative systems (Barnes, Chen, & Hill, 2017), where a mix of adaptive decision-making and action by the IAS is based on the needs of the human and responsive action to requests from the human, may be ideal for reducing the impact of fatigue. Control of task scheduling has been found to moderate the effect of high workload on performance so that individuals with freedom and flexibility to choose when to execute tasks (high control) produced fewer errors and experienced less fatigue (Hockey & Earle, 2006). An IAS teammate could choose to "tag-in" when needed and provide the operator a rest break in order to replan for or, at least, recover from high workload. Similarly, an operator could shift tasking to its IAS partner, which would actively reduce any existing cognitive strain.

Performance when partnered with an IAS may be subject to similar fatigue impacts as human-human partnerships. However, the IAS can be designed to fit the human operator in ways that cannot be controlled in human-human teams and, obviously, the IAS partner will not experience fatigue. Further mitigation of fatigue may be accomplished through anthropomorphic design, which may support appropriate trust. DeVisser et al. (2016) found that having an anthropomorphized partner prevented negative bias that might occur when a highly computerized partner commits an error. Having an avatar may also prevent overreliance on a human partner who may not perform as well as a machine; individuals are almost excessively forgiving of human errors while punishing of machine errors (Merritt, Heimbaugh, LaChapell, & Lee, 2013). With further research aimed at the proper design of machine partners, fatigue impacts on performance may be reduced by the presence of a robotic partner.

The IAS partner may be designed to respond to signs of fatigue with appropriate taskwork assistance but, more uniquely, with teamwork responses that may encourage task engagement and motivation. The IAS partner can be encouraging, alerting, or proactive in replanning based on situational demands.

CONCLUSIONS AND APPLICATIONS

Automation technology provides numerous human factors benefits, but it also has the potential to exacerbate fatigue impacts on the operator. Whereas the operator's neurocognitive state depends on multiple environmental and personal factors, automation that shifts the operator's function toward a monitoring role increases the likelihood of fatigue developing, especially passive fatigue. Performance consequences of automation-induced fatigue partly recapitulate changes in resource availability and effort regulation observed in traditional performance contexts such as vigilance (Warm, Parasuraman et al., 2008) and vehicle driving (Matthews & Desmond, 2002). Sustaining attention is prone to deplete resources, leading to performance deficits on attentionally demanding tasks. The fatigued operator may also lower performance standards or apply other strategies that allow effort-minimization (Hockey, 2012).

Broadly, high workload tasks tend to elicit active fatigue and vulnerability to attentional overload, whereas low workload tasks produce passive fatigue and effort-reduction (Saxby et al., 2013). However, this general principle is simplistic; for example, in driving studies, fatigue effects on manual takeover from automation vary from study to study and appear to be sensitive to contextual factors (e.g., Jamson et al., 2013). From a theoretical standpoint, different mechanisms for fatigue effects vary in context sensitivity. In highly controlled laboratory settings, such as those used for vigilance studies, strategic adjustment to fatigue is limited. Thus, paradigms such as vigilance and dual-task performance are well-suited for identifying fatigue effects on resource availability. Cognitive-psychological explanations can also be integrated with those from neuroscience (Langner & Eickhoff, 2013).

Real-world automated tasks are often more complex than basic laboratory paradigms and provide greater scope for strategies that may hide or compensate for resource shortfalls (Matthews, Wohleber, & Lin, in press). Furthermore, strategy change will depend critically on the operator's appraisal of the significance of the task and available options for coping (Matthews, 2001). Thus, while effort minimization may be a common strategic response to fatigue, it is not an inevitable one, and operators may according to context and personal disposition choose to cope with fatigue in a variety of ways. Fatigued operators may at times even choose to increase effort, perhaps when they perceive a need to compensate for fatigue (Manzey, Reichenbach, & Onnasch, 2009; Reichenbach et al., 2011). General psychological theories of fatigue must be complemented with a contextualized understanding of how fatigued operators interact with automation.

Similar considerations apply to the novel topic of fatigue effects on trust and automation dependence. Attentional resource models (Parasuraman & Manzey, 2010) suggest general trends, such as the likelihood that resource depletion and high workload will increase dependence, provided that the automation is reliable. However, these trends may be overridden by operators' understanding of the task situation and their options for self-regulation. For example, in Wohleber et al.'s (2019) multi-UAS study, operators appear to have perceived the automation as a low-priority task to be shed under fatigue.

The advent of IASs introduces new complexities associated with the novel cognitive, teamwork, and self-regulative demands placed on the operator (Matthews et al., 2016). As for conventional automation, operators remain vulnerable to attentional overload and maladaptive effort regulation, depending on level of cognitive demands and the extent and nature of fatigue. However, the ways in which fatigue might play into features of human-IAS teaming, such as communications, strategic offload of cognitive work, and team management, remain to be explored. As with conventional automation, it is likely that the operator's appraisal and coping will play a critical role. Design solutions for fatigue can develop modes of communication and teamwork that enhance the operator's sense of challenge, control, and task motivation.

Finally, a better understanding of fatigue can support more effective countermeasures. The first line of defense against fatigue is to prevent or attenuate the fatigue response. Traditional remedies such as sleep optimization, rest breaks, and caffeine will be effective across a range of automation contexts. In the current review, we have highlighted the further role of psychological factors that can inform design to mitigate cognitive overload and support sustained task-directed effort. Design solutions can be tailored differently to address active and passive fatigue (May & Baldwin, 2009). These authors also review adaptive automation approaches based on accurate classification of the two types of fatigue that drive appropriate changes to task load. Understanding the role of individual difference factors can also help tailor design to the individual's strengths and vulnerabilities (Szalma, 2009).

ACKNOWLEDGMENT

The authors gratefully acknowledge support from the Air Force Research Laboratory (AFRL). The views and conclusions contained in this document are those of the authors and should not be interpreted as representing the official policies, either express or implied, of the AFRL or the U.S. government.

REFERENCES

Austin, J. R. (2003). Transactive memory in organizational groups: The effects of content, consensus, specialization, and accuracy on group performance. *Journal of Applied Psychology, 88,* 866–878.

Banks, V. A., & Stanton, N. A. (2016). Driver-centred vehicle automation: Using network analysis for agent-based modelling of the driver in highly automated driving systems. *Ergonomics, 59,* 1442–1452.

Banks, V. A., Stanton, N. A., & Harvey, C. (2014). Sub-systems on the road to vehicle automation: Hands and feet free but not "mind" free driving. *Safety Science, 62,* 505–514.

Barnes, C. M., & Hollenbeck, J. R. (2009). Sleep deprivation and decision-making teams: Burning the midnight oil or playing with fire? *Academy of Management Review, 34,* 56–66.

Barnes, C. M., & Van Dyne, L. (2009). "I'm tired": Differential effects of physical and emotional fatigue on workload management strategies. *Human Relations, 62,* 59–92.

Barnes, M. J., Chen, J. Y. C., & Hill, S. (2017). *Humans and autonomy: Implications of shared decision-making for military operations* (ARL-TR-7919). Aberdeen Proving Ground, MD: US Army Research Laboratory.

Billings, C. E. (1996). *Human-centered aviation automation: Principles and guidelines* (Technical memorandum no. 110381). Washington, DC: NASA.

Borghini, G., Astolfi, L., Vecchiato, G., Mattia, D., & Babiloni, F. (2014). Measuring neurophysiological signals in aircraft pilots and car drivers for the assessment of mental workload, fatigue and drowsiness. *Neuroscience & Biobehavioral Reviews, 44*, 58–75.

Boubin, J. G., Rusnock, C. F., & Bindewald, J. M. (2017). Quantifying compliance and reliance trust behaviors to influence trust in human-automation teams. *Proceedings of the Human Factors and Ergonomics Society Annual Meeting, 61*, 750–754.

Bowers, C. A., Oser, R. L., Salas, E., & Cannon-Bowers, J. A. (1996). Team performance in automated systems. In R. Parasuraman & M. Mouloua (Eds.), *Automation and Human Performance: Theory and Applications* (pp. 243–263). Mahwah, NJ: Lawrence Erlbaum.

Burke, C. S., Priest, H. A., Salas, E., Sims, D., & Mayer, K. (2008). Stress and teams: How stress affects decision making at the team level. In P. A. Hancock & J. L. Szalma (Eds.), *Performance under Stress* (pp. 181–208). Abingdon, UK: Ashgate.

Calhoun, G., Funke, G., Matthews, G., Wohleber, R., Lin, J., Chiu, C. Y. P., & Ruff, H. (2016). *Impact of individual differences on reliance optimization* (AFRL-RH-WP-TR-2016-0043). 711 HPW/RHCI. Wright-Patterson AFB, OH: USAF, Air Force Research Laboratory.

Calhoun, G. L., Goodrich, M. A., Dougherty, J. R., & Adams, J. A. (2016). Human-autonomy collaboration and coordination toward multi-RPA missions. In N. J. Cooke, L. J. Rowe, W. Bennett, & D. Q. Joralmon (Eds.), *Remotely piloted aircraft systems: A human systems integration perspective* (pp. 109–136). New York: Wiley.

Calhoun, G. L., Ruff, H. A., Draper, M. H., & Wright, E. J. (2011). Automation-level transference effects in simulated multiple unmanned aerial vehicle control. *Journal of Cognitive Engineering and Decision Making, 5*, 55–82.

Cannon-Bowers, J. A. & Bowers, C. (2011). Team development and functioning. In S. Zedeck (Ed.), *APA Handbook of Industrial and Organizational Psychology* (pp. 597–650). Washington, DC: American Psychological Association.

Chaiken, S., Harville, D. L., Harrison, R., Fischer, J., Fisher, D., & Whitemore, J. (2011). Fatigue impact on teams versus individuals during complex tasks. In P.L. Ackerman (Ed.) *Cognitive Fatigue: Multidisciplinary Perspectives on Current Research and Future Applications*. Washington, DC: American Psychological Association.

Cummings, M. L., Gao, F., & Thornburg, K. M. (2016). Boredom in the workplace: A new look at an old problem. *Human Factors, 58*, 279–300.

Danckert, J., & Merrifield, C. (2017). Boredom, sustained attention and the default mode network. *Experimental Brain Research, 236*(9), 1–12.

Desmond, P. A., & Hancock, P. A. (2001). Active and passive fatigue states. In P. A. Hancock & P. A. Desmond (Eds.), *Stress, Workload, and Fatigue* (pp. 455–465). Mahwah, NJ: Lawrence Erlbaum.

Desmond, P. A., & Matthews, G. (2009). Individual differences in stress and fatigue in two field studies of driving. *Transportation Research Part F: Traffic Psychology and Behaviour, 12*, 265–276.

de Visser, E. J., Monfort, S. S., Mckendrick, R., Smith, M. A., Mcknight, P. E., Krueger, F., & Parasuraman, R. (2016). Almost human: Anthropomorphism increases trust resilience in cognitive agents. *Journal of Experimental Psychology: Applied, 22*, 331–349.

Dewar, A. R., Fraulini, N. W., Claypoole, V. L., & Szalma, J. L. (2016). Performance in vigilance tasks is related to both state and contextual motivation. *Proceedings of the Human Factors and Ergonomics Society Annual Meeting, 60*, 1145–1149.

De Winter, J. C., Happee, R., Martens, M. H., & Stanton, N. A. (2014). Effects of adaptive cruise control and highly automated driving on workload and situation awareness: A review of the empirical evidence. *Transportation Research Part F: Traffic Psychology and Behaviour, 27,* 196–217.

Dixon, S. R., & Wickens, C. D. (2006). Automation reliability in unmanned aerial vehicle control: A reliance-compliance model of automation dependence in high workload. *Human Factors, 48,* 474–486.

Dorneich, M. C., Ververs, P. M., Mathan, S. Whitlow, S., & Hayes, C. C. (2012). Considering etiquette in the design of an adaptive system. *Journal of Cognitive Engineering and Decision Making, 6,* 243–265.

Endsley, M. R. (2018). Situation awareness in future autonomous vehicles: Beware of the unexpected. In S. Bagnara, R. Tartaglia, S. Albolino, T. Alexander and Y. Fujita (Eds.), *Proceedings of the Congress of the International Ergonomics Association* (pp. 303–309). Cham: Springer.

Feldhütter, A., Gold, C., Schneider, S., & Bengler, K. (2017). How the duration of automated driving influences take-over performance and gaze behavior. In C. M. Schlick, S. Duckwitz, F. Flemisch, M. Frenz, S. Kuz, A. Mertens and S. Mütze-Niewöhner (Eds.), *Advances in Ergonomic Design of Systems, Products and Processes* (pp. 309–318). Berlin: Springer.

Flin, R., & Maran, N. (2015). Basic concepts for crew resource management and non-technical skills. *Best Practice & Research Clinical Anaesthesiology, 29,* 27–39.

Gold, C., Happee, R., & Bengler, K. (2018). Modeling take-over performance in level 3 conditionally automated vehicles. *Accident Analysis & Prevention, 116,* 3–13.

Greenlee, E. T., DeLucia, P. R., & Newton, D. C. (2018). Driver vigilance in automated vehicles: Hazard detection failures are a matter of time. *Human Factors, 60,* 465–476.

Harris, W. C., Hancock, P. A., Arthur, E. J., & Caird, J. K. (1995). Performance, workload, and fatigue changes associated with automation. *The International Journal of Aviation Psychology, 5,* 169–185.

Hart, S. G., & Staveland, L. E. (1988). Development of NASA-TLX (Task Load Index): Results of empirical and theoretical research. In P. A. Hancock & N. Meshkati (Eds.), *Advances in Psychology* (Vol. 52, pp. 139–183). Amsterdam, The Netherlands: Elsevier.

Häusser, J. A., Leder, J., Ketturat, C., Dresler, M., & Faber, N. S. (2016). Sleep deprivation and advice taking. *Scientific Reports, 6,* 24386.

Heikoop, D. D., de Winter, J. C., van Arem, B., & Stanton, N. A. (2017). Effects of platooning on signal-detection performance, workload, and stress: A driving simulator study. *Applied Ergonomics, 60,* 116–127.

Hockey, G. R. J. (2012). Challenges in fatigue and performance research. In G. Matthews, P. A. Desmond, C. Neubauer, & P. A. Hancock (Eds.), *Handbook of Operator Fatigue* (pp. 45–60). Aldershot, UK: Ashgate Press.

Hockey, G. R. J., & Earle, F. (2006). Control over the scheduling of simulated office work reduces the impact of workload on mental fatigue and task performance. *Journal of Experimental Psychology: Applied, 12,* 50–65.

Hoeksema-van Orden, C. Y., Gaillard, A. W., & Buunk, B. P. (1998). Social loafing under fatigue. *Journal of Personality and Social Psychology, 75,* 1179–1190.

Hoff, K. A., & Bashir, M. (2015). Trust in automation: Integrating empirical evidence on factors that influence trust. *Human Factors, 57,* 407–434.

Jamson, A. H., Merat, N., Carsten, O. M., & Lai, F. C. (2013). Behavioural changes in drivers experiencing highly-automated vehicle control in varying traffic conditions. *Transportation Research Part C: Emerging Technologies, 30,* 116–125.

Kamzanova, A., Kustubayeva, A. M., & Matthews, G. (2014). Use of EEG workload indices for diagnostic monitoring of vigilance decrement. *Human Factors, 56,* 1136–1149.

Körber, M., Cingel, A., Zimmermann, M., & Bengler, K. (2015). Vigilance decrement and passive fatigue caused by monotony in automated driving. *Procedia Manufacturing, 3,* 2403–2409.

Langner, R., & Eickhoff, S. B. (2013). Sustaining attention to simple tasks: A meta-analytic review of the neural mechanisms of vigilant attention. *Psychological Bulletin, 139,* 870–900.

Lazarus, R. S. (1999). *Emotions and Adaptation.* New York: Oxford University Press.

Lee, J. D., & See, K. A. (2004). Trust in automation: Designing for appropriate reliance. *Human Factors, 46,* 50–80.

Lin, J., Wohleber, R., Matthews, G., Chiu, P., Calhoun, G., Ruff, H., & Funke, G. (2015). Video game experience and gender as predictors of performance and stress during supervisory control of multiple unmanned aerial vehicles. *Proceedings of the Human Factors and Ergonomics Society Annual Meeting, 59,* 746–750.

Manzey, D., Reichenbach, J., & Onnasch, L. (2009). Human performance consequences of automated decision aids in states of fatigue. *Proceedings of the Human Factors and Ergonomics Society Annual Meeting, 53,* 329–333.

Marks, M. A., Mathieu, J. E., & Zaccaro, S. J. (2001). A temporally based framework and taxonomy of team processes. *Academy of Management Review, 26,* 356–376.

Matthews, G. (2001). A transactional model of driver stress. In P. A. Hancock & P. A. Desmond (Eds.), *Stress, Workload and Fatigue* (pp. 133–163). Mahwah, NJ: Lawrence Erlbaum.

Matthews, G., Campbell, S. E., Falconer, S., Joyner, L., Huggins, J., Gilliland, K., Grier, R., & Warm, J. S. (2002). Fundamental dimensions of subjective state in performance settings: Task engagement, distress and worry. *Emotion, 2,* 315–340.

Matthews, G., Davies, D. R., Westerman, S. J., & Stammers, R. B. (2000). *Human Performance: Cognition, Stress and Individual Differences.* London: Psychology Press.

Matthews, G., & Desmond, P. A. (2002). Task-induced fatigue states and simulated driving performance. *Quarterly Journal of Experimental Psychology, 55A,* 659–686.

Matthews, G., Lin, J., & Wohleber, R. (2017). Personality, stress and resilience: A multifactorial cognitive science perspective. *Psychological Topics, 26,* 139–162.

Matthews, G., Neubauer, C., Saxby, D. J., Wohleber, R. W., & Lin, J. (2019). Dangerous intersections? A review of studies of fatigue and distraction in the automated vehicle. *Accident Analysis & Prevention, 126,* 85–94.

Matthews, G., Reinerman-Jones, L., Barber, D., Teo, G., Wohleber, R., Lin, J., & Panganiban, A. R. (2016). Resilient autonomous systems: Challenges and solutions. In *Resilience Week (RWS), 2016* (pp. 208–213). Chicago, IL: IEEE.

Matthews, G., Szalma, J., Panganiban, A. R., Neubauer, C., & Warm, J. S. (2013). Profiling task stress with the Dundee Stress State Questionnaire. In L. Cavalcanti & S. Azevedo (Eds.), *Psychology of Stress: New Research* (pp. 49–90). Hauppage, NY: Nova Science.

Matthews, G., Warm, J. S., Reinerman, L. E., Langheim, L., Washburn, D. A., & Tripp, L. (2010). Task engagement, cerebral blood flow velocity, and diagnostic monitoring for sustained attention. *Journal of Experimental Psychology: Applied, 16,* 187–203.

Matthews, G., Warm, J. S., Shaw, T. H., & Finomore, V. S. (2014). Predicting battlefield vigilance: A multivariate approach to assessment of attentional resources. *Ergonomics, 57,* 856–875.

Matthews, G., Wohleber, R., & Lin, J. (in press). Stress, skilled performance, and expertise: Overload and beyond. In P. Ward, J. M. Schraagen, J. Gore, & E. Roth (Eds.), *Oxford Handbook of Expertise.* New York: Oxford University Press.

May, J. F., & Baldwin, C. L. (2009). Driver fatigue: The importance of identifying causal factors of fatigue when considering detection and countermeasure technologies. *Transportation Research Part F: Traffic Psychology and Behaviour, 12,* 218–224.

Merritt, S. M., Heimbaugh, H., LaChapell, J., & Lee, D. (2013). I trust it, but I don't know why: Effects of implicit attitudes toward automation on trust in an automated system. *Human Factors, 55,* 520–534.

Mouloua, M., Gilson, R., & Hancock, P. (2003). Human-centered design of unmanned aerial vehicles. *Ergonomics in Design, 11*, 6–11.

Neubauer, C., Langheim, L., Matthews, G., & Saxby, D. (2012). Fatigue and voluntary utilization of automation in simulated driving. *Human Factors, 54,* 734–746.

Neuschwander, A., Job, A., Younes, A., Mignon, A., Delgoulet, C., Cabon, P., ... & Tesniere, A. (2017). Impact of sleep deprivation on anaesthesia residents' non-technical skills: A pilot simulation-based prospective randomized trial. *BJA: British Journal of Anaesthesia, 119,* 125–131.

Parasuraman, R., & Manzey, D. H. (2010). Complacency and bias in human use of automation: An attentional integration. *Human Factors, 52,* 381–410.

Parasuraman, R., & Miller, C. A. (2004). Trust and etiquette in high-criticality automated systems. *Communications of the ACM, 47,* 51–55.

Parasuraman, R., Sheridan, T. B., & Wickens, C. D. (2000). A model for types and levels of human interaction with automation. *IEEE Transactions on Systems, Man, and Cybernetics-Part A: Systems and Humans, 30,* 286–297.

Parasuraman, R., & Wickens, C. D. (2008). Humans: Still vital after all these years of automation. *Human Factors, 50,* 511–520.

Parsons, K. S., Warm, J. S., Nelson, W. T., Riley, M., & Matthews, G. (2007). Detection-action linkage in vigilance: Effects on workload and stress. *Proceedings of the Human Factors and Ergonomics Society, 51,* 1291–1295.

Reichenbach, J., Onnasch, L., & Manzey, D. (2011). Human performance consequences of automated decision aids in states of sleep loss. *Human Factors, 53,* 717–728.

Reinerman, L. E., Warm, J. S., Matthews, G., & Langheim, L. K. (2008). Cerebral blood flow velocity and subjective state as indices of resource utilization during sustained driving. *Proceedings of the Human Factors and Ergonomics Society 52,* 1252–1256.

SAE International. (2016). *Taxonomy and Definitions for Terms Related to Driving Automation Systems for On-Road Motor Vehicles* (J3016_201609). Warrendale, PA: SAE International.

Sauer, J., Kao, C. S., & Wastell, D. (2012). A comparison of adaptive and adaptable automation under different levels of environmental stress. *Ergonomics, 55,* 840–853.

Sauer, J., Wastell, D. G., Robert, G., Hockey, J., & Earle, F. (2003). Performance in a complex multiple-task environment during a laboratory-based simulation of occasional night work. *Human Factors, 45,* 657–670.

Saxby, D. J., Matthews, G., Warm, J. S., Hitchcock, E. M., & Neubauer, C. (2013) Active and passive fatigue in simulated driving: Discriminating styles of workload regulation and their safety impacts. *Journal of Experimental Psychology: Applied, 19,* 287–300.

Scerbo, M. W. (2001). Stress, workload, and boredom in vigilance: A problem and an answer. In P. A. Hancock & P. A. Desmond (Eds.), *Stress, Workload, and Fatigue* (pp. 267–278). Mahwah, NJ: Lawrence Erlbaum Associates.

Shaw, T. H., Matthews, G., Warm, J. S., Finomore, V., Silverman, L., & Costa, P. T., Jr. (2010). Individual differences in vigilance: Personality, ability and states of stress. *Journal of Research in Personality, 44,* 297–308.

Solís-Marcos, I., Galvao-Carmona, A., & Kircher, K. (2017). Reduced attention allocation during short periods of partially automated driving: An event-related potentials study. *Frontiers in Human Neuroscience, 11,* 537.

Szalma, J. L. (2009). Individual differences in human–technology interaction: Incorporating variation in human characteristics into human factors and ergonomics research and design. *Theoretical Issues in Ergonomics Science, 10,* 381–397.

Szalma, J. L., Warm, J. S., Matthews, G., Dember, W. N., Weiler, E. M., Meier, A., & Eggemeier, F. T. (2004). Effects of sensory modality and task duration on performance, workload, and stress in sustained attention. *Human Factors, 45,* 349–359.

van der Linden, D., Frese, M., & Sonnentag, S. (2003). The impact of mental fatigue on exploration in a complex computer task: Rigidity and loss of systematic strategies, *Human Factors, 45*, 483–494.

Vogelpohl, T., Kühn, M., Hummel, T., Gehlert, T., & Vollrath, M. (2018). Transitioning to manual driving requires additional time after automation deactivation. *Transportation Research Part F: Traffic Psychology and Behaviour, 55*, 464–482.

Warm, J. S., Dember, W. N., & Hancock, P. A. (1996). Vigilance and workload in automated systems. In R. Parasuraman & M. Mouloua (Eds.), *Automation and human performance: Theory and applications* (pp. 183–200). Mahwah, NJ: Lawrence Erlbaum Associates.

Warm, J. S., Matthews, G., & Finomore, V. S. (2008). Workload and stress in sustained attention. In P. A. Hancock and J. L. Szalma (Eds.), *Performance under Stress* (pp.115–141). Aldershot, UK: Ashgate.

Warm, J. S., Parasuraman, R., & Matthews, G. (2008). Vigilance requires hard mental work and is stressful. *Human Factors, 50*, 433–441.

Warm, J. S., Tripp, L. D., Matthews, G., & Helton, W. S. (2012). Cerebral hemodynamic indices of operator fatigue in vigilance. In G. Matthews, P. A. Desmond, C. Neubauer, & P. A. Hancock (Eds.), *Handbook of Operator Fatigue*, pp. 197–207. Aldershot, UK: Ashgate Press.

Weinbeer, V., Bill, J. S., Baur, C., & Bengler, K. (2018). Automated driving: Subjective assessment of different strategies to manage drowsiness. In D. de Waard, K. Brookhuis, D. Coelho, S. Fairclough, D. Manzey, A. Naumann, L. Onnasch, S. Röttger, A. Toffetti, and R. Wiczorek (Eds.) *Proceedings of the Human Factors and Ergonomics Society Europe Chapter 2018 Annual Conference* (pp. 5–17). http://hfes-europe.org.

Weiss, L. G. (2011). Autonomous robots in the fog of war. *Spectrum, IEEE, 48*, 30–57.

Wickens, C. D., Clegg, B. A., Vieane, A. Z., & Sebok, A. L. (2015). Complacency and automation bias in the use of imperfect automation. *Human Factors, 57*, 728–739.

Wohleber, R. W., Matthews, G., Lin, J., Szalma, J. L., Calhoun, G. L., Funke, G. J., Chiu, C-Y. P., & Ruff, H. A. (2019). Vigilance and automation dependence in operation of multiple unmanned aerial systems (UAS): A simulation study. *Human Factors, 61*(3), 488–505.

8 Human-Automation Interaction and the Challenge of Maintaining Situation Awareness in Future Autonomous Vehicles

Mica R. Endsley

INTRODUCTION

Automation has increasingly become a part of modern system technology in a wide variety of domain areas, from its early uses in manufacturing, aviation systems and power systems to its current implementation in transportation vehicles. Across the past 50 years, considerable evidence has mounted demonstrating benefits from automation but also many challenges involving human interaction with automation that can contribute to catastrophic failures (Bainbridge, 1983; Wiener, 1993; Wiener & Curry, 1980). These challenges include:

1. Loss of manual skills needed for manual performance and decision-making,
2. Decreases in manual workload during low-workload periods, combined with workload spikes in often high-workload periods,
3. Poor operator understanding of system functioning leading to poor expectations of system behavior and inappropriate interactions with the automation, and
4. Loss of operator engagement and low situation awareness (SA) that is required for monitoring automation and intervening appropriately when needed.

A key challenge associated with the operation of automated systems is that it tends to reduce the SA of the human operator (Endsley & Kiris, 1995). SA of both the state of the automation and of the system being automated is critical to the ability of the human operator to effectively oversee it and make appropriate interventions and control inputs as needed. Operators with low SA are said to be "out-of-the loop"

(OOTL). Reduced SA when working with automated systems stems from (Endsley & Kiris, 1995):

1. Changes in information presentation with automation as many times the system developers either accidentally or intentionally remove key cues that operators rely on for successful system operation (Endsley & Kiris, 1995);
2. Poor operator monitoring and vigilance decrements that occur both because of overtrust of automation (Lee & See, 2004; Muir & Moray, 1996) and because people are in general poor at maintaining vigilance in monitoring conditions (Davies & Parasuraman, 1980; Hancock, 2013); and
3. Reduced operator engagement as people move from actively performing a task to watching another entity performing the task (Endsley & Kiris, 1995; Manzey, Reichenbach, & Onnasch, 2012; Metzger & Parasuraman, 2001).

A meta-review of automation research studies revealed a fundamental automation conundrum: "The more automation is added to a system, and the more reliable and robust that automation is, the less likely that human operators overseeing the automation will be aware of critical information and able to take over manual control when needed" (Endsley, 2017b). This indicates that even as automation becomes more technologically capable, with increasing levels of reliability and robustness for performing an ever-widening range of tasks, people will become even more hampered by low SA and fall short in the requirement to oversee the automation and interact with it effectively. Even when system designs are improved and people attempt to be vigilant, the degrading effects of reduced engagement are very difficult to overcome.

Consequently, current automation research and development is focused on introducing new forms of human-automation interaction based on very different concepts of operation (COO) and on designing system interfaces that improve operator SA and ability to interact with automated systems. These concepts will be reviewed and a model of human-autonomous system interaction (HASO) presented (Endsley, 2017b). This model will then be applied to the development of autonomous road vehicles.

HUMAN-AUTONOMY INTERACTION: CONCEPTS OF OPERATION (COO)

Many different human-automation interaction paradigms are possible, including at least four common concepts, Figure 8.1. Whereas most research to date has focused on supervisory control or automation limiter COOs, collaborating teammates and independent autonomous agents are becoming more technologically feasible and are being considered for future systems.

SUPERVISORY CONTROL

The traditional automation interaction paradigm assumes that people are in charge of overseeing the performance of an automated system, setting it up appropriately,

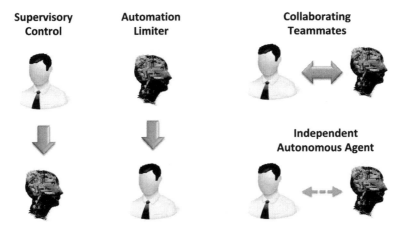

FIGURE 8.1 Common automation concepts of operation (COO).

and intervening as needed to correct its performance. This COO is typical of many of today's automated systems, including flight management systems in aircraft, automated process control systems, and lane-keeping systems in automobiles. The supervisory control COO requires a high degree of human SA to perform basic control tasks and to oversee the automation.

Automation Limiter

In this paradigm, the purpose of the automation is to prevent the human from taking actions that are unsafe. The automation is assumed to know more than the human and have appropriate rules in place to oversee human actions. An example of this COO is the Airbus flight management system that prevents pilots from executing particular flight maneuvers. While the automation limiter COO might be considered as requiring less SA than supervisory control, this assumption has not proven to be the case. Operators of such systems often require significant SA of what the automation is doing to be able to interact with it correctly and can have significant difficulties when that is not the case, such as in the crash of Air France Flight 296 at the Habsheim Airshow (BEA, 1990) and the loss of Air France 447 off the coast of Brazil (BEA, 2012).

Collaborating Teammates

A more collaborative paradigm allows for a broader range of joint human-automation behaviors required for collaboration to achieve shared goals, including coordination, joint planning and replanning, goal reprioritization, and allocation/reallocation of tasks. In this COO, more coordination and cooperation between the human and automation teammates are required to achieve successful performance. An example is joint performance between a manned and unmanned military aircraft as required to patrol dangerous areas and respond to enemy threats.

Significant SA is required of the state of the environment and the system, as well as shared SA between the human-automation teammates for maintaining goal alignment, allocating and reallocating functions, understanding goals, strategies and plans, maintaining awareness of each other's task performance and state, and aligning actions (U.S. Air Force, 2015).

INDEPENDENT AUTONOMOUS AGENT

The autonomous agent paradigm posits the automation as fully capable of assessing the environment and acting independently. No human oversight of the automation or the system environment is expected or required; however, the human may need to interact with the autonomous agent to set appropriate goals. An example of this COO is a fully autonomous automobile in which the passenger sets the desired destination and ride parameters (e.g., scenic or fastest route) but does not have the ability to take control from the vehicle in an emergency. This paradigm likely requires the least SA from the human as minimal interaction is required since the ability for real-time control is not possible. However, it requires very high levels of SA and highly reliable and robust performance from the autonomous system. Some SA will still be required from the human for assessing joint goal achievement (for example: Is it going where I told it to? Have we reached the intended destination?).

THE HUMAN AUTONOMOUS SYSTEM OVERSIGHT (HASO) MODEL

The HASO model, Figure 8.2, details the many factors that have an effect on people's ability to adequately oversee and interact with automation in its many forms (Endsley, 2017b). Performance in the tasks of oversight, intervention, interaction, collaboration, and coordination with the automation are all dependent on the SA of the human operator. The HASO model shows a number of factors that will affect SA in interacting with automation that are summarized here.

TRUST

The ability of the automation to operate accurately (*reliability*) across a wide range of possible conditions (*robustness*) will act to increase the degree to which people trust that automation. While improved automation performance is generally considered a good thing, unfortunately, the increase in trust that accompanies it can act to shift human attention to other competing tasks. Those tasks may be job related (e.g., attending to other parameters or systems) or extraneous (e.g., sightseeing, texting, conversations, or daydreaming). Further, people may increasingly trust automation when there are competing tasks to attend to.

ATTENTION ALLOCATION

How people allocate their attention across competing information and tasks in a given system or environment, of course, has a significant effect on the information they take in and hence their SA. This is partially determined by the characteristics of

Maintaining Situation Awareness in Future Autonomous Vehicles

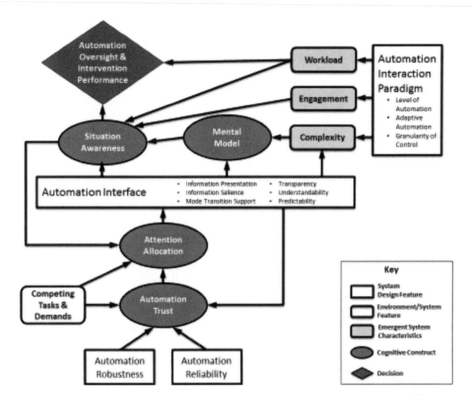

FIGURE 8.2 Human Autonomous System Oversight (HASO) model. (Reprinted with permission from Endsley, 2017b, 5–27.)

that information for grabbing attention (i.e., data-driven processing; Endsley, 1995) as well as how much trust is placed in the automated system. Further, their current SA drives their search for further information from the system interface and external environment (i.e., goal-driven processing; Endsley, 1995).

Automation Interface

A number of features of the automation interface are critical for the formation of accurate SA of the automation within the workload levels and time constraints that are inherent in many operational systems. This includes (a) effective presentation of needed information, (b) salience of cues associated with the state of the automation, including modes and system boundary conditions, (c) support for mode transitions, including from automated to manual, and (d) automation transparency as needed to provide for the understandability of system actions and predictability of future actions.

While many systems have required operators to develop detailed mental models of how automation works in order to understand what it is doing or will do in future, this requires a great deal of training and experience and still may end up incomplete or erroneous, leading to errors (McClumpha & James, 1994; Wiener & Curry,

1980). Further, as future autonomous systems may be based on learning algorithms that are constantly changing, it will be very difficult for human operators to maintain an accurate model of system behavior. In the future, it will become critical to develop more informative and transparent system interfaces that directly portray what the system thinks is happening (e.g., what it sees, how it classifies data) and what it is planning to do next (e.g., projected path or actions) (Endsley & Jones, 2012; Selkowitz, Lakhmani, & Chen, 2017).

Mental Models and Complexity

The ability of people to develop accurate mental models of the systems and automation they are interacting with is critical for correctly interpreting the information provided by the system and for projecting future system states, including those requiring human intervention or interaction (Endsley, 1995). The more complex the system, however, the more difficult it is to acquire an accurate mental model. System complexity is driven up by increases in the number of features, modes, interactions among system components, and by decreases in system predictability that occurs as the system increases its consideration of multiple factors or component states (Endsley & Jones, 2012). System logic, with many branches and mode interactions, and infrequent combinations of events can lead to very complex system behaviors that people may have trouble understanding or remembering. Increased complexity associated with automated systems can directly act to decrease the accuracy of the operator's mental models, unless the system acts very intuitively or the system interface can compensate by making the system behavior more transparent.

Engagement and Workload

High workload is often cited as a rationale for implementing automation. And while automation can in many cases reduce manual workload, it has also been found to increase cognitive workload (required to operate a more complex system; Harris, Goernert, Hancock, & Arthur, 1994; Parasuraman, Mouloua, & Molloy, 1994; Wiener, 1985) and to increase workload at critical times (Bainbridge, 1983). Cognitive engagement has been shown to be critical for maintaining SA when operating in conjunction with an automated system (Endsley & Kiris, 1995). Actively engaging in a task improves task understanding and retention of critical information. Automation that lowers engagement by turning people into passive monitors is highly detrimental to SA and creates the OOTL challenge. A number of automation paradigm approaches have been found to significantly affect the amount of engagement and workload experienced by system operators.

Level of Automation

Considerable research has been directed at the issue of how much automation is applied in terms of a scale of level of automation, from manual at one end to fully automated at the other, beginning with early work by Sheridan and Verplank (1978). Endsley and Kiris (1995) demonstrated that intermediate levels of SA were effective

TABLE 8.1
Effect of Levels of Automation on Human Performance

Taxonomy	Effect of Autonomy Applied to Stage of Task Performance			
	Situation Awareness		Decision	Action
Kaber and Endsley (1997)	Monitoring Information		Option Generation / Action Selection	Implementation
Parasuraman et al. (2000)	Information Filtering	Information Integration	Action Selection	Action Implementation
General Findings	Significant benefit to SA, workload, and performance from systems that present needed information (Level 1 SA)		Significant benefits when system is correct	Significant benefits to performance for routine, repetitive manual labor if reliable
	Significant benefit to SA, workload, and performance from systems that integrate information needed for comprehension (Level 2 SA) and projection (Level 3 SA)		Decreases performance when system is incorrect due to decision biasing	Manual workload may be lower overall
	Better SA and little OOTL problem compared to decision automation		Slower performance due to need to compare recommendations to system information and to other options	Increases in cognitive workload at peak times
			Lowers SA and increases OOTL performance problems	Increases in workload for systems with high false alarm rates and low reliability
Task Specific Findings	Informaton cueing systems create good performance when correct but poor performance when incorrect, similar to decision biasing effects		Automation of selection among alternatives less of a problem for performace than automation that generates options, which affects engagement	Lower SA and significant OOTL problems for automation that employs advanced queuing of tasks
	Information filtering systems can limit Level 3 SA (projection), negatively impacting performance		Decision support based on critiquing systems or what-if reasoning and contingency planning do not create decision biasing problem due to higher engagement	Lower SA and significant OOTL problems for automation of continuous control tasks

(Reprinted with permission from Endsley, 2017b, 5–27.)

at improving SA and reducing the OOTL effect, compared to full automation. Two common levels of automation (LOA) scales used to research the effects of LOA on SA and performance include those by Endsley and Kaber (1997; 1999) and Parasuraman, Sheridan, and Wickens (2000), both of which were developed based on how automation is applied to differing stages of task performance.

Table 8.1 summarizes the research findings on LOA to show its significant effect on SA and performance (Endsley, 2017b). In essence, low LOA that are directed at increasing SA by improving information transmission and integration are highly effective at keeping engagement high, workload low, and improving overall performance. Intermediate LOAs that generate decision options tend to decrease cognitive engagement and can slow performance as people work to compare system options and recommendations to other information. LOAs that select among alternatives or recommend best actions can be helpful if they are correct but can significantly hinder SA if incorrect due to decision biasing. High LOA that act to carry out the results of automated decisions can be beneficial for routine, repetitive tasks but can also increase cognitive workload at peak times, particularly in systems with high false-alarm rates and low reliability. Low SA and OOTL problems are highest for tasks that involve advanced queuing of tasks (such as routes with waypoints set up in advance) and continuous control tasks (such as steering).

Adaptive Automation (AA)

Other research has attempted to increase operator engagement by introducing periods of manual control either by operator action or automatically, based on set time intervals, performance, physiological indicators, or models, (Scerbo, 1996). AA has been found to be primarily effective at reducing workload (Hilburn, Jorna, Bryne, & Parasuraman, 1997; Kaber & Endsley, 2004; Kaber & Riley, 1999) and has also been found to improve operator engagement, which improves SA (Bailey, Scerbo, Freeman, Mikulka, & Scott, 2003; Prinzel, Freeman, Scerbo, Mikulka, & Pope, 2003).

GRANULARITY OF CONTROL

Automation task granularity can range from high (requiring many human actions to carry out a task) to low (requiring only a few), including (a) manual control, (b) programmable control, (c) playbook control, and (d) goal-based control. With lower granularity (e.g., goal-based control), less detailed instructions are required to interact with the automation, thus reducing workload level. Since lower granularity also means that the system may operate across a wider range of functions and a longer period of time without intervention, it may result in lower SA of the system's operations, however.

Taken as a whole, the HASO model shows the key factors that will affect human performance when interacting with autonomous systems and points to key areas of the system design that can be used to improve human interaction with them, including features of the automation interface, the automation interaction paradigm, and the robustness and reliability of the automation itself.

APPLICATION OF THE HASO MODEL TO AUTONOMOUS VEHICLE TECHNOLOGY

SAE LEVELS OF VEHICLE AUTONOMY

The term automobile actually connotes that automation has already been applied to road vehicles, namely in terms of the means of power (as compared to previous horse-powered vehicles). In addition, software is currently being developed to automate the control of those vehicles. SAE International has characterized five broad levels that depict the degree of autonomy of a vehicle: Level 0 — fully manual control with some driver assistance; Level 1 — human must monitor and intervene (1 function); Level 2 — human must monitor and intervene (2 functions); Level 3 — human must intervene with notice; Level 4 — fully autonomous some of the time; and Level 5 — fully autonomous (SAE International, 2018). This taxonomy, therefore, loosely incorporates the degree of automation and the range of functions involved into one scale, along with other characteristics such as temporality and conditions of use. It should also be noted that as the SAE level of vehicle autonomy increases, so too is there a need to increase the overall reliability and robustness of the automation itself. It is based on a supervisory control concept of operation that morphs into that of an independent autonomous agent at very high levels of reliability and robustness.

As shown in Figure 8.3, many current vehicle systems can be classified as SAE Level 0. They provide driver assistance for a very narrowly defined function,

Maintaining Situation Awareness in Future Autonomous Vehicles

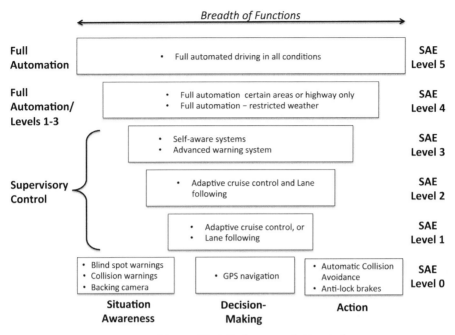

FIGURE 8.3 SAE levels of vehicle autonomy with examples.

generally operating as a system for improving *either* SA, decision-making, or action performance. SAE Level 1 and 2 systems are generally considered supervisory control systems that include automation of SA, decision-making, *and* performance for a limited set of functionality. Current examples include adaptive cruise control (ACC) and/or lane following systems. SAE Level 3 systems are also supervisory control systems but will additionally provide advanced warning of the need to intervene by having some self-awareness of system limitations and upcoming conditions. Whereas SAE Level 5 systems allow for fully autonomous driving by the system across all conditions and situations, SAE Level 4 allows for fully autonomous driving only across a more limited set of circumstances. The rest of the time, the vehicle would be operating at a lower level of automation (e.g., supervisory control or assisted manual control).

Effects of SAE Level of Vehicle Autonomy on SA

Based on the HASO model, it is expected that automation applied to improving SA will largely be beneficial as long as it is reliable and has a low false alarm rate. Automation involving decision aids, such as a global positional system (GPS) navigation system, will be beneficial as long as it gives correct advice but will lead to decision biasing when wrong. For example, people will be drawn to make poor navigational decisions when the GPS map is out of date, uses an inadequate algorithm, or fails to take traffic or road construction into account. Automation involving low-level manual tasks (such as automatic transmission or anti-lock brakes) will also be largely beneficial for both workload and performance, as long as no human intervention is expected. The majority of SAE Level 0 automation will be, therefore, mostly beneficial

As the level of automation increases to that of supervisory control in SAE Level 1 (including aspects of SA, decision-making, *and* task performance), however, SA will start to diminish and incidents of OOTL performance problems will increase. It has also been noted that SA will not be poor all the time but rather will become more variable, with periods of both high and low SA (Endsley, 2017a).

While traditional cruise control is useful in a fairly narrow set of circumstances (such as highway driving), ACC increases its span of operation (by automatically slowing to maintain distance to the vehicle in front), allowing it to be used more often for longer periods of time. This increase in automation robustness will lead to a lower level of SA as periods of manual intervention decline and as the system's behavior becomes more complex and less predictable. For example, when operating under ACC, the vehicle can unexpectedly speed up when it passes from a divided highway to an undivided section, when it moves onto the highway off-ramp, or when the car in front changes lanes (Endsley, 2017a).

The addition of lane following automation (SAE Level 2), which keeps the automobile in-between the lane lines, automates the lateral control of the vehicle just as ACC automates vertical control. In combination, the driver experiences supervisory control across the full set of steering control functionality and is primarily responsible for navigation and responding to events the automation is not robust enough to handle (e.g., lane merges, lane changes, or roadway debris). At this level of automation, as the reliability of the system improves and trust goes up, over time the SA of the driver becomes very low due to low engagement and extended vigilance conditions (Endsley, 2017a). Efforts to keep the driver engaged by requiring "hands on the wheel" or "eyes forward" are largely insufficient to overcome these OOTL challenges, which stem from lack of cognitive engagement and vigilance issues.

SAE Level 3 is expected to improve on the SA problems associated with SAE Level 2 by giving the driver advanced warning of the need to intervene. This will only happen, however, if the automation has (a) a good predictive model of the world that allows it advanced warning of upcoming problems it cannot handle (e.g., road construction) and (b) enough self-awareness to know what it cannot handle. Since most systems that have sufficient sensor systems to provide advanced warnings, and enough intelligence to be aware of its own limitations, will also be easily programmed to deal with those limitations, in practice, it is likely that very few SAE Level 3 systems will ever exist, instead moving directly to SAE Level 4.

But for those that do exist, there is certainly the potential of significantly overcoming the OOTL challenge, as long as the advanced warning is salient enough to grab the driver's attention and gives the driver enough time to gather information and understand the situation to make an intelligent decision and carry it out. Research shows that the OOTL problem is not just due to problems with gaining Level 1 SA but also with having time to integrate that information to correctly understand what is happening (Level 2 SA) when coming back into the loop (Endsley & Kiris, 1995). Given the very short timeframes available for action in driving, this may be unfeasible. As there is the potential for OOTL drivers to actually fall asleep, these systems will also need the ability to default to some "fail safe" condition. So driver SA and engagement will be even lower with SAE Level 3 systems except, hopefully, in the periods where drivers have been alerted to a problem. More research is needed to develop methods to rapidly bring people into the loop and build SA after extended periods of inattention.

SAE Level 4 systems will be fully automated for some predefined set of circumstances such as, in particular, well-understood geographic areas or in certain environmental weather conditions. Thus, they are expected to encompass automation of the full range of driving functionality but not necessarily at a level of robustness that allows implementation in all conditions. Level 4 vehicles, therefore, will operate as fully autonomous (Level 5) some of the time and will revert to a lower level of automation in other conditions. Just as with Level 3 systems, methods will be needed to alert drivers, or to restrict operations, should the allowable conditions of operation be exceeded.

With SAE Level 5 fully autonomous systems, it is expected that the automation will be highly reliable and robust enough to be able to handle all driving conditions and situations. In this case, the system acts as an independent and fully autonomous agent. Human interface issues change from those required for operation to those required for interaction in a human-autonomy teaming COO.

IMPLICATIONS FOR MENTAL MODELS AND SA

As the SAE level of vehicle autonomy increases, the ability of the driver to develop and maintain an accurate mental model of the system needed for understanding and predicting its behavior will be significantly challenged. As the HASO model shows, increases in complexity, as needed to handle an ever-widening set of circumstances associated with system autonomy, will make it harder to develop a mental model of the system.

Figure 8.4 depicts a model of system complexity (Endsley & Jones, 2012). This shows that the actual complexity of the underlying system is not as important as the apparent complexity to the user. While automobiles have become increasingly more complex, their operation is actually simpler than in the early days of driving. For example, while the internal software associated with automatic transmissions or anti-lock brakes may be somewhat complex, the complexity to the user (i.e., the apparent complexity) is really a function of the cognitive challenge of interacting with the system based on its logic (cognitive complexity), the complexity of the displays, and the complexity of the tasks required for interacting with the system. In the case of automatic transmissions and anti-lock brakes, only very simple tasks are required of the driver (shifting into drive or pressing on the brakes), leading to a very low level of apparent complexity.

Many, if not most vehicle features associated with automating task performance to date have led to lower complexity for human drivers (although not necessarily for vehicle mechanics who have very different operational goals and tasks).

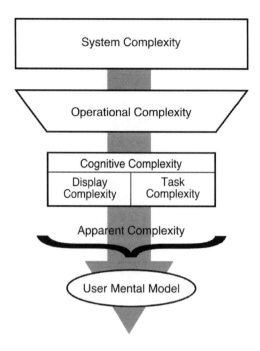

FIGURE 8.4 Layers of complexity. (Reprinted with permission from Endsley & Jones, 2012.)

As higher levels of SAE vehicle autonomy are implemented, however, this situation could change considerably, as shown in Figure 8.5. Rather than automating primarily manual tasks, the cognitive tasks associated with driving are becoming automated, meaning that drivers will need to interact with a cognitive agent (with its inherent logic) to understand what the vehicle is doing. While the cognitive and apparent complexity associated with SAE Level 0 and 1 systems has been fairly low for automation directed at SA and performance, it is considerably higher for systems directed at assisting in decision-making (e.g., GPS navigation). Cognitive complexity will be highest for SAE Level 2 and Level 3 systems, which will involve fairly complex behaviors to accommodate a wide range of situations and behaviors associated with driving. While it is likely that SAE Level 4 and Level 5 vehicles will include even more complex software, the cognitive complexity will likely be much lower for the driver as only very simple interactions with the vehicle will be required, and the need to understand its logic and behaviors will diminish.

Therefore, the need for extensive mental models to understand and predict the behavior of vehicle autonomy will be highest at SAE Levels 2 and 3, where it will also be the most challenging due to its complexity. Further, as the autonomous systems being developed are largely based on deep-learning software techniques, developing an accurate mental model will be quite difficult (Endsley, 2017a; U.S. Air Force, 2015). Deep learning uses approaches such as neural networks to automatically determine key patterns in the environment and map them to appropriate representations or actions. As such, the logic of the system is deeply imbedded and not readily

Maintaining Situation Awareness in Future Autonomous Vehicles

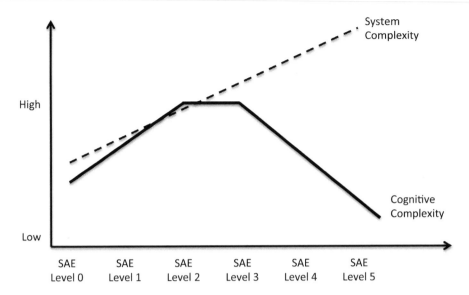

FIGURE 8.5 Complexity associated with SAE levels of vehicle autonomy.

available to inspection or explanation. Therefore, even system developers may not have a full understanding of how the system will behave in all circumstances. This problem will be even further compounded by the advent of frequent system updates, where the system software is updated frequently (as often as monthly or weekly) at the driver's home, changing the internal logic of the system (Endsley, 2017a).

As an additional concern, autonomous vehicle capabilities are being developed by numerous manufacturers and system developers. The capabilities and logic of each vehicle may be very different, even for seemingly similar functions (such as ACC or lane following). In that many drivers move among multiple vehicles, either at home or when driving rental cars, the opportunity for negative transfer or confusion to occur is quite high. And while considerable attention has been paid to the need for the drivers or operators of highly automated vehicles to have good mental models of their behavior, the same is true of those who must share the road with these vehicles, such as other drivers, cyclists, and pedestrians.

TRAINING AND INTERFACES TO SUPPORT VEHICLE AUTONOMY

Given the significant increases in cognitive complexity, the challenges of hidden and changing logic and capabilities associated with deep learning software, and the presence of multiple vehicles with inconsistent logic, behaviors, and interfaces, considerable attention is needed in developing improved driver training and interfaces for addressing this challenge.

At a minimum, vehicle manufacturers will need to spend considerably more time in training drivers on the capabilities, limitations, and behaviors of autonomous vehicle technologies (Endsley, 2017a). As the airlines learned, in-depth automation training is critical, as is refresher training and hands-on practice in operating the automation, along with periods of manual operation without the automation to compensate for loss of manual skills. Road transportation should learn from these experiences. Vehicle manufacturers will need to develop not just training to be delivered at the initial sale of a vehicle but also to accompany any system updates. As the driving population is far more varied in terms of capabilities, age, and experience as compared to pilots, effective training programs will be even more important.

It would be foolish to believe that training alone will solve the complexity problem enough to ensure drivers develop accurate mental models of vehicle autonomy software. The challenges of hidden behaviors and rapid change associated with deep-learning software and the inconsistencies in logic among vehicles makes it an almost certainty that training alone will not be sufficient. The development of effective driver interfaces to support understanding and prediction of vehicle behavior (as well as external interfaces for other drivers, cyclists, and pedestrians), therefore, becomes critical.

Table 8.2 displays some design principals for supporting SA with complex systems (Endsley & Jones, 2012). Of particular importance is the design of the user interface to support SA, as shown in the HASO model. Automation interface designers need

TABLE 8.2
Design Guidelines for Improving SA

Support Human Understanding of Autonomous Systems
Automate only if necessary – avoid out-of-the-loop problems if possible
Use automated assistance for carrying out routine tasks rather than higher-level cognitive functions
Provide SA support rather than decisions
Keep the operator in control and in the loop
Avoid the proliferation of automated modes
Make modes and system states salient
Enforce automation consistency
Avoid advanced queuing of tasks
Avoid the use of information cueing
Decision support should create human/system symbiosis
Provide automation transparency
Minimize Complexity of Autonomous Systems
Ensure logical consistency across features and modes
Minimize logic branches
Map system functions to the goals and mental models of users
Minimize task complexity
Support Situation Awareness
Integrate information to support comprehension of information (Level 2 SA)
Provide assistance for SA projections (Level 3 SA)
Use information filtering carefully
Support assessments of confidence in composite data
Support system reliability assessments

From Endsley & Jones, 2012.

to take care that relevant information for driving and system oversight is presented to the driver and not hidden. This includes both low level data as well as information to support Level 2 SA (comprehension) and Level 3 SA (projection) where possible. Key information, such as that associated with current vehicle mode and mode transitions, needs to be made highly salient. In particular, any time the vehicle changes modes, including unexpected transitions to manual control, or experiences emergent behaviors associated with modes, these changes need to be highly salient to the driver via visualization and auditory alerts. The development of transparent interfaces that make it clear what state the vehicle thinks it is in, and what is driving changes in its behaviors, are needed.

The key to a good mental model and effective interaction with highly autonomous systems rests on predictability. If the system interface can provide an indication of not only what the vehicle automation is currently doing but also what it is going to do next (e.g., speed up, turn at intersection, change lanes to avoid obstacle), this will greatly facilitate successful interaction with the system. Additional research is needed to develop and test effective methods for promoting this type of transparency.

CONCLUSIONS

Autonomous system software is currently being developed for a wide variety of systems including aviation, military systems, power, space, and ground transportation. Over 40 years of research has been conducted on the issue of human-automation interaction that indicates significant hurdles that must be overcome as well as provides guidance for addressing these challenges. The HASO model describes the key factors that are relevant to the performance of human operators who must interact with these systems within various COOs. In particular, HASO includes key features that are implemented during the system design that will substantially impact the effectiveness of the combined human-automated system, including the automation interaction paradigm employed (e.g., level of automation, adaptive automation, and granularity of control), the features of the automation interface, and the reliability and robustness of the automation itself. As the automotive industry rushes toward the implementation of higher levels of autonomy in ground vehicles, it is critical that these factors be carefully addressed during system development in order to meet the stated safety objectives of these systems.

REFERENCES

Bailey, N. R., Scerbo, M. W., Freeman, F. G., Mikulka, P. J., & Scott, L. A. (2003). A brain-based adaptive automation system and situation awareness: The role of complacency potential. In *Proceedings of the Human Factors and Ergonomics Society 47th Annual Meeting* (pp. 1048–1052). Santa Monica, CA: Human Factors and Ergonomics Society.

Bainbridge, L. (1983). Ironies of automation. *Automatica, 19*, 775–779.

BEA. (1990). Commision D'enquête sur l'accident survenu le 26 juin 1988 á Mulhouse-Habsheim (68). Paris: Journal Officiel de la République Française.

BEA. (2012). On the accident on 1st June 2009 to the Airbus A330-203 registered F-FGZCP operated by Air France flight AF 447 Rio de Janeiro–Paris. Cedex, France: Bureau d'Enquétes et d'Analyses pour la sécurité de l'aviation civile.

Davies, D. R., & Parasuraman, R. (1980). *The Psychology of Vigilance*. London: Academic Press.
Endsley, M. R. (1995). Toward a theory of situation awareness in dynamic systems. *Human Factors, 37*(1), 32–64.
Endsley, M. R. (2017a). Autonomous driving systems: A preliminary naturalistic study of the Tesla Model S. *Journal of Cognitive Engineering and Decision Making, 11*(3), 225–238.
Endsley, M. R. (2017b). From here to autonomy: Lessons learned from human-automation research. *Human Factors, 59*(1), 5–27.
Endsley, M. R., & Jones, D. G. (2012). *Designing for Situation Awareness: An Approach to Human-Centered Design* (2nd ed.). London: Taylor & Francis Group.
Endsley, M. R., & Kaber, D. B. (1997). The use of level of automation as a means of alleviating out-of-the-loop performance problems: A taxonomy and empirical analysis. In P. Seppala, T. Luopajarvi, C. H. Nygard, & M. Mattila (Eds.), *13th Triennial Congress of the International Ergonomics Association* (Vol. 1, pp. 168–170). Helsinki, Finland: Finnish Institute of Occupational Health.
Endsley, M. R., & Kaber, D. B. (1999). Level of automation effects on performance, situation awareness and workload in a dynamic control task. *Ergonomics, 42*(3), 462–492.
Endsley, M. R., & Kiris, E. O. (1995). The out-of-the-loop performance problem and level of control in automation. *Human Factors, 37*(2), 381–394.
Hancock, P. (2013). In search of vigilance: The problem of iatrogenically created psychological phenomena. *American Psychologist, 68*(2), 97–109.
Harris, W. C., Goernert, P. N., Hancock, P. A., & Arthur, E. J. (1994). The comparative effectiveness of adaptive automation and operator initiated automation during anticipated and unanticipated taskload increases. In M. Mouloua & R. Parasuraman (Eds.), *Human Performance in Automated Systems: Current Research and Trends* (pp. 40–44). Hillsdale, NJ: LEA.
Hilburn, B., Jorna, P. G., Bryne, E. A., & Parasuraman, R. (1997). The effect of adaptive air traffic control decision aiding on controller mental workload. In M. Mouloua & J. M. Koonce (Eds.), *Human Automation Interaction: Research and Practice* (pp. 84–91). Mahwah, NJ: LEA.
Kaber, D. B., & Endsley, M. R. (2004). The effects of level of automation and adaptive automation on human performance, situation awareness and workload in a dynamic control task. *Theoretical Issues in Ergonomic Science, 5*(2), 113–153.
Kaber, D. B., & Riley, J. (1999). Adaptive automation of a dynamic control task based on secondary task workload measurement. *International Journal of Cognitive Ergonomics, 3*(3), 169–187.
Lee, J. D., & See, K. A. (2004). Trust in automation: Designing for appropriate reliance. *Human Factors, 46*(1), 50–80.
Manzey, D., Reichenbach, J., & Onnasch, L. (2012). Human performance consequences of automated decision aids: The impact of degree of automation and system experience. *Journal of Cognitive Engineering and Decision Making, 6*, 57–87.
McClumpha, A., & James, M. (1994). Understanding automated aircraft. In M. Mouloua & R. Parasuraman (Eds.), *Human Performance in Automated Systems: Current Research and Trends* (pp. 183–190). Hillsdale, NJ: LEA.
Metzger, U., & Parasuraman, R. (2001). Automation-related "complacency": Theory, empirical data and design implications. In *Proceedings of the Human Factors and Ergonomics Society 45th Annual Meeting* (pp. 463–467). Santa Monica, CA: Human Factors and Ergonomics Society.
Muir, B. M., & Moray, N. (1996). Trust in automation: Part 2. Experimental studies of trust and human intervention in a process control simulation. *Ergonomics, 39*, 429–460.
Parasuraman, R., Mouloua, M., & Molloy, R. (1994). Monitoring automation failures in human-machine systems. In M. Mouloua & R. Parasuraman (Eds.), *Human Performance in Automated Systems: Current Research and Trends* (pp. 45–49). Hillsdale, NJ: LEA.

Parasuraman, R., Sheridan, T. B., & Wickens, C. D. (2000). A model of types and levels of human interaction with automation. *IEEE Transactions on Systems, Man and Cybernetics, 30*(3), 286–297.

Prinzel, L. J., Freeman, F. G., Scerbo, M. W., Mikulka, P. J., & Pope, A. T. (2003). Effects of a psychophysiological system for adaptive automation on performance, workload, and the event-related potential P300 component. *Human Factors, 45*(4), 601–613.

SAE International. (2018). *Taxonomy and Definitions for Terms Related to Driving Automation Systems for On-Road Motor Vehicles*. Warrendale, PA: SAE.

Scerbo, M. W. (1996). Theoretical perspectives on adaptive automation. In R. Parasuraman & M. Mouloua (Eds.), *Automation and Human Performance: Theory and Application* (pp. 37–63). Mahwah, NJ: Lawrence Erlbaum.

Selkowitz, A. R., Lakhmani, S. G., & Chen, J. Y. C. (2017). Using agent transparency to support situation awareness of the autonomous squad member. *Cognitive Systems Research, 46*, 13–25.

Sheridan, T. B., & Verplank, W. L. (1978). *Human and computer control of undersea teleoperators*. Cambridge, MA: MIT Man-Machine Laboratory.

U.S. Air Force. (2015). *Autonomous Horizons*. Washington, DC: United States Air Force Office of the Chief Scientist.

Wiener, E. L. (1985). Cockpit automation: In need of a philosophy. In *Proceedings of the 1985 Behavioral Engineering Conference* (pp. 369–375). Warrendale, PA: Society of Automotive Engineers.

Wiener, E. L. (1993). Life in the second decade of the glass cockpit. *In Proceedings of the Seventh International Symposium on Aviation Psychology* (pp. 1–11). Columbus, OH: Department of Aviation, The Ohio State University.

Wiener, E. L., & Curry, R. E. (1980). Flight deck automation: Promises and problems. *Ergonomics, 23*(10), 995–1011.

9 Human Factors Issues Regarding Automation Trust in UAS Operation, Selection, and Training

Mustapha Mouloua, James C. Ferraro, Alexandra D. Kaplan, Phillip Mangos, & Peter A. Hancock

The proliferation of unmanned autonomous systems (UASs) in both civilian and military settings has dramatically increased in recent years. A recent report from BIS Research (2016) foresaw that the market volume of unmanned aerial vehicles alone would increase exponentially over the following two years—and growth does not appear to be slowing down. This is just one market trend relevant to the use of unmanned systems that highlights the current and potential growth of these technologies across domains. As a result of this expansion, research has revealed multiple challenges and threats involved in UAS operation from a human factors/ergonomics standpoint. The U.S. Air Force (USAF), along with other military agencies, have increasingly adapted and used UAS technologies in the last 20 years. This rapid deployment of these remotely operated vehicles has changed the nature of modern warfare. UAS superiority ensures air power control, combat efficiency, and increased safety. However, such progress has also resulted in numerous pressing issues related to human performance and human-automation interaction (Gilson, Richardson, & Mouloua, 1998; Hancock, 2014; Mouloua et al., 2001b; Mouloua, Gilson, & Hancock, 2003). As the automation implemented in these unmanned systems increases, and humans are tasked with monitoring and operating multiple systems, it is critical to address and attempt to mitigate these human factors and engineering concerns related to UAS design (Hancock, 2013; Mouloua et al., 2001a). The present chapter provides a synthesis of relevant literature on trust and its influence on unmanned vehicle operations (and see Hancock, Billings, & Schaefer, 2011) and emphasizes the need to develop effective training and selection measures for future operators of these highly automated systems.

The next generation of USAF personnel and other military operators will experience an increasing level of challenge due to the proliferation of UASs, evolving mission requirements, task load, personnel selection and training, and rapid deployment of military personnel (Mouloua et al., 2001a). Undoubtedly, this will also place significant demands on operators' performance, workload, and safety. As a result, the USAF will need to select and train operators who show higher levels of adaptability and resilience to these new challenges in order to ensure optimal performance, improved safety margins, higher levels of retention, better levels of job satisfaction, and better physical and mental fitness-for-duty (Hoffman & Hancock, 2017). Therefore, there is an evident need to systematically examine these emerging challenges within the context of human performance evaluation, selection, training, and systems design of current and future UAS operations. This challenge is that which our chapter looks to address.

PERFORMANCE DEMANDS IN UAS OPERATION

The fact that UASs are considered "unmanned" is largely a misnomer. In this context, "uninhabited" may represent a more fitting term. It is unlikely that humans will be completely removed from any such system as they are required in order to at least initiate these devices and, while remote operation may be safer, the workload associated with such a task is not insignificant (see Hancock & Warm, 1989). The particular performance demand placed on the operator relies largely on the level of control. Sheridan (1997) has described three general levels of automation control: manual, supervisory, and autonomous. When automation functions under direct/manual control, it means the human operator has total control over the actions of the system and it does not act independently or without input from the operator. When operating these unmanned vehicles, manual control of the vehicle is accomplished remotely, and spatial orientation and navigation skills become essential for safe operations. Supervisory control occurs when the human operator acts by providing occasional feedback or direction to the autonomous system. Without any input from the operator, an autonomous system will operate independently to accomplish its task. In the last mode, the system operates completely independently of the human operator under autonomous control. Full manual control generates the highest workload for human operators, requiring complete control and attention on behalf of the pilot. Fully automated control imposes less demand in general on the operator, but this comes with significantly less ability to intervene. Placing the human operator in a supervisory role allows such operators to more easily monitor multiple aircraft simultaneously. After a targeted review by Mouloua, Gilson, and Hancock (2003), it was determined to be the ideal level of automation for UAS operators. This hybrid model of control allows one operator to control many vehicles without necessarily creating work overload.

However, just because a system allows a human to simultaneously operate multiple vehicles at once, does not mean that workload is no longer an issue. Supervisory control of UASs would require constant monitoring and observation. A large influence in aviation accidents is frequently considered due to human error and, as automation becomes more reliable, a higher percentage of accidents is often attributed to humans (Wiegmann & Shappell, 2001). This holds true also in the arena of driving. To assume that simply adding distance between the human and the vehicle will lessen the risk of accident is almost certainly flawed. In fact, the need for attention most probably increases with distance as well as with the growing number of vehicles under one operator's control.

UASs do not require a single driver in the same way that a manned vehicle does. That is, it would be impossible for one person to physically operate two vehicles at once or to be in two separate spaces at the same moment in time. With the advent of

UASs, the spatial demands are mitigated; however, the temporal ones are not. While UAS controls allow operators to control multiple vehicles at once, they do nothing to increase a human's attentional capacity nor do they allow any diminishing of the attentional resources demanded by such vehicle operation. According to queuing theory, humans completing multiple concurrent tasks have to switch their attention between tasks; thus, even with the technology granting the opportunity to complete concurrent tasks, the ability to multitask is limited by the simple fact that a person's eyes can only fixate on one area at a time (Senders, 1964).

A model of the impact of workload on the performance of multiple unmanned vehicles and supervisory control has found that higher workload negatively affected performance (Donmez, Nehme, & Cumming, 2010). In a simulation study, these researchers measured the time between the emergence of a threat and participant response to that threat. This response time was higher when participants were required to monitor multiple UASs, particularly when the UASs were of different sorts (e.g., aerial versus underwater). The authors attributed this increased response time to attentional inability, or a participant's difficulty switching from one task to another.

Dixon and Wickens (2006) found that in the cases of high workload, unreliable automation serves to harm performance. In their simulation study, participants directed a UAS to several locations, while conducting the secondary task of looking for target objects. Additionally, while they were completing these tasks, participants were required to monitor for system failures. They did this with the help of automated aids that alerted them to system failures—though only with a rated 67% accuracy. The automated aids were designed to be either omission prone (not alerting the participant to a system failure) or false alarm prone (causing an alert when there was in fact no system failure). Results indicated that in a high workload condition, false alarms severely degraded performance on the system failure monitoring task. This was most likely due to the increased workload and distraction required of double-checking the system.

A further study by Dixon, Wickens, and Chang (2005) examined licensed pilots in a multitasking environment. While piloting either one or two unmanned aerial vehicles, the pilots were also required to perform a system monitoring task. These findings showed that reliable automation decreases workload. However, the multitask nature of the job showed overall performance decrement with increased workload. Even when the automated aspects of the piloting task were without flaw, performance on the system monitoring task still degraded with increased workload. When piloting only one UAS, participants suffered from a 7% loss in accuracy on the system monitoring task when they were given one concurrent monitoring task and a 15% loss in accuracy when they were given two concurrent tasks. While decrements in the single-UAS condition were small, they doubled when pilots were required to control two UASs. This showed that monitoring a second unmanned vehicle is no easier than monitoring the first; thus, workload increases do not dissipate with the addition of each similar task. Such increments remain the same, meaning that control of a fleet of UASs still poses an incredible workload challenge for pilots. Dixon, Wickens, and Chang (2005) concluded that in the task of monitoring multiple displays, while navigating UASs, workload demand is great enough that decrement in performance may be unavoidable.

The number of UASs controlled by a single person is perhaps the largest factor in inducing overload. According to Dixon and Wickens (2003), performance is degraded when a single operator is responsible for two UASs, but other than registering a performance decrement, it still remains feasible for one person to control two vehicles. The Army Research Laboratory (ARL) conducted a study to determine the amount of time spent in work overload while operating multiple UASs. They found that while it was well within the ability of a single operator to control one UAS, when they controlled two, an operator's cognitive workload was above the threshold of overload—being about 80% to 90% of the time (Pomranky & Wojciechowski, 2007). This shows that it may not be ideal for a single operator to control multiple vehicles; though, of course, the importance of the task, its difficulty, and the consequence of performance decrement could each enable an operator to control multiple vehicles simultaneously if marginal decrement performance is not a critical issue.

The use of UASs, and other highly automated systems, has prompted examinations of the relationships among human performance, workload, and situational awareness (SA) as a component of the system. Research investigating situational awareness in the presence of high levels of automation has found a trade-off between situational awareness and workload (Onnasch, Wickens, Li, & Manzey, 2014). Higher levels of automation decrease workload, taking certain tasks out of the hands of a human operator. However, it is postulated that this decrease in workload will be accompanied by a loss of SA. Designing for appropriate workload proves to be difficult as we must ensure that workload is neither too great nor too small (Mouloua, Gilson, Kring, & Hancock, 2001b). Excessive demands, and the operator will produce degraded performance; too little demand, and the operator's performance is similarly affected (Hancock & Warm, 1989). However, the true challenge lies in managing the workload of a dynamic system. In UAS operating situations, demands do not remain stable. Operators can face long periods of boredom, interspersed with moments of higher workload (Hancock & Krueger, 2010). This occurs when spotting targets or dealing with system malfunctions or overload. For this reason, there is no specific, optimal level of workload to be specified for an unmanned system. Considerations should be taken that any risk of work overload be minimized if not eliminated.

Recent advances in automated technologies have contributed to the proliferation of UASs across a wide variety of application domains. These include military, aviation, homeland security, and entertainment, among others (Mouloua & Koonce, 1997; Mouloua & Parasuraman, 1994; Parasuraman & Mouloua, 1996; Scerbo & Mouloua, 1999; Vincenzi, Mouloua, & Hancock, 2004). As a result of these advancements, there have been noted benefits and costs to overall human-machine system safety and performance (Ferraro et al., 2018; Mouloua, Gilson, & Hancock, 2003; Mouloua et al., 2001a). In aviation, in both civilian and military contexts, flying a UAS requires a variety of skills. These are somewhat similar to various skills required in manual flight and driving tasks (Ferraro et al., 2018; Mouloua, Gilson, & Koonce, 1997; Mouloua et al., 2004; Pavlas et al., 2009; Smither et al., 2004). Such tasks include manual navigation of the UAS, monitoring of various flight- and craft-related parameters, and searching for potential points of interest (Dixon & Wickens, 2003). Previous research on human interaction with highly automated systems has also shown that operators/pilots may become complacent (Ferraro et al., 2018; Mouloua, Parasuraman, & Molloy, 1993; Parasuraman, Molloy, & Singh, 1993). They may also become overreliant or underreliant on automation (Lee & See,

2004; Muir & Moray, 1996; Parasuraman & Riley, 1997), and they may lose their SA (Endsley, 1995; Smith & Hancock, 1995). Consequently, automated systems can fail because of any of a variety of human, environmental, design, and engineering factors (Mouloua, Hancock, Jones, & Vincenzi, 2010; Mouloua, Smither, & Kennedy, 2009). Although these types of failures have occurred in a variety of human-machine systems, users and operators of highly autonomous systems continue to rely on them. This is mainly due to the level of trust that is established between user and machine (Hancock, Billings, & Schaefer, 2011). This trust level can also mediate operators' levels of reliance on these autonomous systems.

INDIVIDUAL DIFFERENCES IN AUTOMATION TRUST

Trust in automation is such an important facet in understanding automation use. Several meta-analyses have already been conducted on the subject (Hancock, Billings, Schaefer, Chen et al., 2011; Hoff & Bashir, 2015; Schaefer, Chen, Szalma, & Hancock, 2016). Hoff and Bashir (2015) analyzed 127 studies in which humans interacted with automation. Of these studies, the automation was most often categorized first as a decision aid, second as an information analysist, third as an action implementation agent, and finally as an information acquisition element. The review found three main precursors to human-automation trust: 1) human-related antecedents, 2) environmental factors, and 3) automation factors. Considerations were made of dispositional trust—that is, a person's trust in automation without a specific system in mind. The authors listed factors such as age, culture, gender, and personality as important antecedents of trust. In terms of environmental or situational trust factors, these authors divided trust antecedents into two groups: the external, physical environment and the internal, state-dependent environment of the operator. For automation-related trust antecedents, Hoff and Bashir defined learned trust as the type of trust that relies on an operator's previous knowledge of a particular system. For example, an individual's experience with the system or the system's reputation is most pertinent here. Additionally, anthropomorphism, transparency, and ease of use consistent with the earlier meta-analysis by Hancock et al. (2011) can all influence whether or not an operator trusts the automated system.

Schaefer and colleagues (2016) similarly examined empirical studies and divided the antecedents of trust along comparable lines, into 1) human aspects, 2) technology aspects, and 3) environmental aspects. In terms of individual differences, such aspects were divided into categories of 1) traits, 2) states, 3) cognitive factors, and 4) emotive factors. Trait factors include a) elements of demographics, b) personality, and c) the propensity to trust. State factors encompass a) attentional, b) control, c) fatigue, and d) stress. Cognitive factors were identified as understanding, ability, and expectancy. Finally, the emotive factors explored were 1) confidence in automation, 2) attitudes, 3) satisfaction, and 4) comfort. Automation factors included features of the automation, such as the level of automation, mode of automation, appearance, and communication method and/or the capabilities of the automation such as its performance. The meta-analysis also focused on shared environmental factors such as team collaboration.

Before these synthetic assessments, Lee and Moray (1994) examined some of the precursors to automation reliance. Participants here were given a task that was based on the process used in an orange juice pasteurization plant with the help of a semi-automatic system. The system, which controlled several important functions such as the feedstock pump, steam pump, and steam heater, could operate automatically, and operators needed only intervene when necessary. Participants were asked to perform in a supervisory capacity and to intervene when and if needed. Normally, however, they could leave all systems in the automatic setting and not switch to manual control. At their discretion, they could operate aspects of the system manually if they chose to do so. Participants were measured for their trust in the automation and their own level of self-confidence. Results showed, unsurprisingly, that those who trusted the automation tended to rely on it more. However, those participants who scored higher on self-confidence, or a belief in their own capability in performing well on the task, tended to rely on the automation less. When self-confidence was higher than trust, participants did not rely on the automation. That is, the distance between their level of self-confidence and their level of automation trust predicted whether or not they would use the automation (Lee & Moray, 1994). Others have also evaluated these trust issues.

For example, in an X-ray screening task, participants were instructed to search luggage for fake weapons. This task was similar to the job performed by the Transportation Security Administration (TSA); it was shown that individual differences in the propensity to trust automation accounted for 52% of the variance in trust for equally performing automation (Merritt & Ilgen, 2008). Participant perception of automation had a strong mediating effect on history-based trust, though the definition would suggest that history-based trust should depend largely on performance of the automation and not on user attitudes.

Hergeth and colleagues (2015) examined trust in a highly automated driving system using a simulator. In this study, participants from either German or Chinese cultural backgrounds were instructed to drive along a set route while the automation was engaged. At certain intervals, the automation relayed a "take-over request." This required the participants to drive for a set period of time before reengaging the automation. Using single-item questions, these researchers found that trust in the automation was temporarily decreased, for a short while, following the take-over request, but was later repaired after the automated system had been used for a period after the take-over request. Automation and mistrust were measured both before and after the experiment, and each of these factors were examined separately. While culture exerted no impact on initial trust in automation ratings, Chinese participants exhibited higher levels of mistrust in automation compared to German participants. These results showed that mistrust was not simply the opposite of trust or a lower level of trust, but rather was its own separate component of what goes into an individual's decision to rely on automation.

AUTOMATION-INDUCED COMPLACENCY, RELIANCE, COMPLIANCE, AND TRUST

The use of UASs evidently serves to mitigate physical danger to pilots (Mouloua, Gilson, & Hancock, 2003). However, this is simply because a vehicle is unmanned, but that does not mean that it is without risk. The most danger, perhaps, exists to those on the ground in the immediate proximity to UASs (Clothier, Walker, Fulton, & Campbell, 2007). It is necessary that operators understand these inherent dangers to civilians that are potentially greater due to the unmanned nature of the vehicles. Operators must exercise caution and remember that, though they themselves might be far from harm's way, they are indeed piloting what may be a very large and often expensive vehicle through unfamiliar territory. While it has been argued that the human factor is the most unpredictable element, the proper use of reliable automation is what makes it safer. To that end, appropriate trust must be calibrated appropriately while complacency is avoided.

Lee and See (2004) produced a benchmark examination of trust in automation. They define trust as, "The attitude that an agent will help achieve an individual's goals in a situation characterized by uncertainty and vulnerability" (p. 54). Additionally, they examine the consequences of incorrectly calibrated trust such as the misuse, disuse, and/or abuse of automation. Misuse of automation relates to situations in which a person relies inappropriately or too much on automation. Disuse refers to a situation in which a person underuses automation. Lastly, abuse is the automatization of a task without appropriate consideration of the effects of such automatization (Parasuraman & Riley, 1997). Lee and See (2004) note that overtrust is a cause of misuse, while undertrust or distrust can cause disuse. In order to avoid both of these negative outcomes, it is important that trust is calibrated for appropriate reliance. Reliance on, as well as compliance with, automation depends on levels of trust. The authors found that when operators could appropriately evaluate automation's performance compared to human performance, the relationship between trust and reliance was stronger. Additionally, the ability to decide how and when to use the automation also leads to a stronger relationship between trust and reliance.

While appropriate reliance proves valuable, complacency is detrimental to performance. Automation-induced complacency is a phenomenon that occurs when an operator becomes inattentive and fails to recognize when it is necessary to intervene. This can result from levels of trust that are higher than desired, based on the capabilities and reliability of the system at hand. Parasuraman, Molloy, and Singh (1993) required participants to multitask in a flight simulator. One of these tasks was to monitor the system for malfunctions with automated alerts. The alerts remained either at constantly high or low levels or varied between high and low performance. Results showed that the participants were less likely to detect system malfunctions in the constant-reliability conditions. This showed a complacency effect when "aided" by the highly reliable automation. Participants were less likely to spot a malfunction that they would normally have been able to detect on their own. The researchers made the point that too much trust or overreliance, even on near-perfect automation, could have significant consequences.

Reliance on automation refers to a certain level of dependency or trust that is placed in a person or, in the present example, an autonomous vehicle. Reliance and trust are related in that, in order to rely on someone, you must first place some degree of trust in them. Similarly, there are situations where trust must be placed in an individual or autonomous agent because reliance is required. These situations appear necessary in order to properly calibrate trust. Researchers believe that, in order to properly calibrate how trustworthy the automation is, it is first necessary to rely on that automation and adjust levels of trust appropriately (Lee & See, 2004; Muir & Moray, 1996). Compliance with an automated system also is associated with levels of trust placed in the automation. Compliance with an alarm or action requested by the automation would indicate that an operator trusts the decision to be the right one. This trust fluctuates, however, according to experiences of false alarms, which make the operator less likely to comply (Rice, 2009).

Parasuraman and Manzey (2010) later examined the dual examples of automation complacency and bias, which had previously been seen as different phenomena. Their definition of complacency focused on situations in which operators did not sufficiently check their automation, and their examples of bias highlighted those times when operators used decision aids to replace, rather than enhance, their own abilities to secure information. Their thorough review found that complacency is most likely to occur when there are multiple concurrent tasks, particularly when the operator is focused on the manual as opposed to the automated task. Previously, the two instances of overreliance had been considered to be separate. However, Parasuraman and Manzey postulated that both depended on similar attentional processes. Both complacency and bias were problems, for even expert participants, and could not be mitigated through training. These separate phenomena were both a result of overtrust and frequently could be problematic.

TRUST IN UAS OPERATION

According to Parasuraman and Riley (1997), "Occasional failures of automation do not seem to be a deterrent to future use of the automation" (p. 237). Similarly, humans often calibrate their reliance on other humans (e.g., friends, spouse, neighbors, teammates, etc.). These two constructs are interchangeably related. We trust someone because we can rely on them, but likewise, we can also rely on someone because we trust them. This bidirectional relationship between people and their interpersonal relations is sometimes very ambiguous. Lee and See (2004) defined trust as, "The attitude that an agent will help achieve an individual's goals in a situation characterized by uncertainty and vulnerability" (p. 54). Research investigating trust between humans and automation has consistently supported the importance of trust as a determinant of appropriate automation use (e.g., Lee & Moray, 1992; Muir & Moray, 1996). The effects of incorrectly calibrated trust on system performance have been well documented, and using inappropriate levels of trust may also result in misuse, disuse, or abuse of automation (Parasuraman & Riley, 1997). With the development of fully autonomous vehicles proceeding at a rapid pace, understanding the various human factors issues that affect trust will be crucial for ensuring appropriate interaction and trust development between future UAS operators and the systems they control. Lee and See (2004) proposed a cyclical model of trust development in which individual differences influence users' evaluation of machine characteristics, forming an initial baseline of trust. Such trust evolves as users are exposed to, and evaluate, machine characteristics and behaviors. In such a model, the role of individual differences is of significant importance; as such, differences influencing an operator's propensity to trust automated systems could affect his or her initial use decision and the manner in which they interpret system behavior.

Previous studies have also examined a variety of system characteristics that influence the development of trust in autonomous systems (Schaefer et al., 2016) including anthropomorphism (Pak, Fink, Price, Bass, & Sturre, 2012; Stedmon et al., 2007; Waytz, Heafner, & Epley, 2014), alarm bias (Lees & Lee, 2007), and the inclusion of adaptive task allocation (Mouloua, Parasuraman, & Molloy, 1993; Parasuraman, Mouloua, & Molloy, 1996). However, more research is still needed to understand how individual differences in cognitive, perceptual, and emotional characteristics influence operators' trust development during interaction with UAS systems. A recent meta-analysis by Schaefer and colleagues (2016) examined a number of human characteristics that have been shown to influence operators' willingness to trust and rely on autonomous systems. These factors range from traits and states to cognitive, emotional, and environmental variables.

While research examining the influence of these characteristics on trust in UAS systems is still in its infancy, trust development and attitudes toward automation in other domains have been found to vary significantly as a function of age (Donmez, Boyle, Lee, & McGehee, 2006; Ezer, Fisk, & Rogers, 2008; Ho, Wheatley, & Scialfa, 2005; Steinke, Fritsch, & Silbermann, 2012), gender (Hohenberger, Spörrle, & Welpe, 2016; Tung, 2011), culture (Huerta, Glandon, & Petrides, 2012; Li, Rau, & Li, 2010), and personality variables (Merritt & Ilgen, 2008; Szalma & Taylor, 2011). In addition, attitudes toward automation (Ardern-Jones et al., 2009; Bailey et al., 2006; Gao, Lee, & Zhang, 2006; Merritt, Heimbaugh, LaChapell, & Lee, 2013) and satisfaction with automation performance (Donmez et al., 2006, Wang, Jamieson, & Hollands, 2011) have also been shown to influence the process of trust development. In a similar vein, self-reported comfort with autonomous systems, defined in terms of familiarity (van den Broek & Westerink, 2009), perceived similarity of intent (Verberne, Ham, & Midden, 2012), and degree of automaton control (Ward, 2000) can be strong determinants of system use. Finally, research examining the influence of cognitive and environmental factors has found differences in trust development and automation reliance between operators with low and high capacities for attentional control (Chen & Barnes, 2012; Chen & Terrence, 2009), low and high spatial ability (Chen, 2010; Endsley & Bolstad, 1994), with varying levels of executive functioning (Panganiban & Matthews, 2014), working memory capacity (de Visser et al., 2010), and among operators experiencing varying states of fatigue (Reichenbach, Onnasch, & Manzey, 2011).

Significant differences in trust and reliance have also been observed between operators under low and high workload (McBride, Rogers, & Fisk, 2011) and as a function of operators' expectancies concerning the perceived benefits of system use (Lee, Kim, & Kim, 2007) as well as the reputation of a system for performing a task successfully (De Vries & Midden, 2008; Madhavan & Weigman, 2007. Together, these findings from varying performance domains suggest that individual differences in key human characteristics will likely have a strong influence on trust and reliance during UAS system operation. As such, selection and training methods for UAS operators must acknowledge these influences in addition to the other cognitive, sensory, and perceptual factors on UAS operator trust and performance.

TRUST MEASUREMENT APPROACHES

As trust has been shown to be integral to the use of automation, it is important to understand the various ways in which it has measured the construct. For instance, cross-cultural studies of highly automated driving systems have found that trust and mistrust are two related but separate concepts (Hergeth et al., 2015). So, it is necessary to examine and compare the various ways in which trust and mistrust in automation are measured prior to, during, and following any interactions.

There are two main ways in which researchers determine the level of trust that an individual puts in any system or machine. The first method uses scales completed by the individual indicating levels of trust, while the second uses behavioral measures that indicate reliance on a system. A trust scale may refer to any method of trust measurement where a person says whether or not (and to what extent) they trust a piece of automation. These are normally highly validated; however, with the use of a scale come the associated pitfalls of any self-report. The companion method of measuring trust relies not upon surveys but on usage. The greater the potential for damage if the automation should fail, the greater is the trust required to use it. This method can be compared to a "trust fall," in which people express their faith in one another by falling backward and relying on their partners to catch them. Trust in automation can be expressed in similar ways; those who do not trust an automated machine will not place their safety in its hands. This comparison among experiences of trust may apply to self-driving cars. A driver surrenders the control of a heavy, and deadly, vehicle and allows him- or herself to become only a passenger. It may apply to automation uses in military settings also, where failures can be life threatening for both the operator and their teammates. While "use" does not always provide the type of numerical values useful for statistics, it often tells us more than surveys since it records actions and not perceptions.

That is not to say that surveys and scales cannot provide valuable insight into the degree to which people trust automation. Perhaps the most commonly used scale for measuring trust in automation was developed by Jian, Bisantz, and Drury (1998). This Checklist for Trust between People and Automation uses a Likert scale rating system (where 1 means "not at all" and 7 means "extremely"). It asks that participants rate their perception of an automated system in order to measure their trust and distrust in that system. Several surveys exist that measure more specific elements of the trust relationship. Singh, Molloy, and Parasuraman (1993) developed the complacency potential rating scale to measure potential trust-related complacency. This scale measures the extent to which a person risks becoming complacent with automation predicated on their level of trust in the system. A further scale, focused specifically on robotics, measured trust both before and after an interaction in order to determine how that interaction had affected trust (Schaefer, 2013). All of these scales are useful in determining the effects that various factors can have on trust in automation.

TRAINING AND SELECTION

The operation of the UASs puts the human operator in cognitively demanding environments where failures of any component could have tragic consequences. These operators experience combat and occupational stressors, making the effective

Automation Trust in UAS Operation, Selection, and Training 179

recruitment, training, and selection of these operators essential to overall UAS performance and safety. The unique cognitive demands of UAS operations create special challenges for the assessment, recruitment, selection, and training of qualified operators. Performance demands in UAS operation include several key components including teamwork, technical proficiency, vigilance, stress reliance, and effective training, and selection methods for UAS operators must accurately and efficiently assess and predict each of these factors.

Historically, military UAS testing platforms have borrowed heavily from existing assessment systems designed originally for predicting performance in manned flight operations. The dominant mindset has been that skilled aviators would be the best candidates for unmanned operations given the common skill and ability requirements shared across the two domains. For example, both manned and unmanned flights require common visual and spatial orientation abilities, teamwork, communication skills, navigation skills, and a variety of higher-level mental abilities related to attention and action planning.

However, there are several key differences between manned and unmanned aviation. These include the evident, increased reliance on autonomy, human–machine teaming, and trust in automation within unmanned aviation. Such differences render existing selection methods deficient. Unmanned operations not only put a unique pressure on skill domains shared with manned aviation but require several unique cognitive and noncognitive attributes. For example, as the pilot is not physically located in the aircraft that is under their control, they need to engage in unique mental gymnastics with respect to spatial orientation skills to maintain SA as to the orientation and direction of the aircraft. This will often diverge from their own perspective and field of view.

The deficiencies of current selection methods translate to suboptimal outcomes for UAS operators who are hired using these systems. This is manifest in critical work effectiveness metrics such as job performance, workplace safety, selection, fairness, mission fratricide and collateral damages, and return on investment (ROI) related to improved job tenure, satisfaction, and engagement. There is currently no single, comprehensive technology that meets the military's current and projected requirements for UAS operator selection. Current efforts to select UAS operators are piecemeal and still borrow heavily from existing assessment systems designed originally to support manned aviation.

The absence of a single, comprehensive technology to meet military UAS operator selection needs highlights the following characteristics that will be required for a dedicated UAS operator selection system:

1. Accurate performance prediction enabled by efficient, holistic assessment of a range of personal attributes designed to collectively predict overall UAS operator effectiveness, with particular emphasis on adaptive management of operational stress
2. Reduction in assessment time
3. Measures to ensure content protection and counteract intentional response distortion
4. Realistic, immersive, performance-based assessments of latent cognitive traits while affording realistic performance of simulated UAS mission tasks

5. Flexible scoring algorithms to boost the operational validity of assessments developed for UAS operator personnel selection that provide the basis for custom scoring for selection of different UAS platforms
6. Ability to objectively evaluate criterion validity, adverse impact, predictive bias, and test utility

One example of an emerging personnel selection technique designed to meet these criteria for the U.S. military is the "Stealth Adapt" assessment system (Mangos, 2016). The Stealth Adapt technology solution has been developed for the U.S. Department of Defense (DoD) as a means of effectively predicting UAS operator performance. This comprehensive measure takes into account factors of executive level cognition, biographical history, and personality in order to create an adaptive assessment measure. The Stealth Adapt program and measurement process includes a series of self-report questions and a simulation-based search-and-rescue mission during which individuals pilot a UAS through a realistic game environment.

The assessment targets three general domains related to effective UAS operations: 1) cognitive attributes, 2) nonability (personality) attributes, and 3) biographical history. Collectively, these assessments form the foundation for an integrated performance prediction architecture for UAS operator selection. The cognitive assessments are embedded within a realistic UAS mission simulation. The basic scenario for the simulation represents a UAS search and rescue (SAR) task. The simulation consists of two phases. During a mission planning phase, the player-operator is tasked with planning an SAR mission and then launching one or more drones to execute this mission. The objective is simply to perform visual identification of the stranded friendly forces (rescue targets) in the order of their priority while avoiding hostile forces that might initiate aggression against either the UAS or the rescue targets. The rules of engagement, designed to measure advanced cognitive skills during gameplay, focus the player's attention on the targets' priority level, location, and lethality of hostile forces, hazardous weather systems, battery and fuel levels of the drone, no-fly zones, and the enemies' policies for aggressive action. The core set of cognitive skills measured include mental simulation, task prioritization, mission replanning, multitasking, working memory, mental rotation, time estimation, problem-solving, prioritizing, attention allocation, and real-time mission replanning.

The second assessment category includes a comprehensive suite of personality assessment administered in an adaptive forced choice format. This format, designed to preemptively mitigate the possibility of applicant faking, combines item stems, each measuring a different personality attribute, into a series of adaptively administered dyads. The respondent is instructed to select the item stem that best describes his or herself. Collectively, the items measure 11 unique personality attributes deemed essential to the accurate prediction of UAS operator performance. These include factors such as teamwork orientation, stress resilience, trust in automation, assertiveness, confidence, and attention to detail.

The final category of assessment content is biographical data. These items assess multiple aspects of a candidate's biographical history with respect to a number of content areas, including interest in robotics, military interest and experience, science, technology, engineering, and math (STEM) academic history, gaming, sports and teamwork, computers and cyber history, and history of emotional fitness (problem-focused coping style).

The benefit of this technology, as an initial example of a stand-alone UAS operator selection system, is its potential to enhance the effectiveness of any military operations using UAS platforms through measurable enhancements in job performance, workplace safety, and fair selection practices; reduced mission fratricide and collateral damages; and ROI related to improved job tenure, satisfaction, and engagement. Table 9.1 summarizes the assessment strategies, constructs, and associated performance criteria of the Stealth Adapt system.

The advantages associated with each critical feature of the solution are summarized in the table below:

TABLE 9.1
Summary of Advantages Associated with Each Critical Feature of the Solution

Selection Phase	Domain	Measured Attributes	Assessment Format	Target Criteria (Outcome to be Improved)
1	Executive-level cognition	- Mental simulation - Task prioritization - Dynamic mission replanning - Working memory - Multitasking - Mental rotation - Prospective/retrospective time estimation	Immersive simulation with embedded performance-based tasks	- Technical UAS operator performance - Probability of catastrophic performance failures - Mission outcomes/target accuracy
2	Nonability/ Normal personality	- Operational stress resilience - Emotional stability and control - Disengagement/compartmentalization - Assertiveness - Polychronicity - Initiative - Dependability - Teamwork orientation - Adaptability - Resourcefulness - Trust in automation	Adaptive, forced-choice personality items	- Symptom-free survival (time to onset of stress symptomology) - Teamwork effectiveness
3	Biographical history	- Interest in robotics - Military interest and experience - STEM academic history - Sports leadership and teamwork - Gaming, computers, & cyber history - Emotional fitness; problem-focused coping style	Self-report assessment items	- Mission productivity - Tenure/termination probability - Work engagement

TRUST IN AUTOMATION DEVELOPMENT

Levels of trust in a system will waiver or strengthen with time and as the operator gains experience with the automation. However, it is often the case that a certain level of trust and reliance must be placed in the automation with the operator relegated to other, often critical, tasks. The development of trust in automation is certainly dynamic and may be, based on the evidence provided in the literature mentioned in this chapter, cyclical in its calibration. The amount of trust placed in an unmanned system is shaped and altered through a series of subjective experiences as well as deliberate efforts of training for proper calibration.

Figure 9.1 represents the dynamic and often cyclical nature of the development of trust in an automated agent. The first stage is the first formation of any judgment about the trustworthiness of the automation. This pre-exposure judgment is often influenced by any prior knowledge the operator has about the capabilities and limitations of the system, as well as any visible qualities that may make an individual more likely to trust automation, such as anthropomorphic qualities. The second stage is the initial compliance and reliance during operation of the system. This initial compliance/reliance is necessary to gather information about the activity of the system, including its capabilities and limitations, and make a proper judgment when calibrating trust in the system. The degree to which a person will initially rely on

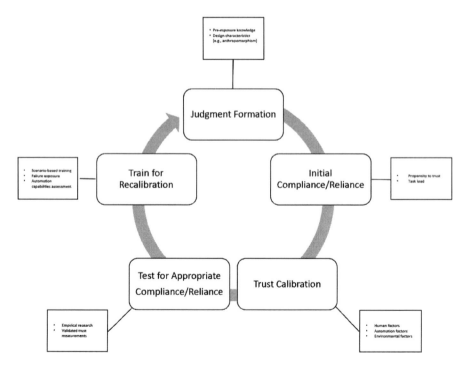

FIGURE 9.1 Depicting the dynamic nature of automation trust and its calibration/development.

automation is strongly impacted by their own personal propensity to trust. After gathering information during an initial exposure phase, trust calibration takes places and is influenced by human factors, automation factors, and environmental factors alike. Human factors include individual state variables such as workload and stress, while automation factors such as reliability and environmental factors such as the physical environment also contribute during this calibration stage. Following trust calibration and exposure, it is necessary to examine the levels of trust now exhibited by the individual and how those levels impact performance. The most effective way to do that is through empirical research utilizing validated measures of trust in automation. Examples of these measures include those described previously. Finally, the last step in this process would be to recalibrate trust to optimal or even appropriate levels to ensure system safety and efficiency. Researchers have found scenario-based training that exposes operators to a multitude of automation failures may effectively calibrate trust. Ultimately, through this recalibration, a new judgment about the system and its trustworthiness/reliability is formed and the cycle begins again.

CONCLUSIONS

The role of trust has been examined in contexts as diverse as health-care systems, nuclear power plant operation, cyber-security, and even inter-personal relationships. Such results support the need to properly calibrate trust in order to maximize safety and efficiency. The research that has been described here provides evidence that the level of trust an operator places in automation plays a significant role in their interactions with UASs. Automation provides an aid to unmanned system operators, particularly in stressful, multitasking environments common to autonomous vehicle operation. However, it is important to acknowledge that even at this stage and in spite of how much technology has advanced, automation still remains fallible. The nature of such unreliable automation can result in either overreliance or underreliance, and thus underutilization of the system. The study previously mentioned by Dixon and Wickens (2006) is one example of how experience of false alarms or misses by an automated system have differential effects on operator trust. Placing too much trust in automation, which surpasses the abilities of the system itself, can result in misuse of the automation (Lee & See, 2004; Parasuraman & Riley, 1997). Applications of trust in the operation of UASs have become more relevant, for example, as the number of self-driving cars and military drone operations have increased.

The investigations into the role of trust in human-automation interaction have implications for unmanned system design and operator training/selection. The relationship between an operator and an automated agent appears to be fragile, and operators will be less likely to rely on the automation after a false alarm or miss. Unreliable automation impacts levels of trust operators place in automation have been shown to impact operator complacency and reliance, with complacency indicating the likelihood of responding to an alarm and reliance the operator's tendency to let the automation go unchecked in the presence of no alarm. Both complacency and reliance, as they relate to trust, impact the operator's ability to perform concurrent tasks and utilize the automation. Future efforts should emphasize ways to moderate the effects of operator trust and avoid instances of misuse and disuse of automation (Parasuraman & Riley, 1997). Theories of adaptive automation have improved human monitoring performance of such automated and autonomous systems and contributed to better adaptive and adaptable system design. Parasuraman, Molloy, and Singh (1993) found a performance drop-off in a monitoring task following 20 minutes of consecutive monitoring under static automation reliability. Interestingly, by varying the reliability of the automation, this performance decrement, attributed to the previously mentioned complacency, was mitigated. These results suggest a form of adaptive automation may work to eliminate the bias that trust introduces and prevent the complacency that tends to result in missed signals that can impact system performance.

Over the last 25 years, a significant body of scientific knowledge, both theoretical and applied, has focused on examining the critical role of human-monitoring efficiency in automated and autonomous systems (Mouloua & Koonce, 1997; Mouloua & Parasuraman, 1994; Parasuraman & Mouloua, 1996; Scerbo & Mouloua, 1999; Vincenzi, Mouloua, & Hancock, 2004). As drivers' and pilots' roles shift to more passive, monitoring ones, research dedicated to investigating the impact operator trust has on system performance has revealed several risks and potential consequences of autonomous systems. Such research highlights the importance of investigating these problems in the context of effective human-system design to mitigate the consequences. As the U.S. military will continue to use autonomous systems for various operational tasks and missions, further research into system design is needed to address how to enhance UAS operator performance, workload, mission success, and safety in the context of trust calibration. Training methods should emphasize a thorough understanding of the system's capabilities and limitations by the operator. A calibration of trust is most appropriate when it matches the usability and functionality of the system.

Ergonomists and human factors engineers face specific challenges when attempting to keep a human engaged in the presence of high levels of automation. There is a tendency for humans to become overreliant on automation, resulting in a complacency that impairs their ability to stay engaged and attentive. This reliance and complacency can again be tied in large part to the construct of trust, and much research has been conducted investigating its impact on the ergonomic design of automated products or systems. The more trust persons place in the automation, the more reliant and/or complacent they may become, counting on the automation to flawlessly execute the tasks it is designated to. However, inappropriate levels of trust can have varying impacts on how operators handle and monitor automation. A proper calibration of trust has proven to be a significant factor in the effective monitoring of automated systems, with too much or too little trust having unique and potential catastrophic consequences. The role of trust must be seriously considered when researchers and engineers work toward optimizing the performance of complex human-machine systems. Future research efforts should be devoted to establishing effective training and selection measures for operators of unmanned systems that ensure a proper calibration of trust in the system, especially considering the impact research has consistently shown it has on how humans operate automated systems and vehicles.

REFERENCES

Ardern-Jones, J., Hughes, D., Rowe, P. H., Mottram, D. R., & Green, C. F. (2009). Attitudes and opinions of nursing and medical staff regarding the supply and storage of medicinal products before and after the installation of a drawer-based automated stock-control system. *International Journal of Pharmacy Practice, 17*(2), 95–99.

Bailey, N. R., Scerbo, M. W., Freeman, F. G., Mikulka, P. J., & Scott, L. A. (2006). Comparison of a brain-based adaptive system and a manual adaptable system for invoking automation. *Human Factors, 48*(4), 693–709.

BIS Research.com. (2016). Unmanned aerial vehicles (UAV) market, by value and volume: Analysis and forecast, 2015–2020. Retrieved November 1, 2018 from: https://bisresearch.com/industry-report/uav-industry-by-type-volume-analysis.html?gclid=Cj0KCQiAzKnjBRDPARIsAKxfTRBdx53k3bH-oxTvGz63unD306ahCCrNVGsXfbrbIAxXEvJV6_sI7x4aAgbAEALw_wcB

Chen, J. Y. C. (2010, March). Effects of operator spatial ability on uav-guided ground navigation. In *Proceedings of the 5th ACM/IEEE International Conference on Human-Robot Interaction,* 139–140. Osaka: IEEE Press.

Chen, J. Y. C., & Barnes, M. J. (2012). Supervisory control of multiple robots in dynamic tasking environments. *Ergonomics, 55*(9), 1043–1058.

Chen, J. Y. C., & Terrence, P. I. (2009). Effects of imperfect automation and individual differences on concurrent performance of military and robotics tasks in a simulated multitasking environment. *Ergonomics, 52*(8), 907–920.

Clothier, R. A., Walker, R. A., Fulton, N., & Campbell, D. A. (2007). A casualty risk analysis for unmanned aerial system (UAS) operations over inhabited areas. In *Proceedings of the AIAC12 – Twelfth Australian International Aerospace Congress, 2nd Australasian Unmanned Air Vehicles Conference,* 42–57. Melbourne, Australia: QUT Queensland University of Technology.

de Visser, E., Shaw, T., Mohamed-Ameen, A., & Parasuraman, R. (2010, September). Modeling human-automation team performance in networked systems: Individual differences in working memory count. In *Proceedings of the Human Factors and Ergonomics Society Annual Meeting, 54*(14), 1087–1091. Los Angeles, CA: SAGE Publications.

De Vries, P., & Midden, C. (2008). Effect of indirect information on system trust and control allocation. *Behaviour & Information Technology, 27*(1), 17–29.

Dixon, S. R., & Wickens, C. D. (2003). Control of multiple-UAVs: A workload analysis. *Proceedings of the 12th International Symposium on Aviation Psychology,* pp. 1–5. Dayton, Ohio: Wright State University.

Dixon, S. R., & Wickens, C. D. (2006). Automation reliability in unmanned aerial vehicle control: A reliance-compliance model of automation dependence in high workload. *Human Factors, 48*(3), 474–486.

Dixon, S. R., Wickens, C. D., & Chang, D. (2005). Mission control of multiple unmanned aerial vehicles: A workload analysis. *Human Factors, 47*(3), 479–487.

Donmez, B., Boyle, L. N., Lee, J. D., & McGehee, D. V. (2006). Drivers' attitudes toward imperfect distraction mitigation strategies. *Transportation Research Part F: Traffic Psychology and Behaviour, 9*(6), 387–398.

Donmez, B., Nehme, C., & Cummings, M. L. (2010). Modeling workload impact in multiple unmanned vehicle supervisory control. *IEEE Transactions on Systems, Man, and Cybernetics—Part A: Systems and Humans, 40*(6), 1180–1190.

Endsley, M. R. (1995). Toward a theory of situation awareness in dynamic systems. *Human Factors, 37*(1), 32–64.

Endsley, M. R., & Bolstad, C. A. (1994). Individual differences in pilot situation awareness. *The International Journal of Aviation Psychology, 4*(3), 241–264.

Ezer, N., Fisk, A. D., & Rogers, W. A. (2008). Age-related differences in reliance behavior attributable to costs within a human-decision aid system. *Human Factors, 50*(6), 853–863.

Ferraro, J., Clark, L., Christy, N., & Mouloua, M. (2018). Effects of automation reliability and trust on system monitoring performance in simulated flight tasks. In *Proceedings of the Human Factors and Ergonomics Society Annual Meeting 62*(1), 1232–1236. Los Angeles, CA: SAGE Publications.

Gao, J., Lee, J. D., & Zhang, Y. (2006). A dynamic model of interaction between reliance on automation and cooperation in multi-operator multi-automation situations. *International Journal of Industrial Ergonomics, 36*(5), 511–526.

Gilson, R., Richardson, C., & Mouloua, M. (1998). Key human factors issues for UAV/UCAV mission success. In *Proceedings of the Annual Meeting of the Association for Unmanned Vehicle Systems International, AUVSI '98*, 477–484). Washington, DC: Association for Unmanned Vehicle Systems International.

Hancock, P. A. (2013). Task partitioning effects in semi-automated human–machine system performance. *Ergonomics, 56*(9), 1387–1399.

Hancock, P. A. (2014). Automation: how much is too much? *Ergonomics, 57*(3), 449–454.

Hancock, P. A., Billings, D. R., & Schaefer, K. E. (2011). Can you trust your robot? *Ergonomics in Design, 19*(3), 24–29.

Hancock, P. A., Billings, D. R., Schaefer, K. E., Chen, J. Y., De Visser, E. J., & Parasuraman, R. (2011). A meta-analysis of factors affecting trust in human-robot interaction. *Human Factors, 53*(5), 517–527.

Hancock, P.A. & Krueger, G.P. (2010) *Hours of Boredom, Moments of Terror: Temporal Desynchrony in military and Security Force Operations*. Report, Center for Technology and National Security Policy, Washington, DC: National Defense University.

Hancock, P. A. & Warm, J.S. (1989). A dynamic model of stress and sustained attention. *Human Factors, 31*(5), 519–537.

Hergeth, S., Lorenz, L., Krems, J. F., & Toenert, L. (2015). Effects of take-over requests and cultural background on automation trust in highly automated driving. In *Proceedings of the Eighth International Driving Symposium on Human Factors in Driver Assessment, Training and Vehicle Design in Salt Lake City, UT,* 331–333. Iowa City: Iowa Research Online.

Ho, G., Wheatley, D., & Scialfa, C. T. (2005). Age differences in trust and reliance of a medication management system. *Interacting with Computers, 17*(6), 690–710.

Hoff, K. A., & Bashir, M. (2015). Trust in automation: Integrating empirical evidence on factors that influence trust. *Human Factors, 57*(3), 407–434.

Hoffman, R. R., & Hancock, P. A. (2017). Measuring resilience. *Human Factors, 59*(4), 564–581.

Hohenberger, C., Spörrle, M., & Welpe, I. M. (2016). How and why do men and women differ in their willingness to use automated cars? The influence of emotions across different age groups. *Transportation Research Part A: Policy and Practice, 94*, 374–385.

Huerta, E., Glandon, T., & Petrides, Y. (2012). Framing, decision-aid systems, and culture: Exploring influences on fraud investigations. *International Journal of Accounting Information Systems, 13*(4), 316–333.

Jian, J. Y., Bisantz, A. M., & Drury, C. G. (1998). Towards an empirically determined scale of trust in computerized systems: Distinguishing concepts and types of trust. In *Proceedings of the Human Factors and Ergonomics Society Annual Meeting, 42*(5), 501–505. Los Angeles, CA: SAGE Publications.

Lee, H., Kim, J., & Kim, J. (2007). Determinants of success for application service provider: An empirical test in small businesses. *International Journal of Human-Computer Studies, 65*(9), 796–815.

Lee, J., & Moray, N. (1992). Trust, control strategies and allocation of function in human-machine systems. *Ergonomics, 35*(10), 1243–1270.

Lee, J. D., & Moray, N. (1994). Trust, self-confidence, and operators' adaptation to automation. *International Journal of Human-Computer Studies, 40*(1), 153–184.

Lee, J. D., & See, K. A. (2004). Trust in automation: Designing for appropriate reliance. *Human Factors, 46*(1), 50–80.

Lees, M. N., & Lee, J. D. (2007). The influence of distraction and driving context on driver response to imperfect collision warning systems. *Ergonomics, 50*(8), 1264–1286.

Li, D., Rau, P. P., & Li, Y. (2010). A cross-cultural study: Effect of robot appearance and task. *International Journal of Social Robotics, 2*(2), 175–186.

Madhavan, P., & Wiegmann, D. A. (2007). Similarities and differences between human–human and human–automation trust: an integrative review. *Theoretical Issues in Ergonomics Science, 8*(4), 277–301.

Mangos, P. (2016). *Stealth Adapt [Computer Software]*. Tampa, FL: Adaptive Immersion Technologies.

McBride, S. E., Rogers, W. A., & Fisk, A. D. (2011). Understanding the effect of workload on automation use for younger and older adults. *Human Factors, 53*(6), 672–686.

Merritt, S. M., Heimbaugh, H., LaChapell, J., & Lee, D. (2013). I trust it, but I don't know why: Effects of implicit attitudes toward automation on trust in an automated system. *Human Factors, 55*(3), 520–534.

Merritt, S. M., & Ilgen, D. R. (2008). Not all trust is created equal: Dispositional and history-based trust in human-automation interactions. *Human Factors, 50*(2), 194–210.

Mouloua, M., Gilson, R., Daskarolis-Kring, E., Kring, J., & Hancock, P. (2001a, October). Ergonomics of UAV/UCAV mission success: Considerations for data link, control, and display issues. In *Proceedings of the Human Factors and Ergonomics Society Annual Meeting, 45*(2), 144–148. Los Angeles, CA: SAGE Publications.

Mouloua, M., Gilson, R., Kring, J. A., & Hancock, P. A. (2001b). Workload, situation awareness, and teaming issues for UAV/UCAV operations. In *Proceedings of the Human Factors and Ergonomics Society Annual Meeting 45*(2), 162–165. Los Angeles, CA: SAGE Publications.

Mouloua, M., Gilson, R., & Hancock, P.A. (2003). Human-centered design of unmanned aerial vehicles. *Ergonomics in Design 11*(1), 6–11.

Mouloua, M., Gilson, R. D., & Koonce, J. (1997). Automation, flight management and pilot training- issues and considerations. *Aviation Training: Learners, Instruction and Organization.* 78–86.

Mouloua, M., Hancock, P., Jones, L., & Vincenzi, D. (2010). Automation in aviation systems: Issues and considerations. In J. Wise., D. Garland., & D. V. Hopkin (Eds.). *Handbook of Aviation Human Factors*. Boca Raton, FL: CRC Press (Taylor & Francis Group).

Mouloua, M. E., & Koonce, J. M. (1997). *Human–Automation Interaction: Research and Practice*. Mahwah, NJ: Lawrence Erlbaum Associates, Inc.

Mouloua, M., & Parasuraman, R. (1994). *Human Performance in Automated Systems: Current Research and Trends* (pp. 264–269). Hillsdale, NJ: Erlbaum.

Mouloua, M., Parasuraman, R., & Molloy, R. (1993, October). Monitoring automation failures: Effects of single and multi-adaptive function allocation. In *Proceedings of the Human Factors and Ergonomics Society Annual Meeting, 37*(1), 1–5. Los Angeles, CA: SAGE Publications.

Mouloua, M., Smither, J., Hancock, P., Duley, J., Adams, R., & Latorella, K. (2004). Aging and driving II: Implications of cognitive changes. In D. A. Vincenzi., M. Mouloua, & P. A. Hancock (Eds.), *Human Performance, Situation Awareness and Automation: Current Research and Trends* (pp. 320–323). Mahwah, NJ: Lawrence Erlbaum Associates.

Mouloua, M., Smither, J., & Kennedy, R. S. (2009). Space adaptation syndrome and perceptual training. In D. Vincenzi, J. Wise, M. Mouloua, & P. A. Hancock (Eds.). *Human Factors in Simulation and Training* (pp. 239–255). Boca Raton, FL: CRC Press (Taylor & Francis Group).

Muir, B. M., & Moray, N. (1996). Trust in automation. Part II. Experimental studies of trust and human intervention in a process control simulation. *Ergonomics, 39*(3), 429–460.

Onnasch, L., Wickens, C. D., Li, H., & Manzey, D. (2014). Human performance consequences of stages and levels of automation an integrated meta-analysis. *Human Factors, 56*(3), 476–488.

Pak, R., Fink, N., Price, M., Bass, B., & Sturre, L. (2012). Decision support aids with anthropomorphic characteristics influence trust and performance in younger and older adults. *Ergonomics, 55*(9), 1059–1072.

Panganiban, A. R., & Matthews, G. (2014, September). Executive functioning protects against stress in UAV simulation. In *Proceedings of the Human Factors and Ergonomics Society Annual Meeting, 58*(1), 994–998. Los Angeles, CA: SAGE Publications.

Parasuraman, R., & Manzey, D. H. (2010). Complacency and bias in human use of automation: An attentional integration. *Human Factors, 52*(3), 381–410.

Parasuraman, R., Molloy, R., & Singh, I. L. (1993). Performance consequences of automation-induced "complacency." *The International Journal of Aviation Psychology, 3*(1), 1–23.

Parasuraman, R. E., & Mouloua, M. E. (1996). *Automation and Human Performance: Theory and Applications*. Mahwah, NJ: Lawrence Erlbaum Associates, Inc.

Parasuraman, R., Mouloua, M., & Molloy, R. (1996). Effects of adaptive task allocation on monitoring of automated systems. *Human Factors, 38*(4), 665–679.

Parasuraman, R., & Riley, V. (1997). Humans and automation: Use, misuse, disuse, abuse. *Human Factors, 39*(2), 230–253.

Pavlas, D., Burke, C. S., Fiore, S. M., Salas, E., Jensen, R., & Fu, D. (2009). Enhancing unmanned aerial system training: A taxonomy of knowledge, skills, attitudes, and methods. In *Proceedings of the Human Factors and Ergonomics Society Annual Meeting, 53*, 1903–1907.

Pomranky, R. A., & Wojciechowski, J. Q. (2007). *Determination of mental workload during operation of multiple unmanned systems* (No. ARL-TR-4309). Army Research Lab. Aberdeen Proving Ground, MD: Human Research and Engineering Directorate.

Reichenbach, J., Onnasch, L., & Manzey, D. (2011). Human performance consequences of automated decision aids in states of sleep loss. *Human Factors, 53*(6), 717–728.

Rice, S. (2009). Examining single-and multiple-process theories of trust in automation. *The Journal of General Psychology, 136*(3), 303–322.

Scerbo, M., & Mouloua, M. (1999) (Eds.). *Automation Technology and Human Performance: Current Research and Future Trends*. Mahwah, NJ: Lawrence Erlbaum Associates.

Schaefer, K. E. (2013). The perception and measurement of human–robot trust (Doctoral dissertation). Orlando: University of Central Florida.

Schaefer, K. E., Chen, J. Y., Szalma, J. L., & Hancock, P. A. (2016). A meta-analysis of factors influencing the development of trust in automation: Implications for understanding autonomy in future systems. *Human Factors, 58*(3), 377–400.

Senders, J. W. (1964). The human operator as a monitor and controller of multidegree of freedom systems. *IEEE Transactions on Human Factors in Electronics, 5*(1), 2–5.

Sheridan, T. B. (1997). Supervisory control. In G. Salvendy (Ed.). *Handbook of Human Factors and Ergonomics* (pp. 1295–1327). New York: Wiley.

Singh, I. L., Molloy, R., & Parasuraman, R. (1993). Automation-induced complacency: Development of the complacency-potential rating scale. *The International Journal of Aviation Psychology, 3*(2), 111–122.

Smith, K., & Hancock, P. A. (1995). Situation awareness is adaptive, externally directed consciousness. *Human Factors, 37*(1), 137–148.

Smither, J., Mouloua, M., Hancock, P., Duley, J., Adams, R., & Latorella, K. (2004). Aging and driving I: Implications of physical and perceptual changes. In D. A. Vincenzi., M. Mouloua, & P. A. Hancock (Eds.), *Human Performance, Situation Awareness and Automation: Current Research and Trends* (pp. 315–319). Mahwah, NJ: Lawrence Erlbaum Associates.

Stedmon, A. W., Sharples, S., Littlewood, R., Cox, G., Patel, H., & Wilson, J. R. (2007). Datalink in air traffic management: Human factors issues in communications. *Applied Ergonomics, 38*(4), 473–480.

Steinke, F., Fritsch, T., & Silbermann, L. (2012). A systematic review of trust in automation and assistance systems for older persons' overall requirements. In *eTELEMED 2012, the Fourth International Conference on eHealth, Telemedicine, and Social Medicine,* 155–163. Valencia, Spain: IARIA.

Szalma, J. L., & Taylor, G. S. (2011). Individual differences in response to automation: The five factor model of personality. *Journal of Experimental Psychology: Applied, 17*(2), 71–96.

Tung, F. W. (2011). Influence of gender and age on the attitudes of children towards humanoid robots. *Human-Computer Interaction. Users and Applications* (pp. 637–646). Berlin: Springer.

van den Broek, E. L., & Westerink, J. H. (2009). Considerations for emotion-aware consumer products. *Applied Ergonomics, 40*(6), 1055–1064.

Verberne, F. M., Ham, J., & Midden, C. J. (2012). Trust in smart systems: Sharing driving goals and giving information to increase trustworthiness and acceptability of smart systems in cars. *Human Factors, 54*(5), 799–810.

Vincenzi, D. A., Mouloua, M., & Hancock, P. A. (Eds.). (2004). *Human Performance, Situation Awareness and Automation: Current Research and Trends: HPSAA II* (Vol. 2). London: Psychology Press.

Wang, L., Jamieson, G. A., & Hollands, J. G. (2011, September). The effects of design features on users' trust in and reliance on a combat identification system. In *Proceedings of the Human Factors and Ergonomics Society Annual Meeting, 55*(1), 375–379. Los Angeles, CA: SAGE Publications.

Waytz, A., Heafner, J., & Epley, N. (2014). The mind in the machine: Anthropomorphism increases trust in an autonomous vehicle. *Journal of Experimental Social Psychology, 52*, 113–117.

Wiegmann, D. A., & Shappell, S. A. (2001). Human error analysis of commercial aviation accidents: Application of the Human Factors Analysis and Classification System (HFACS). *Aviation, Space, and Environmental Medicine, 72*(11), 1006–1016.

10 Autonomous Systems Theory and Design and a Paradox of Automation for Safety

David Kaber

CHALLENGES IN AUTOMATED AND AUTONOMOUS SYSTEMS DESIGN

There has been a recent explosion of development in the area of autonomous systems, or systems that are considered to self-sufficient in an operating context and self-directed in function. These systems can range from independent mechanical and biological agents (or "things") in an environment to teams of both mechanized/computerized systems with humans acting together to deliver specific functions to task performance. Examples of mechanized and computerized autonomous systems include some unmanned aerial vehicles (UAVs; e.g., Global Hawk). The Federal Aviation Administration (FAA) predicts the UAV industry will be worth $90 billion in a decade (Gent, 2015). Other autonomous systems can be found in space applications, such as the next generation of rover technology being developed by the National Aeronautics and Space Administration (NASA) for space missions, including lunar exploration (e.g., KRex2). NASA is focused on creating human and autonomous agent systems for space applications (NASA ASCLT, 2018), which involves defining taxonomies of autonomous system capabilities. Still other autonomous systems can be identified for ground transportation. Ford Motor Company recently announced development of a high-production autonomous vehicle by 2021 (Ford, 2016).

Despite these advances, there are several challenges that exist in the design and engineering of autonomous systems including: (1) identification of characteristics of autonomous (versus automated) agents; (2) identification of roles, viability, and impact of autonomous (versus automated) agents; (3) identifying constraints on design for autonomy; and (4) definition of approaches for designing autonomous versus automated systems.

CHARACTERISTICS OF AUTONOMOUS AGENTS

Recent research in this area (Bradshaw, Hoffman, Woods, & Johnson, 2013; Johnson et al., 2011; Kaber, 2017) has identified a number of unique characteristics of autonomous agents including: (1) robustness to an environment or "viability";

(2) independence in action/function; and (3) self-determination of goals and resource allocation. By identifying these characteristics, including differentiating viability from independence in function, we are able to define many different agent states, such as survival in an environment, existence with self-governance, self-control, existence with the capability to function independently under certain circumstances, self-governance plus independent function, automatic behavior, and autonomy (see Figure 10.1 from Kaber, 2017). All of these states come with different functional capacities and limitations. For example, an agent that is viable in an environment may not necessary be capable of independent action. Agents that are free from external control are also not necessarily self-governing but require specific "cognitive" capacities, such as learning and strategizing, to achieve goals in order to exhibit self-determination in context.

This framework of autonomy also provides some insight into the relationship between autonomous and automated agents as well as the concept of levels of automation (LOAs). The state of automation, or an extension of agent survival, is achieved through machine or agent programming for forms of automatic behavior in context. Depending upon the degree or level of agent programming for specific task functions, an agent may realize different degrees or levels of automation. However, in order for an agent to be truly autonomous, there is a need for some capability of an automated agent to learn relationships with, and among, other agents in an environment and how to negotiate and acquire resources to achieve goals under such circumstances. Unlike the state of automation with various levels, the state of autonomy is conceptualized as being discrete for a specific context and mission; that is, an agent is viable, independent, and self-governing or not!

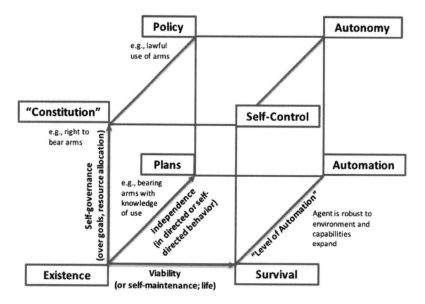

FIGURE 10.1 A conceptual framework of facets of autonomous agents and the relation to levels of automation and automated agents (from Kaber, 2017).

Conceptually speaking, there are no levels of autonomy, but there may be types of autonomy based on the role or mission an agent is to achieve in a particular environment. For example, an autonomous agent may be developed to map the role of piloting a UAV, but this same agent may not have sufficient programming or learning capabilities to fulfill the role of air traffic control for an unmanned systems operating space.

Roles, Viability, and Impact of Automated and Autonomous Agents

Automation provides for a partial mapping of human activity in human-automated systems. Automated agents demand human operation and coordination for task success. Automation also requires human articulation of command input relative to "data entry" capabilities as well as human interpretation of output relative to system "display" capabilities. With respect to operational reliability, automation is assumed to be "imperfect"; that is, functions are limited to specific operating conditions in the absence of disturbances. In the event of off-nominal operating conditions, there must be additional programming of the agent or plans to address such conditions; otherwise, errors and failures are likely. For such reasons, automated agents require human monitoring and supervision. This situation means humans support automation in function with the hope that the technology will lead to greater overall system performance. This expectation, however, typically comes with the cost of alternate forms of workload, such as cognitive load versus visual motor demands.

On the other hand, autonomous agents are typically designed to fulfill human roles in a mission context and to adapt performance to various forms of input as well as presentation of output for human observation and comprehension in the case of a collaborative mission. Unlike automated agents, if there are fluctuations in environmental demands or changes in functional demands, autonomous agents have the capability to adapt to or accommodate these changes, unless the agent has been "misapplied" to a particular domain (i.e., a role for which the agent cannot demonstrate self-direction and independent function). In this case, failure of the autonomy relegates the agent to a form of automation. By fulfilling a human role in a collaborative mission, autonomous agents are intended to support human partners in achieving a broader mission; for example, autonomous agents for search, discovery, and manipulation tasks may support humans in data analysis and inferences on states of remote environments. If an autonomous agent is truly a partner to the human, the agent should not pose additional workload demands for the human in terms of communication, command, and control.

Constraints on Design for Autonomy and Agent Design Processes

Considering this comparison of automated and autonomous agents, and the potential benefits of autonomous systems, there are some steep constraints on the process of designing for agent autonomy. In general, engineers seek to develop autonomous systems for "unknown" and dynamic environments. Unfortunately, the reality of what we often actually achieve in "autonomous" systems is typically limited to application in known and static environments. This situation is due to the fact that technological viability, independence of action, and self-control can only be achieved through design for known states, potential actions (efferent channels), and outcomes under relatively constant conditions. Any design is limited by our knowledge and stability of an application environment. Any design of an autonomous agent must account for context in order to support agent viability. Agent viability can support some system operation even if there is poor design or projection of necessary independent behaviors/functions and needs of self-governance. A key problem here is that agent viability (alone) is not sufficient to support human goals or off-load workload. Consequently, autonomous systems design demands technological advancement and detailed knowledge of mission contexts.

Our conventional concept of design for automated systems is that a human occupies an environment and defines a goal for self-fulfillment. Automation is designed to address tasks in order to support human achievement of the target goal. This concept includes the human as the motivation for the design process and identifies human needs/desires as central design objectives. The reality of design for automated systems is that any limitations or demands of existing technology pose design or configuration requirements for the application environment or training the human user to operate automation to effectively achieve task objectives. Once the type of automated agent is conceptualized, as humans, we begin to look for flexibility in other aspects of the overall system, or ways in which to transform the system (aside from the characteristics of the technology), to support use of the automation. Although all human-machine systems design processes start with the objectives of promoting human performance, reducing workload and supporting situation awareness, my contention is that such processes often degrade to accommodate the limitations of "weak" automated technologies in terms of application to a particular operating context. The upshot of these situations can be technology that: (1) makes work and activity more complicated for humans; (2) increases actual user workload in ensuring appropriate application of the technology to the task; and (3) degrades user situation awareness as a result of the demands the automation places on user perceptual and cognitive resources relative to the actual task at hand.

Opposite to this design scenario, autonomous agent design represents a complete accounting for human needs in terms of situation awareness, workload, and task performance. Based on the definition of autonomy, agents should not require human partner assistance for operation within a defined environment or for delivering specific task functions. Furthermore, autonomous agents should not require human assistance in formulating objectives to facilitate the human partner's goal and in managing resources (e.g., battery life) for viability over the course of a mission. More specifically, autonomous agents should not require human adaptation/accommodation of specific forms of agent input and output capabilities and any agent implementation should support collaboration with a human partner or other autonomous agent acting in other coordinated roles to achieve a common mission. In this way, the agent serves as a resource for the human to achieve broader goals as part of the mission; it reduces overall human work complexity, decreases human workload, and supports human awareness of other aspects of the common mission.

Multiple "Autonomous" Agent Networks

Finally, the recent research on autonomous systems design (Kaber, 2017) has roughly explored the development of multiple agent networks as a basis for achieving autonomous system functionality or operations. For example, in situations in which independent human work may not be possible due to the remote nature of a target task environment, or danger in the environment, mechanized/computerized systems may be developed for teleoperation or robustness to the environment. However, such systems may be weak in terms of the other dimensions of autonomy, including independent function and self-determination. In this situation, designers may enlist a team and/or network of multiple humans to support the "weak" automation through various cognitive and control activities, including information acquisition, information analysis, decision-making, and action implementation. In these scenarios, the automation may provide some benefit of access to uncharted domains from a performance perspective. It may also ensure human safety in the case of dangerous environment exposures. However, the system configuration places demands of the automation technology on humans, increases human workload, and fails to represent a force-multiplier in terms of goal attainment. Unless all agents in a network are of a similar capability level, or bring unique capabilities to an application that can be independently delivered, dependencies develop among agents that most commonly involve the human supporting automation. This situation is akin to the outcomes of the automation design process versus design for autonomous agents.

Summary

In summary, the main contribution of this research is identification of underlying issues in the design of automated systems that are revealed by the definition of autonomous agents and characterization of their capabilities. A comparison of the roles, viability, and impact of various agents in human-defined missions makes clear the demands of automated versus autonomous agents on human performance. In addition, identification of the facets of autonomy reveals the requirements for achieving autonomous systems through design processes.

RECENT ADVANCES IN HUMAN-AUTOMATION INTERACTION MODELING

Given the previously identified issues with the design of autonomous agents, there is a need and interest within the human factors research community to attempt to improve design processes. One way to do this is by developing more accurate and applicable models of human-automation interaction (HAI). Some recent research has been published on HAI models for supporting systems design processes. The collection of methods that have been developed for solving HAI design problems can be considered as a subfield of human factors science. A special issue of the *Journal of Cognitive Engineering and Decision Making* (*JCEDM*, Vol. 12, Issue 1) was recently devoted to this topic and revealed many new and important ideas for how we might approach automation design in the future and promote the effectiveness and efficiency of HAI. Prior research on the definition of levels of human-machine system automation, and the development of taxonomies of LOAs as a basis for conceptual systems design, was reviewed and critiqued.

The LOAs approach basically involves: (1) identifying a set of information processing functions to be addressed by a human-machine system; (2) enumerating the capabilities of the human operator and automation for addressing functions; (3) formulating combinations of human and machine assignment to the functions; and (4) empirically assessing the impact of various function allocation schemes on human and overall system performance. Many examples of the LOAs approach currently exist in the literature (e.g., Endsley and Kaber, 1999; Parasuraman, Sheridan, & Wickens, 2000; Sheridan and Verplank, 1978). The new ideas evolving from the *JCEDM* special issue include one piece that I contributed on developing descriptive accounts of human and system responses to LOAs in order to detail automation frameworks with outcomes that engineers can expect when designing for a specific LOA. This idea was motivated by the Society of Automotive Engineers' (SAE, 2014) recent development of a new taxonomy of automation for automated vehicles with a focus on task-oriented forms of driver assistance mechanisms. I suggested that there is a need to make clear the driving task performance, driver workload, and situation awareness implications of various function allocations as part of the SAE taxonomy under both nominal and off-nominal operating conditions. This and the other ideas presented in the *JCEDM* special issue need to be carefully studied and potentially synergized.

It may be possible to determine an integration of multiple conceptual and detailed automation design approaches to create better human-machine systems. Many of the new approaches to HAI modeling share similar inputs, and a single method may yield outputs that can support other methods applied in a sequential manner during the systems design and engineering cycle. Having said this, a starting point for identification of any such integration of methods is to: (1) review the literature and classify methods in terms of contributions to existing automation design practice and then (2) make classification of methods in terms of applicability to various stages of the systems design and engineering cycle. On this basis, it may be possible to determine an optimal timing with which each HAI design approach can and should be delivered in the cycle in order to promote effective system implementation and application. The following passages present reviews of the new HAI modeling ideas captured in the recent *JCEDM* special issue on the same topic.

Literature Review

In her recent article on automation and "the race" for human factors research to catch up with the pace of development, Burns (2018) questioned whether existing methods of automation design, such as LOAs taxonomies, allow for rapid response to new developments in technology in order to prevent potential accidents. This time-pressure situation may encourage the use of historical automation "look-up" tables, such as Sheridan and Verplank's (1978) and levels of decision automation, as a basis for conceptual system design. However, the situation also emphasizes the need for any design references to reflect current and elaborate ("big") data on real systems. Burns argues that there is a need in the systems design process to conduct "worst-case" scenario analysis for a conservative understanding of the implications of various LOAs on human performance. She also suggests the integration of cognitive work analysis (CWA) with the degrees of automation concept, as developed by Wickens and colleagues (Onnasch, Wickens, Li, and Manzey, 2014; Wickens, 2018), for simultaneously analyzing effects of function allocations on HAI. This combination of techniques may identify where cognitive demands exist and whether and how automation can be applied to address demands. Burns also advises that the application of such approaches should be coupled with more empirical testing on the performance implications of how and when LOAs occur, types of automation interface design, use of LOAs after failures, etc. In general, there is a current need to adapt automation design approaches to the availability of "big data" for promoting the accuracy and timeliness of any projections of human or system performance and safety outcomes. Burns' research extends existing approaches to automation design and may be applicable to system conceptual and detailed design.

In another recent study on HAI modeling, Cummings (2018) also expressed concern with the recent explosion and rapid development of sophisticated autonomous systems and associated risks. She considers historical LOAs look-up tables to be useful for generally categorizing types of systems but to be inexact and coarse as a basis for design. Cummings also contended that there may be confusion on how such general references may be applied in real design practice. On this basis, Cummings developed the skills, rules, knowledge, and expertise (SRKE) framework (as an extension of Rasmussen's [1983] seminal work) to help engineers understand how integrated system functions could be distributed exclusively or collaboratively to human and automated agents and the resulting level of complexity of automation development. In presenting this new approach, Cummings highlights the fact that collaboration of human and automated agents may be critical to ensuring system safety and that there is a need to identify how agents might support one another under off-nominal conditions. This is similar to Burns' (2018) call for collecting data on system performance under automation failures or worst-case automation use scenarios. Cummings also contends that empirically characterizing uncertainty in automated agent behavior under different environmental conditions is critical to effectively determining function allocations. This uncertainty must be accounted for in the automation design process. Cummings' research mainly focuses on how agent design may account for environmental uncertainty. She offers that application of the SRKE approach needs to be coupled with engineering knowledge of technological capabilities to determine what LOAs may be feasible for new systems and operating conditions. This research also represents an extension of existing approaches to automation design and may be applicable to system conceptual and detailed design.

Endsley (2017; 2018) also published recent studies on approaches to design of human and autonomous systems. Endsley (2017) presented a new model of human-autonomy system oversight (HASO), including a comprehensive layout of system, human, and environmental factors that are considered influential in human monitoring and control of automated agents. Based on prior empirical work, LOAs are identified as a predictor of operator workload, degree of system engagement, and situation awareness. Endsley (2018) considers the concept of LOAs to be a fundamental characteristic of human-automation systems that determines demands in monitoring and control. In her research, Endsley summarizes common LOAs effects on human performance when applied to various stages of task performance. Endsley also observes that prior research on LOAs has revealed features of automation that are most relevant to cognitive load and situation awareness (SA). Endsley also raises the automation conundrum that the stronger the form of automation for human support, the more opaque and complex the automation may be for takeover. She emphasizes that this issue is even more urgent today with the explosion of human-autonomous systems. Endsley's new HASO model may be important for providing engineers with a broader sense of human factors and automation design features in system performance outcomes. Endsley's (2017; 2018) research represents both an extension of the LOAs approach to systems design and a new overarching concept for HAI. Endsley's (2017) new HASO model could be used to support the system conceptual design phase.

Johnson, Bradshaw, and Feltovich (2018) have also conducted recent research on identification of the needs of automated systems designers and the development of a design tool that addresses: (1) human and machine capabilities, (2) representation of work processes, and (3) identification of functional and information-sharing relationships that might exist between humans and machines. Johnson et al. suggest that shared (human-machine) control and collaborative control of complex systems are alternative approaches to defining LOAs for systems design. They also contend that any design tool needs to support the development of automation as a "team player" with a focus on agent presentation or interface design. Johnson et al. put forth the approach of coactive design, which is essentially a method for sorting out human-automation collaboration for specific joint activities, that is, how functional collaborations may occur. The approach involves consideration of agent observability, predictability, and directability or the nature of HAI. In the same work, they propose a method for human-agent interdependence analysis to support designers in addressing the question of what functions/tasks should be automated, given agent capabilities. For Johnson et al. the interaction between the human and the machine is a priority in the design process that ultimately leads to a definition of appropriate forms of system automation. However, the current state of technology fundamentally limits the types of interaction that can occur between a human and a machine system. Johnson et al.'s coactive design is a new HAI modeling approach with utility for detailed system design or test and evaluation. It may be an effective follow-on to the application of other approaches, such as designer consideration of possible degrees of system automation in the conceptual design phase. Interdependency analysis may also be revisited at the end of a system life cycle when there are detailed data on patterns of human performance and tool use that could be used as a basis for revisions to legacy system design.

In other recent research on HAI modeling, Kirlik (2018) calls out a need for researchers and designers to account for adaptive human cognition and behavior in HAI analysis, modeling, and design. Kirlik points out that adaptive cognition and behavior, like satisficing, influences situation awareness, judgment, decision-making, and projection and that the design of automation interfaces is a mediating factor in the occurrence of such behavior. For example, if insufficient information is delivered to system users to support adaptive cognition, they may simply not make use of automation for performance. Kirlik offers that adaptive cognition and behavior explains operator use of minimal sets of cues for projection of system states counter to designer expectations, which can result in mismatches of expectations for HAI relative to actual system performance. Given particular task demands, such as diverting attention in the use of automation, interface design that does not support specific adaptive or "rational" behaviors can lead to automation misuse. Kirlik contends that unless automation interfaces are structured to account for biases in judgment and projection as well as satisficing in decision-making, we can continue to expect gaps between HAI model predictions and system outcomes. Kirlik provides a brief review of some empirical work demonstrating this very issue. Kirlik's identification of the need for adaptive behavior analysis in the design and development of automation is an extension of existing automation design methods for human factors and could have application to the operations phase of the systems engineering cycle in order to ensure effective human use of automation technology.

Like Johnson et al. (2018), Lee (2018) has also conducted recent research focusing on the need for greater precision in describing automation and how agents relate to humans. Lee suggested focusing on the question of who or "what is connected" in a system and combining this with other perspectives in automation design. Lee also identifies the general utility of prior LOAs research as a basis for design and applies some of the findings of the prior empirical work to project implications of vehicle design at various levels of the new SAE taxonomy. He suggests that a useful extension of using LOAs look-up tables, like the SAE publication, would be to focus on the capabilities of automation in context. Lee also observes that some of the earlier approaches to identification of LOAs (e.g., Sheridan & Verplank, 1978) led to the interpretation of automation as a single element. These approaches do not account for how networks of humans and automated agents could be modeled. Lee's research demonstrates the importance of contextual considerations in automation design and that a lack of such perspective could lead to errors in system performance predictions and failures to reveal trade-offs in human information processing, depending upon the nature of automation as applied to vehicle subsystems. Lee's network perspective on automated systems design highlights opportunities for how to effectively structure modes of automation of interconnected subsystems interacting with humans, which may not be achieved with historical LOAs look-up tables. Network analysis has the benefit of revealing broader constraints on agent function, given various operating conditions. Lee's research represents an extension of historical LOAs design approaches and provides a new modeling method involving consideration of networks of agents. This new approach has applicability to detailed system design and test and evaluation of complex integrated networks of human and automated agents. In addition, designers may revisit network work analysis for HAI design at the end of a system life cycle (disposal) for characterizing actual patterns of system operation and providing an additional basis for iterations of vehicle automation for new roadway network situations.

Miller (2018) recently presented research on limitations of existing approaches to defining LOAs for human-machine systems and aspects of design that such methods overlook. Miller contended that prior schemes of LOAs had been used for representing entire systems and that this approach may fail to completely represent how humans interact with automation. Miller said that the reduced "vocabulary set" of LOAs taxonomies can lead to gross classification of specific variations on automation function and interface designs and may reduce the accuracy of predicted LOAs implications for specific systems design. He did, however, allow that discrete approaches to systems representation, like the historical taxonomies of LOAs, can sometimes remove distractions from the design process and distill major causal relationships between automation features and human performance. Miller identified the expansion of definitions of LOAs to multiple stages of information processing, and potentially different subsystems, to be an improvement addressed by the Parasuraman et al. (2000) approach. Like Lee's (2018) research in the same volume, Miller advocated for describing LOAs of each and every subsystem in an entire system in order to more precisely identify overall task LOAs and any relationships that might exist among multiple components. This type of approach may allow for identification of common information needs among agents and provide insight into how to design agents that can communicate in very precise ways for achieving shared "mental models" with humans and effective teamwork. Miller's proposal represents an extension to existing approaches to LOAs design and could be useful in the test and evaluation stage of the systems engineering cycle as well as the system operations phase in which performance and agent interaction data are being generated as a basis for determining how shared "mental models" can be achieved through agent information sharing.

Like Miller (2018), Naikar (2018) recently published research with a focus on how to design automation for complex sociotechnical systems that involve multiple humans working with multiple technological agents and for which there is a requirement for shared mental models to achieve effective performance. Naikar identified such systems as being capable of dynamic self-organization under different demand profiles but with organizational structure constrained in part by individual agent capacities. For these types of problems, Naikar recommends the use of work-organization possibility (WOP) diagramming as an extension of the CWA framework to identify organizational possibilities for multiple agents/actors in a system. Fundamentally speaking, this approach involves identifying specific work demands (from among a broad set of information processing functions) that each agent/actor can address. Any flexibility in organization is based on overlap among actors in terms of demand capabilities. This approach is actually akin to use of historical LOAs taxonomies for identifying multiple information processing demands and the possibility of shared agent responsibility. Any variations in WOP diagrams due to contextual factors is similar to identifying various LOAs that might be used for a specific set of agent capabilities and system operating conditions. Naikar contends that HAI design should focus on identifying the limits of each agent/actor capability or "who can do what" in addition to providing constrained operational flexibility in the workplace. Naikar's work is really an extension of the CWA framework as a basis for automation design. The approach may have applicability during system test and evaluation when some information becomes available on exactly how specific agents field various work demands and what forms of system organizational flexibility might be achievable for refining teamwork and actual operations.

Finally, Wickens' (2018) recent research on HAI modeling identifies the origins of historical LOAs look-up tables and how they might be extended and applied in contemporary systems design. Wickens offers that the Parasuraman et al. (2000) model of types and levels of automation emerged from an attempt to apply the Sheridan and Verplank (1978) taxonomy of LOAs (with a focus on decision-making) to the problem of air traffic control system automation design, involving multiple information processing functions, including filtering, integration, decision aiding, and action implementation. Wickens identifies how different automation hierarchies may apply to various stages of information processing. His recent work provides a helpful decomposition of components of the LOAs approach in systems design, including identification of a taxonomy, description of the modeling approach, and identification of an underlying theory. Wickens contends that the LOAs approach may not have been entirely accurate in describing or predicting various system outcomes as a result of some levels of automation not being as critical to human behavior or performance as we might have previously expected. Therefore, certain LOAs could possibly be lumped together to more accurately describe real forms of automation and system performance. Wickens also contends that some deviations between previous HAI model predictions of system performance and actual outcomes might actually be due to poor automation interface design in terms of supporting operator SA. Wickens identifies a need to ensure that the manner of presentation of automation supports good operator SA in order for us to make clear assessments of how various LOAs effect performance. This work clearly represents an extension of prior LOAs taxonomies for automation design and likely has applicability to the conceptual and detailed design phases of the systems engineering cycle.

Summary

Table 10.1 presents a brief summary of the types of methods for HAI modeling and design that have been proposed by the aforementioned researchers, as well as a classification of the various studies according to whether an extension of the historical LOAs approach was proposed or if the research defined a new modeling approach based on another type of earlier method, or if the study defines an entirely new method. The majority of studies suggest some extension of the LOA approach or define new extensions of other types of existing models. In general, the review of the current state of HAI modeling research reveals that many researchers perceive a need for advancing the state of design methods. There are also new ideas on how to account for various factors influencing human performance with automation, including adaptive human cognition and behavior, networks of agents, dynamic work demands, etc.

Organizing Design Methods According to the System Life Cycle

As I mentioned previously, with this review and classification of contribution of various ideas, it may be possible to come to some schedule of methods for application during the system design and life cycle. I identified some phases of the life cycle to which specific methods may have applicability. There are many different

TABLE 10.1
Summary of HAI Modeling Methods Suggested by Contemporary Research with General Classification of Contribution

Authors	Method Descriptor	Extension of LOA	New Approach (within Existing Type of Model)	New Modeling Approach
1. Burns (2018)	Identifies a novel integration of CWA with degrees of automation analysis.	X	X (CWA)	
2. Cummings (2018)	Extends Rasmussen's (1983) SRK model to consider human expertise and uncertainty in agent and environment states.		X (SRK)	
3. Endsley (2018)	New HASO model revealing range of automation design factors and identifying LOAs as a variable in monitoring and control.	X		X (HASO model)
4. Johnson et al. (2018)	Proposes coactive design for defining automation based on interaction with human as well as interdependence analysis for what "should" be automated.	X		X (Coactive design)
5. Kirlik (2018)	Classification of satisficing as one form of adaptive cognition and behavior; identification of other adaptive behaviors and need to address in HAI models.	X	X (Adaptive behavior analysis)	
6. Lee (2018)	Network analysis as a basis for defining the HAI context and constraints on human and automation behavior; not viewing automation as single independent element.	X		X (Network analysis)
7. Miller (2018)	Discretization of complex HAI problems can distill some causal relationships but also leads to missing details for design of interaction; need to characterize shared agent mental models for negotiating task performance.		X (Using mental models and common ground for task negotiation)	

TABLE 10.1 (*Continued*)
Summary of HAI Modeling Methods Suggested by Contemporary Research with General Classification of Contribution

Authors	Method Descriptor	Extension of LOA	New Approach (within Existing Type of Model)	New Modeling Approach
8. Naikar (2018)	Work organization process diagramming to identify specific limits of agent capability and exploit operational flexibility (classification assumes LOAs are not prescriptive).	X	X (Extension of CWA)	
9. Wickens (2018)	Degrees of automation feature emerging for application of stages and levels of automation model; simplification and elaboration of model to account for discontinuities in human responses to LOAs.	X		

conceptualizations of the system life cycle; however, the ISO 9000 (2015) quality management standard presents one of the most commonly referenced cycles in engineering practice. The life cycle spans from conceptual design of the system to detailed design (requiring identification tasks and system functions), development, test and verification, production, operation, and disposal (see Figure 10.2). The cycle

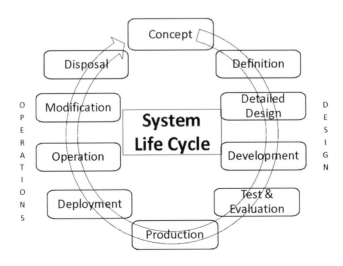

FIGURE 10.2 Design and operations system life cycle (similar to ISO 9000 standard content).

is divided into the stages of design and operations. In general, any approach to injecting HAI knowledge in the system life cycle should be both proactive and systematic in nature. That is, HAI considerations (and potential outcomes of various LOA) should be made in the concept and design phases in order to potentially mitigate the need for retrofitting systems or relying on operating procedures and training (administrative controls) in an attempt to account for "bad" design. Furthermore, continued consideration of HAI knowledge should be made throughout the life cycle, including during test and evaluation as well as system production and deployment, in order to refine detailed design and ensure optimal agent collaboration under both nominal and off-nominal operating circumstances.

In terms of organizing specific HAI modeling methods by stages and phases of the life cycle, given that there are so many different system life cycle representations in the engineering literature, in Figure 10.3, I present a basic "timeline," including system concept, design, development (test and evaluation), operation, and disposal. The key input and outputs to the timeline are also identified, including human "generation of a system idea" to human "recording and documenting performance outcomes and lessons learned." Of course, this timeline represents an iterative process for most complex human-automation systems with legacy and revision. Figure 10.2 identifies those phases at which specific HAI modeling approaches may be most impactful and effective in terms of system outcomes. Here, it is important to note that some of the methods I have included in the timeline (based on the literature review) may still be undergoing development, and the timeline certainly does not capture all existing HAI design and modeling

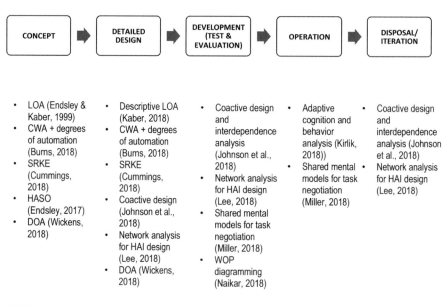

FIGURE 10.3 New HAI modeling methods organized by applicability to phases of system life cycle.

tools. The diagram is intended as an example of how the aforementioned reviewed methods might be organized.

System Concept. In this phase, the original LOAs method (e.g., Endsley & Kaber, 1999) may be useful for identifying potential function allocation approaches and projecting associated system outcomes. The CWA and degrees of automation approach, conceptualized by Burns (2018), may be possible if sufficient information on standard operating procedures is available for analysis. This approach can also be carried forward into the detailed design phase when more information on the system and "means-end" connections (goals to functions/components) may exist. Cummings' (2018) SRKE model may also be appropriate for the conceptual design phase when engineers are considering whether to develop a partially or fully automated system. Endsley's (2017) HASO method also has utility at the concept phase for providing engineers with the "big picture" of all human, task, system, and environment factors that should be considered in designing automated systems. Lastly, Wickens' (2018) degrees of automation (DOA) approach can be used in conceptual design to facilitate designer function allocation decisions relative to the potential for automation failures and human out-of-the-loop performance issues.

System Design. In the detailed design phase, the new descriptive LOA approach (Kaber, 2018) may have utility, provided that detailed system outcomes are identified for each LOAs based on system observations. Cummings' (2018) SRKE model could also be extended to this phase, provided that detailed information is available on system sensor and actuator capabilities with the objective of highlighting limitations of technology for addressing "rule" and "knowledge" based functions. Johnson et al.'s (2018) coactive design method may also be applicable here provided that detailed information on automated agent capabilities as well as the proposed system workflow is available. Lee's (2019) network analysis may also be applicable at this stage when more information may be available on the range of operating environment conditions and the range of agents competing for environmental resources, as well as the interconnections among those agents to ensure safety and performance. Wickens' (2018) DOA approach could also be applied in this phase to more specifically identify when a designer should consider alternate functions allocation schemes to account for failure performance as well as the stages and levels at which automation should be applied.

Development (Test and Evaluation). In this phase, coactive design (Johnson et al., 2018) could be further applied to fine-tune HAI through interactions with system prototypes. The approach could also be extended with applied interdependence analysis for verifying human-machine teaming or automation function as a "team player." It is also possible that Lee's (2018) network analysis for HAI design could be applied in this phase based on data from prototype testing and assessments of interactions of agents within the target task environment. Beyond this, Miller's (2018) concept of agent sharing of mental models as a basis for negotiation and allocation of tasks in the work mix could possibly be applied as well. Furthermore, Naikar's (2018) suggestion for use of WOP diagrams to identify work demands fielded by various actors and the potential for constrained system organizational flexibility relative to demand states could be applied at this phase, when a prototype system exists and agent adaptive behaviors may be observable.

Operation. Once the system is in operation, there is a need for verification of the accuracy of HAI model predictions of human use, misuse, or disuse of automation based on model accounting for forms of behavior. This phase is likely where Kirlik's (2018) concept of interface design for adaptive cognition and behavior needs to be assessed. Operators often develop heuristics during actual system use. Such patterns of adaptive behavior need to be observed and addressed through interface design revisions as well as refinements of models used as a basis for original design. Miller's (2018) concept of agent communication and sharing of mental models would likely extend to this phase wherein real-time task/function allocation negotiations might occur based on operational experiences.

Disposal and Iteration. In disposal of the technology (and preparation for iteration of the design), a return to Johnson et al.'s (2018) coactive design approach and interdependence analysis may be helpful. Detailed operational data on patterns of user task performance (prioritization, sequencing) and tool use can be used as inputs to refine concepts of how HAI should occur in revisions of a legacy system. Lee's (2018) network analysis for HAI design may also have utility at this stage for re-characterizing the system's operating environment as well as multi-agent resource competition (e.g., a driving environment with escalating automation and physical parameterization). These demands could then be translated to redefine human driver and automated assistance system relationships in vehicle iterations for new roadway "network" situations.

CURRENT RESEARCH NEEDS

The majority of this writing on new approaches to HAI is largely targeted at promoting effective human and machine interaction in terms of situation awareness and task performance. Little work on conceptual approaches to HAI design has been motivated by systems' safety considerations. It is important for us to consider what "the march" of advanced automation for road vehicles means for driver safety. The intent for advanced automation in the aviation domain was generally twofold, including: (1) greater trajectory control and efficiency in fuel use and travel times and (2) greater safety by reducing the potential for pilot errors influencing flight operations. The general concept that many advanced automated automobile manufacturers have is that increasing levels of automation in driving tasks and/or vehicle control functions will lead to greater driver and vehicle safety. This concept is generally achieved by taking inventory of demand characteristics of various driving scenarios and potential hazard exposures and developing automated vehicle control technology to account for the elements and dynamics of such events.

To some extent, automotive manufacturers have been successful in developing automated technologies for longitudinal and lateral vehicle control as well as collision avoidance. These technologies include adaptive cruise control (ACC) systems, lane-keeping systems (LKSs), and emergency braking systems (EBSs). As a simplistic perspective, the majority of these systems integrate some form of radar-sensing technology with onboard computing hardware to identify vehicle-to-object distance threshold violations. These systems also integrate vehicle controls, including accelerator, brake, and steering, in order to modify performance parameters and maintain safe driving conditions. ACC systems are based on constant time-headway logic; that is, the vehicle maintains a stable headway distance to a lead vehicle. It is important to note that for some systems, the threshold for triggering changes in vehicle performance are set by the driver (e.g., two to six car lengths). Some systems, such as the EBS, also take into account time-to-contact or time-to-collision as a basis for triggering vehicle control functions (i.e., a vehicle-to-object distance is considered critical and ownship is accelerating). The functionality of these systems is primarily dependent upon the object detection algorithms, which are often based on machine learning technology (reinforcement learning) with capability and accuracy improving over time. These systems may be introduced into vehicles for independent operation or in conjunction with one another creating integrated capabilities, such as "autopilots."

The overall objective of the ACC, LKS, and EBS technologies is to decrease the likelihood or occurrence of crashes. This objective is critically dependent on the capabilities of manufacturers to identify complex road conditions and driving scenarios that may lead to hazard exposures as well as the possibility for crashes. Similarly, there is a need for automation of evasive maneuvering in order to prevent crashes. As with any automation technology, ACC and EBS, for example, must start with simple input conditions, such as high-speed braking by a vehicle directly in the lane in front of ownship. For this circumstance, the ACC may trigger emergency braking in order to prevent a collision. However, other more complex roadway conditions may develop, such as cross traffic illegally entering an intersection directly in front of ownship, an event for which EBS sensors may not be capable of detecting the offending vehicle in time to effectively trigger the automated emergency braking response. For such a complex hazard exposure, there is a need for further development of advanced vehicle automation systems in order to ensure driver safety.

The evolution of vehicle automation from an accounting for frequent but less severe hazard exposures to addressing low frequency but high severity hazards creates a situation in which drivers rely on the technology to manage minor infractions of their vehicle trajectory but they remain responsible, at a moment's notice, for highly complex and often high criticality events. Unfortunately, driver reliance on automation for simple hazard avoidance actually limits regular manual skill use and may undermine driver capabilities to address less frequent but more critical hazard exposures when they do occur. This is the paradox of the design and incremental development of advanced vehicle automation for safety. As manufacturers target simpler roadway hazards through automation design, and drivers exploit such early technology, this progression of events degrades driver manual take-over skills in the event of automation failure and may increase the likelihood of driver failures to effectively negotiate complex roadway hazards and avoid crashes.

Let us assume that there is some distribution of roadway hazards (see Figure 10.4) in terms of the complexity of vehicle automation or driver control responses. Let us also assume that the distribution is skewed with the likelihood of lower complexity events (the left side of the distribution in Figure 10.4) being more frequent than the higher complexity events (the right side of the distribution). Furthermore, let us

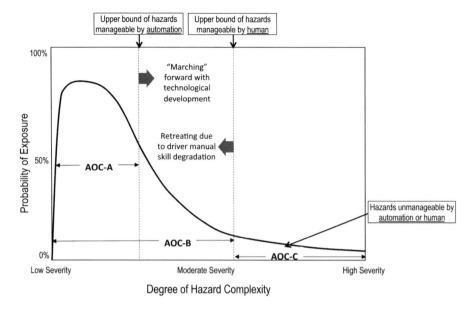

FIGURE 10.4 The paradox of automation for vehicle safety.

assume that the distribution is "heavy-tailed" with some rather extremely complex roadway hazards but with low probabilities of occurrence. The set of hazards to the left of the distribution (area under the curve (AOC)-A) can be considered those that are the target of early stage vehicle automation (independent operation of ACC, LKS, or EBS [SAE Level 1]), which can be used to manage vehicle longitudinal and lateral position. The hazards to the far right of the distribution (AOC-C) can be considered as those that currently cannot be managed by either vehicle automation technology or the human driver in take-over scenarios. AOC-B is the set of hazards for which drivers can effectively deliver manual take-over skills. AOC-B overlaps AOC-A and "contacts" the lower bound of AOC-C. The region above the upper bound of AOC-A and below the lower bound of AOC-C represents a set of hazards that extends beyond the design capabilities of existing vehicle automation and is the design target for the next generation of automation.

Having defined these AOCs of roadway hazards, the paradox of advanced vehicle automaton occurs when driver skills for addressing complex hazards (AOC-B) degrade faster than the rate at which new automation is developed for addressing the same set of hazards (AOC-B). That is, the upper bound of AOC-B (the set of hazards manageable by humans) moves leftward along the distribution of roadway hazards at a faster rate than the upper bound of AOC-A (the set of hazards manageable by current automation). The paradox may not necessarily lead to a greater total incident rate for personal vehicles but may lead to a higher severity or loss level associated with vehicular accidents during the period of development of automation.

This paradox of automation for safety in the rapidly developing technology of automated vehicles is nothing new relative to the same paradox that historically emerged in advanced aircraft development and expert human use. The key difference in the development of new "autonomous" vehicles is that we are now in a position to apply the many lessons learned from aviation and more mature industries. This knowledge should allow us to better deliver design processes to prevent major automation "pitfalls" that the human factors community has worked so hard to identify and characterize.

Implications. One approach to preventing the paradox of automation for safety is to return to the novel schedule of application of HAI modeling and design methods throughout the system's life cycle. In order to promote the capability of automaton for dealing with more complex roadway hazards, and supporting human driver manual takeover under circumstances that extend beyond automation capabilities, there is a need to ensure the decision-making framework of a driver is adequately captured in the automated vehicle system design. This problem may be addressed by applying Burns' (2018) approach of using CWA for characterizing work demands in specific driving situations as well as driver decision response strategies. Coupling this insight with the concept of DOAs (Wickens, 2018) may support design of types and stages of automation that better reflect human decision-making in vehicle control and promote compatibility with driver mental models. As Miller (2018) discussed, in such complex systems, the sharing of mental models among agents and shared information is critical to system performance. This approach may generally serve to accelerate the rate at which advanced vehicle automation accounts for a greater subset of those roadway hazards that are currently manageable only by human drivers (AOC-B).

Beyond accounting for driver decision-making strategies, advanced vehicle automation also needs to be designed around specific driver behaviors occurring under particular roadway conditions. As Kirlik (2018) noted, humans exhibit adaptive cognition and behavior, and "shortcuts," like satisficing, apply not only in decision-making but perception, judgment, and achieving SA. Driver behaviors depend on the value that is placed on particular events (e.g., crashes) and actions (e.g., acceleration, deceleration) from moment to moment. Drivers make "trade-offs" among control actions in terms of safety and efficiency outcomes. There is a need to ensure that the design of automation presentation (e.g., in-vehicle interfaces) accounts for driver adaptive cognitive behaviors and addresses SA needs when posed with critical trade-off scenarios. This approach may serve to reduce the negative effects of driver manual skill degradations due to low-level automation use over time and promote readiness for effective takeover under demanding hazard conditions (AOC-B).

Finally, in designing advanced vehicle technology, there is also a need for automated or autonomous agents to account for specific environmental circumstances, including the presence of other autonomous or automated vehicles that might serve as information sources for ownship or, otherwise, compete with ownship for roadway "resources" (e.g., lane, position, etc.). The application of Lee's (2018) network analysis approach to this problem may allow for a broader accounting of constraints or "crutches" to agent function, which are not addressed by existing LOAs design approaches. Lee's (2018) method can support effective structuring of modes of automation of interconnected systems in order to allow for humans and machines to address the current set of roadway hazards that are currently unmanageable by both automation and the driver (AOC-C).

REFERENCES

Bradshaw, J. M., Hoffman, R. R., Woods, D. D., and Johnson, M. (2013) The Seven Deadly Myths of "Autonomous Systems." *IEEE Intelligent Systems*, 28(3), 54–61.

Burns, C. (2018). Automation and the human factors race to catch up. *Journal of Cognitive Engineering & Decision Making*, 12(1), 83–85.

Cummings, M. (2018). Informing autonomous system design through the lens of skill-, rule- and knowledge-based behaviors. *Journal of Cognitive Engineering & Decision Making*, 12(1), 58–61.

Endsley, M. R. (2017). From here to autonomy. *Human Factors*, 59(1), 5–27.

Endsley, M. R. (2018). Level of automation forms a key aspect of autonomy design. *Journal of Cognitive Engineering & Decision Making*, 12(1), 29–34.

Endsley, M. R., and Kaber, D. B. (1999). Level of automation effects on performance, situation awareness and workload in a dynamic control task. *Ergonomics*, 42(3), 462–492.

Ford Motor Company. (2016). Ford targets fully autonomous vehicle for ride sharing in 2021; invests in new tech companies, doubles Silicon Valley team. Retrieved June 13, 2019 from: https://media.ford.com/content/fordmedia/fna/us/en/news/2016/08/16/ford-targets-fully-autonomous-vehicle-for-ride-sharing-in-2021.html

Gent, E. (2015). The future of drones: Uncertain, promising and pretty awesome. LiveScience. Retrieved June 13, 2019 from: http://www.livescience.com/52701-future-of-drones-uncertain-but-promising.html

International Organization for Standardization (ISO; 2015). ISO 9000 - Quality management systems: Fundamentals and vocabulary (ISO 9000:2015). Geneva: ISO.

Johnson, M., Bradshaw, J. & Feltovich, P. (2018). Tomorrow's human-machine design tools: From levels of automation to interdependencies. *Journal of Cognitive Engineering & Decision Making*, 12(1), 77–82.

Johnson, M., Bradshaw, J. M., Feltovich, P. J., Hoffman, R. R., Jonker, C., van Riemsdijk, B., and Sierhuis, M. (2011). Beyond cooperative robotics: The central role of interdependence in coactive design. *IEEE Intelligent Systems*, 26(3), 81–88.

Kaber, D. B. (2017). A conceptual framework of autonomous and automated agents. *Theo. Issues in Ergo. Sci.*, 19(4), 406–430.

Kaber, D. B. (2018). Issues in human-automation interaction modeling: Presumptive aspects of frameworks of types and levels of automation. *Journal of Cognitive Engineering & Decision Making*, 12(1), 7–24.

Kirlik, A. (2018). Automation and adaptive behavior. *Journal of Cognitive Engineering & Decision Making*, 12(1), 70–73.

Lee, J. (2018). Perspectives on automotive automation and autonomy. *Journal of Cognitive Engineering & Decision Making*, 12(1), 53–57.

Miller, C. (2018). The risks of discretization: What is lost in (even good) "levels of automation" schemes. *Journal of Cognitive Engineering & Decision Making*, 12(1), 74–76.

Naikar, N. (2018). Human-automation interaction in self-organizing sociotechnical systems. *Journal of Cognitive Engineering & Decision Making*, 12(1), 62–66.

NASA Autonomous Systems Capability Leadership Team (NASA ASCLT; April, 2018). *Autonomous Systems Taxonomy* (Tech. Doc. 2018-04-26). NASA Ames, Moffet Field, CA: NASA ASCLT.

Onnasch, L., Wickens, C. D., Li, H., & Manzey, D. (2014). Human performance consequences of stages and levels of automation: An integrated meta-analysis. *Human Factors*, 56(3), 476–488.

Parasuraman, R., Sheridan, T. B. & Wickens, C. D. (2000). A model for types and levels of human interaction with automation. *IEEE Transactions on Systems, Man, and Cybernetics*, Part A, 30(3), 286–297.

Rasmussen, J. (1983). Skills, rules, and knowledge; signals, signs, and symbols, and other distinctions in human performance models. *IEEE Transactions on Systems, Man and Cybernetics*, SMC-13(3), 257–266.

SAE On-Road Automated Vehicle Standards Committee (2014). Taxonomy and definitions for terms related to on-road motor vehicle automated driving systems. SAE Standard J3016, 01–16.

Sheridan, T. B., & Verplank, W. L. (1978). *Human and Computer Control of Undersea Teleoperators* (Technical Report). Cambridge, MA: MIT Man-Machine Laboratory.

Wickens, C. D. (2018). Automation stages and levels: 20 years after. *Journal of Cognitive Engineering & Decision Making*, 12(1), 35–41.

11 Workload and Attention Management in Automated Vehicles

Joonbum Lee, Vindhya Venkatraman, John L. Campbell, & Christian M. Richard

INTRODUCTION

Over the past decade, driver assistance technologies have evolved from merely assisting the driver in specific situations (e.g., cruise control systems) to replacing the driver's vehicle control in limited situations. Recent demonstrations of automated vehicles by automotive companies, technology developers, and ride-sharing services reinforce the notion that even limited automation in the vehicle can provide benefits compared to fully manual driving. These benefits include increased access to employment and greater levels of independence for millions of people, including those that are disadvantaged due to age- and disability-related declines (National Highway Traffic Safety Administration, 2017). Vehicle automation can also reduce traffic congestion since these vehicles can follow others more closely, leading to smoother traffic flow, lower emissions, and optimized fuel use (Fagnant & Kockelman, 2015). Most importantly, transitioning from a manual to an automated vehicle fleet will substantially improve overall traffic safety by reducing the involvement of fallible human drivers in the driving task.

Changes in the nature and timing of driver workload are at the core of automation's promise of increased safety. Automation shifts the primary responsibility of controlling the vehicle away from the driver to a complex system of sensors, algorithms, and vehicle controllers. This shift reflects some important road safety realities. Drivers contribute to almost 94% of crashes (National Highway Traffic Safety Administration, 2017) and, while most crashes are the result of multiple contributing factors, driver behaviors in the form of speeding, inattention, distraction, and impaired driving play a significant role in the crash statistics. The logic of automation is that by shifting tasks and workload away from drivers, these unsafe behaviors can be eliminated or rendered irrelevant. The introduction of vehicle automation fundamentally changes the nature of driving by reducing or eliminating the moment-to-moment task demands and workload that driving imposes on individuals (Bainbridge, 1983). However, if drivers are not engaged with the driving task, they may engage in other activities (e.g., reading, texting, watching videos, or even sleeping), which can lead to difficulties when they are required to take back control of the vehicle, especially in time-critical situations.

This chapter examines driver workload during vehicle automation and the management of drivers' attention and awareness of the situation when automation fails. The following discussion describes the sources of workload associated with manual driving and how different levels of automation can impact driver workload. We describe specific challenges associated with potential driver underload and overload conditions, including take-over (or handoff) situations where drivers' workload and attention levels can impact their ability to safely and quickly reengage in the driving task. Finally, we present future opportunities for managing attention and workload through innovative approaches to the conceptualization and design of the human-machine interface (HMI).

BACKGROUND: TASK DEMANDS IN MANUAL DRIVING

Manual driving is a multifaceted activity that consists of many subtasks, including the visual scanning of the environment, identification of potential hazards, speed control, and the control of lateral/longitudinal position with respect to the roadway and relative speeds of other vehicles (Senders, 2017; Wierwille, 1993; Young, Regan, & Hammer, 2007). The driver's vehicle control actions are predominantly responses to the demands of the roadway; thus, the driver must perceptually sample the environment for critical control-related cues (Gibson & Crooks, 1938; Morgan & Hancock, 2009). Drivers are continuously performing these tasks concurrently, and some of them are more effortful than others.

Hale et al. (1990) mapped Rasmussen's three levels of performance (Rasmussen, 1987) to Michon's three levels of driving task classification (Michon, 1985) to characterize driving tasks based on their complexity and cognitive requirements (Table 11.1). According to this framework, tasks primarily vary along two key dimensions. The first dimension reflects the complexity of the tasks and the degree of thinking

TABLE 11.1
Framework of Driving Tasks Based on Task Complexity and Cognitive Requirements (Adapted from Hale et al., 1990)

		Task Complexity			
		Strategic/Navigational level (Planning)	Tactical/Maneuvering level (Normal driving actions)	Operational/Control level (Vehicle control)	
Cognitive Requirements	Knowledge-based	Navigating in an unfamiliar town	Driving through a fog bank	Student driver	Attention Demanding & Effortful ↕ Fast & Effortless Performance
	Rule-based	Estimating travel times	Speed selection around a school zone	Test driving a new car	
	Skill-based	Traveling between work and home	Negotiating freeway interchanges	Driving through a work zone	

(Example tasks are provided for each combination of these dimensions.)

required to conduct them. These tasks are generally divided into strategic, tactical, and operational tasks, which correspond to planning, normal driving, and vehicle control tasks. The second dimension describes the nature of the cognitive requirements. These tasks are categorized as knowledge-based, rule-based, and skill-based. Knowledge-based tasks are relatively slow, cognitively effortful, and require attention to conduct successfully. Rule-based tasks require a moderate degree of mental effort, whereas skill-based tasks are typically highly practiced actions that can be performed fast, effortlessly, and without deliberately attending to them. The particular combination of tasks that drivers must perform in a specific situation imposes mental workload on a driver based on how much attention and cognitive effort the tasks require. If the demands of the driving tasks exceed what individuals can manage, it can lead to driver errors in decision-making and task execution. One of the promising aspects of vehicle automation is that it will eliminate driver errors caused by task overload in many situations; however, as will be discussed later in this chapter, automation may also lead to unsafe driver behaviors when they are required to quickly transition to manual vehicle control.

WORKLOAD IN AUTOMATED VEHICLES

Driver Workload in Manual Driving and Vehicle Automation

Driver workload is directly related to the basic demands of the driving task discussed earlier; it is a psychological concept that represents the proportion or amount of a driver's mental and physical capacity (i.e., perceptual, cognitive, and psychomotor) that is used to complete the driving task (Campbell et al., 2016). Workload is complex and has been conceptualized in a variety of ways—as the time demands of a task, the number of competing activities, or the complexity of activities. Primary driving tasks, such as controlling the vehicle, scanning for hazards, navigating, etc., impose workload on the driver. Workload increases or decreases based on the driving conditions (e.g., roadway complexity, weather, traffic flow, etc.) or driver state (e.g., fatigued, alert, etc.). Importantly, task demands, driver characteristics, and situational factors will influence a driver's perception of the moment-to-moment demands of the driving task. This perception affects how drivers develop and implement strategies for coping with the demands based on their own capabilities and available resources. Note that some of this driver coping may happen unconsciously, particularly with highly practiced actions, such as speed decisions or lane maintenance. These types of "automatic" behaviors typically require less attention or deliberation on the driver's part (Shinar, Meir, & Ben-Shoham, 1998), but workload is always present to some degree during manual driving.

Unlike manual driving, vehicle automation disrupts the nature and allocation of driving-related workload between the driver and automation. Current implementations of automated vehicles span a general range of capabilities. The Society of Automotive Engineers (SAE) describes six levels of driving automation systems based on the allocation of driving subtasks (SAE International, 2014). Level 0 automation specifies manual driving, though emergency collision avoidance systems may be present. Level 1 vehicles automate *either* lateral or longitudinal control and the driver performs the other control function, for example, when the driver steers, and the adaptive cruise control (ACC) is activated for speed control. The driver is relieved of the psychomotor work of operating the pedals, while still engaged in the steering task. Level 2 vehicles present an increase in automation capabilities and a dramatic reduction in driver workload as both lateral and longitudinal controls are automated. During normal operation of a Level 2 vehicle, drivers are still held responsible for perceptual and cognitive (monitoring) functions; however, they are relieved of psychomotor (operational) actions. A Level 3 vehicle is similar to Level 2, except that the automation is responsible for monitoring the driving environment, and drivers are given advance notice of the need for interventions. At Level 3, the need for continuous perceptual and cognitive engagement in the driving task is further reduced compared to Level 2 vehicles. Level 4 and Level 5 vehicles are highly automated, with Level 4 vehicles being restricted to a subset of driving situations. Generally, occupants of these vehicles are treated as "users" or "passengers" and are not expected to assume the responsibilities of a driver; Level 4 vehicles are expected to be capable of performing a safe stop in case of emergencies. Level 5 vehicles are fully automated with all occupants considered as passengers throughout a trip, and the role of the driver is eliminated. Driver workload considerations do not generally apply to Levels 4 and 5 vehicles.

Hence, an increase in the level of driving automation specifies an increase in the automation of vehicle control and can correspond to a decrease in driver engagement with the driving task and lower overall workload. However, the management of driver workload and attention is a concern in vehicles with partial/conditional automation (SAE Levels 2 and 3 vehicles). The safety benefits of these vehicles are realized only through appropriate interactions between the driver and automation. This interdependency poses human factors challenges; drivers will likely adapt to and interact with automation in ways that are unintended by system designers. Such interactions introduce uncertainties in the moment-to-moment workload experienced by drivers in partially/conditionally automated vehicles.

Driver Workload in Imperfect Automation

Despite the purpose and potential benefits of automation, automated vehicles can exacerbate existing errors or cause different types of errors than those caused by drivers for several reasons. First, sensor technology can be unreliable and impediments such as object recognition/classification and all-weather performance is often in the way of robust and reliable performance. Recent fatal crashes of Tesla Model X and S vehicles (e.g., Allen, 2018; Valdes-Dapena, 2018) demonstrated imperfections of sensors and algorithms in the automation. Second, from a human factors perspective, the design of the interdependent driving performance (e.g., seamless transfers of control and the HMI interactions) between the driver and vehicle automation determines overall safety outcomes (Endsley, 2018; Lee, 2018). Finally—and more relevant to our purposes here—a specific barrier to interdependent performance is the effect of the automation on drivers' workload. As we describe in more detail to follow, partial/conditional automation can introduce issues related to both "underload" and "overload" conditions within the same take-over scenario. Critically, the design and implementation of vehicle automation and the HMI will impact driver workload, performance, and safety.

UNDERLOAD DRIVEN BY THE AUTOMATION

As discussed previously, vehicle automation will reduce the tasks that drivers must perform, and the corresponding reduction in workload can have beneficial effects, especially in challenging, complex driving scenarios. However, under normal driving conditions, underload can also occur due to the reduction or elimination of manual tasks (including sensing the environment, making decisions, and controlling the vehicle). Reduced workload due to automation can lead to a loss of situation awareness if the driver is disengaged from driving (Lee, Wickens, Liu, & Boyle, 2017). In situations of normal operation, as automation level increases, both workload and situation awareness decreases (Endsley & Kiris, 1995).

Safety in a Level 2 automated vehicle depends on the driver's monitoring performance and their ability to quickly reengage with the driving task. In this respect, the reduced workload may encourage driver inattention to the driving task. In the underload condition, the main problem is not only that drivers can be disengaged from driving tasks, but also that they can actively engage in other nondriving tasks. This pattern is commonly observed and even expected; even under fully manual driving conditions, drivers use mobile phones and tablet computers, interact with in-vehicle infotainment systems, and chat with passengers. It is also consistent with trends toward automation that we have seen for decades. For example, automatic transmission and power steering reduces a driver's physical workload (and cognitive workload, especially for novice drivers) but provides additional resources for multitasking, such as using a mobile phone with one hand while driving. Despite their source, the key consequences of underload occur when drivers disengage with all or some of the ongoing demands of fully manual driving. This can cause them to lose track of the important visual cues that provide information on what is happening around them, thus diminishing their situation awareness.

Potential underload problems in automated driving can vary in terms of severity of outcomes compared to manual driving. An example is the self-driving Uber crash in Tempe, Arizona, in 2018. Tempe police reported that, "The vehicle was in motion for 21 minutes, 48 seconds. Of that time, the total amount of time the driver's eyes were averted from the roadway was 6 minutes, 47.2 seconds, or approximately 32% of the time" (Buono, 2018). Also a glance region classification algorithm was applied to the driver's face video before the accident, and it estimated glance location and duration (Fridman, 2018). The results showed that the Uber driver glanced at her lap/knee area multiple times, and the duration of the off-road glances ranged between 3 to 5 seconds (and sometimes exceeded 5 seconds). Driver distraction is not a novel safety issue; however, these glances are much longer than what is considered unsafe in manual driving (e.g., over 1.7 to 2.0 seconds; Liang, Lee, & Horrey, 2014; Victor et al., 2015). This suggests that a vehicle with partial/conditional automation can contribute to specific types of underload associated with low levels of engagement with the driving environment.

Overload Driven by Transfer of Control

The driver workload and related safety concerns introduced by automation include overload as well as underload, though these states are really two different sides of the same coin. According to SAE's Levels 2 and 3 automation descriptions, the driver must be prepared to step in and resume control and/or respond as needed. Examples of "take-over" scenarios could include situations where lane markings are undetectable by the system, the vehicle is moving outside a specific geo-fenced area that defines its zone of allowable operation in automated mode, or the system detects some other "out-of-boundary" conditions (e.g., Gold, Naujoks, Radlmayr, Bellem, & Jarosch, 2017). Levels of workload associated with such take-over scenarios will clearly vary. Resuming vehicle control after a period of disengagement can be as easy and simple as starting driving after parking (e.g., van der Meulen, Kun, & Janssen, 2016). However, if the take-over situation requires prompt reaction from the driver due to an imminent conflict, and if the required vehicle control and the driver's intervention(s) requires a long time to recover situation awareness, the workload can be high and exceed the driver's capacity to execute the necessary response. The timeline workload analysis (Parks & Boucek, 1989) estimates workload as the ratio of the "time required" for task completion to the "time available," and can be applied as a parsimonious estimate of driver workload in take-over scenarios.

Specifically, a drivers' readiness to quickly respond to high-workload situations will be compromised as they shift their attention to secondary or nondriving tasks. In this case, drivers are not updating their understanding of the driving situation as frequently (or at all), and they may become oblivious to information that becomes important for performing a takeover. When the actual take-over request occurs, the driver has to catch up by quickly scanning and processing the driving scene to regain their understanding of the situation and key information elements so that they can make decisions and take actions that are sufficiently safe. Trying to do all of this within a short time budget (i.e., period from when the automation requests take-over to when the automation predicts it will reach its own performance limit; see Figure 11.1) will lead to elevated workload—sometimes to the point where drivers are unable to sufficiently regain situation awareness and perform safe vehicle control actions within the time budget.

With imperfect automation, the potential for sudden take-over situations undermines the benefits of lower workload because the reduced workload can make recovery of situation awareness challenging when the driver must intervene. Figure 11.1 illustrates a hypothetical driver's attention and demand level under partial/conditional automation and manual driving in a take-over scenario. The automation provides a "benefit" by reducing workload compared to manual driving, and this may encourage drivers to disengage from the driving task and engage in

Workload and Attention Management

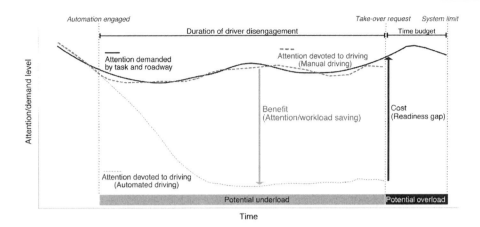

FIGURE 11.1 Hypothetical driver's attention and demand level under partial/conditional automation and manual driving.

activities unrelated to driving. Automation would also reduce the likelihood that drivers will make safety-related driving errors. However, when the driver has to take back vehicle control, this benefit can actually become a "cost." This is because a disengaged driver in the automated vehicle may have a poor understanding of what has happened, what is happening, and what needs to happen in order to maintain safety. This impoverished situation awareness before/during takeover can lead to high workload if drivers must obtain and interpret a lot of information within a short time budget.

The "gap" between the attention demanded by the take-over situation and the attention devoted by the driver during the period of automation engagement represents a deficit in the driver's readiness to respond (Figure 11.1, "Readiness gap"). Based on the timeline workload analysis (Parks & Boucek, 1989), "time available" is likely determined by the automation capabilities, and "time required" can be affected by multiple factors (e.g., driving experience, driver state, HMI support, etc.) but mostly by the readiness gap. Therefore, the workload level that drivers experience during a takeover reflects this readiness gap, with the possibility that the gap is too great for the driver to sufficiently overcome it in the time available to safely respond. One approach to facilitating a successful takeover could be by balancing the readiness gap with the duration of the time budget. With long time budgets, overload can be avoided because the tasks needed to "get ready" for regaining manual control are performed over a longer duration. However, when the duration is shorter and more constraining, certain strategies can be adopted to keep workload at manageable levels. One strategy is to reduce the time required for drivers to be ready. This can be done by using an HMI to assist drivers during takeovers. Another strategy is to keep the readiness gap from becoming substantial in the first place. This can be done by implementing automation in a way that requires drivers to periodically reengage with the driving task, thus refreshing their understanding of what is going on around them. Both these approaches rely on the driver's *attention management*.

ATTENTION MANAGEMENT

Attention management is the on-going process by which a driver allocates and focuses attention among competing objectives, activities, decisions, and information elements to meet specific goals within a driving context As shown in Figure 11.2, attention management has strategic, tactical, and operational elements that vary based on how engaged the individual is in act of driving.

The strategic aspect captures the driver's high-level objectives for deploying attention. Individuals that prioritize driving will attend to information that is important to driving and that builds up their understanding of the situation. Their deployment of attention would be conducted in a manner that supports this activity (e.g., active scanning of locations where hazards may occur). In contrast, if an individual chooses to attend to other information—perhaps because they do not have responsibility for vehicle control—they will spend a greater proportion of their time attending to nondriving information and their understanding will correspondingly suffer.

The tactical aspect of attention management captures how individuals allocate their attention. An engaged driver would be focused on attending to driving-relevant information, whereas a less engaged driver may spend time multitasking and switching between driving and nondriving information sources.

Finally, the operational aspect of attention management captures the specific information to which individuals attend. Engaged drivers that have a clear understanding of what is going on around them will anticipate and seek out the most meaningful information (MMI) at a particular moment (Campbell et al., 2012). In contrast, disengaged drivers would probably lack the understanding to know what they should be looking at, or more likely, they might simply be more interested in the information related to their nondriving activity. However, characteristics of the driving scene and properties of key information can still guide a disengaged driver to important information. For example, salient and compelling visual signals, such as illuminating taillights and visual looming can draw a driver's attention to a

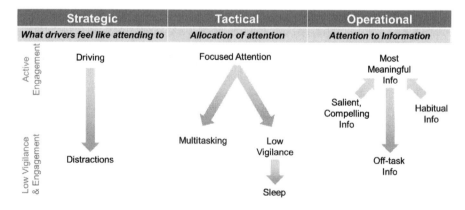

FIGURE 11.2 Strategic, tactical, and operational aspects of attention management under active driver engagement and low engagement.

hard-braking lead vehicle. Similarly, familiarity with common roadway scenarios can guide a driver's attention to key information sources out of habit (e.g., traffic signals at intersections). These processes would only provide disengaged drivers with a limited understanding of the situation; however, it could still be enough for them to execute simple emergency actions, such as hard braking.

Attention management is a cognitive process that is internal to the driver; however, it can be influenced and even controlled by external elements, such as the vehicle HMI. In particular, the status of a driver's attention management is observable. Engaged drivers will look at driving-relevant information, while disengaged drivers will do so less frequently, if at all. This can be quantified by recording eyes-off road-time, calculating gaze concentration, or more precisely by recording what information elements drivers actually look at. A driver monitoring system could assess the level of an individual's engagement in the driving task. Another aspect of attention management is that attention can be captured by external stimuli in a stimulus-driven manner. Auditory and haptic signals can generally alert drivers, and salient visual messages can direct a driver's attention to specific information (Campbell et al., 2016). Another relevant aspect is that some important information that drivers missed while they were distracted can be filled in with longer, narrative messages that describe a situation or problem (i.e., a required takeover at an upcoming work zone). This type of supplemental information can assist drivers in regaining their understanding and in identifying the MMI to develop a clearer sense of what they need to do and what choices they should make. These aspects of attention management provide a conceptual foundation for designing HMIs that can help drivers reengage with the driving task through information that is tailored to their observed level of disengagement from driving.

CHALLENGES TO ATTENTION MANAGEMENT IN AUTOMATED DRIVING

The dynamic allocation of attention during partial/conditional automation can be a challenge to drivers. Drivers are expected to monitor the driving task and be ready to intervene as necessary. Such sustained visual attention or vigilance in monitoring the roadway can be compromised in at least two ways. First, vigilance is mentally effortful (Bainbridge, 1983; Warm, Parasuraman, & Matthews, 2008). A prolonged and monotonous automated drive can lead to a reduction in the useful field of view (Rogé, Kielbasa, & Muzet, 2002). Drivers may focus attention within a small "cone" in the forward roadway, rather than actively attending to different roadway and in-vehicle information elements; subsequently, this leads to a failure to recognize peripheral hazards and drowsiness (Casner & Schooler, 2015; Körber, Cingel, & Zimmermann, 2015; Vogelpohl, Kühn, Hummel, & Vollrath, 2018). Another way vigilance decrements can occur is when drivers choose to engage in other activities due to task underload, which can result in a failure to notice key information and loss of situation awareness (Casner & Schooler, 2015).

Even when drivers are highly motivated to be vigilant, their attention management strategies may be insufficient to perform safe takeovers in some situations. The perceptual cycle (Lee et al., 2017, p. 81; Neisser, 1977) describes how drivers' psychomotor actions are coupled with perceptual attention and an internal model of the

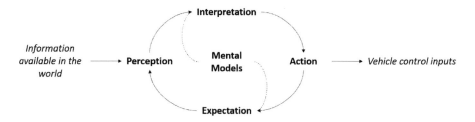

FIGURE 11.3 Perceptual cycle applied to the driving task (adapted from Lee et al., 2017).

requirements of the task ("mental models" in Figure 11.3). A mental model encompasses the user's knowledge of an automated system's purpose, how it functions, and how it is likely to function in the future (Goodrich & Boer, 2003). Mental models are developed over longer periods of time than a single trip and are updated as drivers gather information ("experience") through driving in different situations. Mental models are crucial in directing dynamic attention and may affect decisions at every stage of the perceptual cycle. When driving manually, a driver's mental model influences their expectation of events, which then directs perceptual sampling of available information and subsequent interpretation. The driver then acts on the vehicle controls to adjust the error between the perceived information and the intended state; this error adjustment results in a new set of expectations about the world, motivating further visual sampling (Stanton, Chambers, & Piggott, 2001). This process is represented in Figure 11.3.

During automated driving, drivers may monitor the roadway to identify hazards, but they do not perform vehicle control. Consequently, drivers' motor readiness is not coupled to their sensory system because perceived information is not processed with the goal of influencing action. When required to take over, this latency in motor readiness can negatively influence both the timing of actions (Louw, Merat, & Jamson, 2015; Merat, Jamson, Lai, Daly, & Carsten, 2014) and the subsequent quality of control maneuvers (DinparastDjadid et al., 2017). In addition, monitoring automated driving, without understanding automation capabilities and limitations, can impair the driver's expectations of hazards. The driver should understand what elements the automation can sense and interpret, how it acts upon this information, how this capability changes with the situation, and how the driver's own visual sampling should bridge gaps in automation capability. This understanding contributes to drivers' mental models of the interdependencies between the automation and themselves in performing the driving task (Victor et al., 2018). Thus, the driver's awareness of the automation's "attention" should be a crucial component of their own attention management and visual sampling strategies. However, the driver's understanding of automation capabilities and their own roles are seldom well-calibrated.

Indeed, even dealerships, with the opportunity to provide in-person interactions, may do a poor job of educating drivers of the limits of automation (Abraham, McAnulty, & Mehler, 2017; Endsley, 2017). As such, drivers may not have appropriate knowledge about the interaction requirements or the limits of automation, which can lead to a mismatch between what the drivers think the system can

Workload and Attention Management

do and what the system actually does. Drivers will have to shift from being an inattentive passenger to an active driver very quickly, with insufficient situation awareness, when automation behaves unexpectedly. Such takeovers will seldom be safe; rather, the driver is likely to react suboptimally. In addition to decrements in immediate performance, these automation surprises (Sarter, Woods, & Billings, 1997) can lead to long-term changes in driver-automation interaction, including loss of trust and disuse.

Overall, attention management can be compromised due to unclear driver expectations of their role when automation is engaged. In partially/conditionally automated vehicles, the driver's role can vary from "monitor" to "driver" based on the level and operational limits of automation. The convenience of "free time" is a key attraction that influences decisions of purchasing a vehicle with automated capabilities, and marketing materials often promote this pitch. For example, the branding of a Level 2 system as an "autopilot" instead of a "super cruise" may lead to a person attributing more automated capabilities to the former than the latter (Abraham, Seppelt, Mehler, & Reimer, 2017), even though the level of automation and the designer-intended attention requirements of both systems may be similar. Drivers may come to expect their own roles to be that of passengers in a piloted vehicle when the "autopilot" is active, instead of considering themselves as the monitor of an advanced "cruise" system. This misperception can lead to misuse due to overtrust in the automation and complacency in monitoring behaviors, particularly during the initial stages of use of the vehicle automation. Support for appropriate management of driver attention should be provided at different timescales—from the design of marketing material to ongoing in-vehicle feedback on automation performance. One intervention is through the inclusion of a proactive HMI, as will be discussed in the subsequent section.

IMPLICATIONS FOR HUMAN-MACHINE INTERFACE DESIGNS

The vehicle HMI can play an important role in automation takeovers. During takeovers with short time budgets, the HMI can facilitate driver readiness either after a takeover becomes imminent, using appropriate messages and alerts, or in advance of the takeover by keeping drivers partially engaged in driving and maintaining their readiness at elevated levels. For some failures, take-over events may be so uncommon that drivers are uncertain about how to respond. In these cases, the HMI can guide the driver through the takeover with the use of appropriate messages and alerts.

There are multiple ways to use the HMI to effectively communicate relevant information. Several aspects of HMI messages can influence a driver's take-over performance. The related design considerations include questions such as:

- *What information should drivers receive about the transition in level of automation?*
- *When should drivers be alerted about the transition between one level of automation and another level of automation?*
- *How should drivers be informed about the transition in levels of automation?*

Designers responsible for HMI messages must consider how automation can change the driver's role and the nature of the driving task. The effects of automation on driver workload can vary and will depend greatly on the extent to which key human factors issues are considered during the design of the automated system. As discussed, the gap between the task demands and the driver's readiness level is a key determinant of the driver's take-over workload. One of the ways to address this gap is to improve the technological capabilities of the system through better sensors, better predictive algorithms, and improved access to real-time data from the surrounding environment. This will reduce the severity of the problem through improvements to the capabilities of the technology and will reduce the number of driving situations where these problems can occur. Until we achieve Level 5 vehicles, however, we are only likely to mitigate this readiness gap, not eliminate it. Another way is to use the HMI proactively as a means to help drivers manage their attention more effectively, and to improve drivers' detection and response performance to ongoing and emerging hazardous situations. Effectively, a proactive HMI, designed to support drivers' attention management strategies, can not only reduce the number of situations associated with "peak workload," it can also improve driver trust, improve the driver's understanding of the automated system's capabilities, support a proactive approach to improving situation awareness, and improve the driver's ability to detect and respond to hazardous situations.

Campbell et al. (2018) identified a number of well-supported strategies for attention management and workload reduction that can be applied to the design of HMIs in automated vehicles. These include:

- *Presentation of system status information.* Information about the automation system status, on/off status, system mode (i.e., the type or level of automation that is currently active), transition confirmations (i.e., changes in mode), alerts, or information about a projected future state of the automation would greatly improve situation awareness. A considerable body of research (Arroyo, Sullivan, & Selker, 2006; Parasuraman & Riley, 1997; Seppelt & Lee, 2007; Stanton et al., 2001) has addressed the need for such information and strategies for presenting it, including questions about presentation modality, the availability of the information, and the needed level of complexity of the information.
- *Provide information about system transitions.* Although well-examined in other domains, control transitions have not been extensively examined for Level 2 or Level 3 vehicles. Basic human factors research, combined with the (largely) exploratory vehicle automation research suggests that designers should: (a) support the manual-to-automated transitions by acknowledging drivers' requests to engage/disengage the automation and providing information about the various stages of the transition of the transition and (b) support automated-to-manual transitions by telling the driver when and how they should take control and by providing information about any conditions that may threaten safe transitions. Table 11.2 (adapted from Campbell et al., 2018) summarizes some key design principles.

TABLE 11.2
Design Principles for Supporting Transitions Between Manual and Automated Driving

For Transitions from Manual to Automated Driving	For Transitions from Automated to Manual Driving
• Provide the current system status at all times (Toffetti et al., 2009). • The system acknowledges automation transition requests within 250 milliseconds of input (Alliance of Automobile Manufacturers, 2006; International Organization for Standardization, 2017) to prevent duplicate or conflicting inputs from the driver, and to prevent the driver from releasing control without knowing that the automation is active. • If the transfer was successful, provide the driver with a notification and update the displayed status of the automation. • If the transfer was unsuccessful, provide the driver with a notification of the failure of the automation to engage and why the automation did not engage (Merat & Jamson, 2009; Tsao, Hall, & Shladover, 1993). • The use of unimodal or multimodal notifications and messages should reflect the context of the situation. • Use distinctive messages for successful and unsuccessful transfer of control events.	• Provide the driver with information on when to take back control (Blanco et al., 2015; Gold, Damböck, Lorenz, & Bengler, 2013). • Provide the driver with information on how to take control if a specific control input is required (Toffetti et al., 2009). • Provide the driver with information on why the driver needs to take control. For time-critical situations, this may be a simplified "take control" message. For less time-critical situations, more information may be provided. • The current system status is always provided, allowing the driver to validate the disengagement (Sheridan & Parasuraman, 2005). • Notifications and messages related to a "take control" message are multimodal (Blanco et al., 2015; Brookhuis, van Driel, Hof, van Arem, & Hoedemaeker, 2009; Toffetti et al., 2009). • If applicable, provide the driver with any collision avoidance system messages. • Provide the driver with information about the situation that the system may have been monitoring at the time of transfer (e.g., position of a lead vehicle).

- *Improve situation awareness for both normative and time-critical driving situations.* In particular, use the HMI to directly provide critical information to the driver or to help direct the driver's attention to critical information and/or information elements. This could include timely presentation of attentional cues, alerts, or critical roadway information (e.g., missed guide signs or temporary roadside messages). (See also principles for designing to support SA developed by Endsley, 2016.)
- *Support an improved match between the drivers' expectations and experience, and the system's real capabilities.* Design an automation system that supports the development and maintenance of a functionally accurate "mental model" of how the automation system operates. In particular:
    - Instructions, training, and actual use of the system are the best opportunities to develop and maintain functionally accurate mental models.

- Provide information about the system at a level of detail that can be understood by a representative driver. The automation should be described as tasks or functions (e.g., "the system looks for lane markings," NOT in terms of complex technology, e.g., "near-infrared camera and machine vision algorithms are used to detect lane boundaries"). The information's level of specificity/detail should be geared toward relevant details for the Driver-Vehicle Interface (DVI) and the driver's direct experience with the system without the inclusion of unnecessary detail (see also Pollock, Chandler, & Sweller, 2002) or technical jargon.
- Use error training as a strategy to reduce driver overconfidence in using an automated system and to influence drivers to generate their own coping strategies for novel situations not covered in training (Ivancic & Hesketh, 2000).
- Carefully prioritize and organize training and instructional materials. Training on the use and function of in-vehicle systems should occur after novice drivers have obtained the basic and rudimentary skills needed for safe driving (Panou, Bekiaris, & Touliou, 2010).
- Avoid exaggerating the capabilities of the system. The level of trust that drivers place in the automation should be commensurate with the capabilities of the automation (Lee & See, 2004).

SUMMARY AND DISCUSSION

It seems that each day brings news, discussion, and debate concerning technological advances, corporate mergers, promises, predictions, or even tragic failures associated with driving automation. Driving automation is generally viewed as a means to reduce and even eliminate the many traffic fatalities that we have become normalized to through living in a society so dependent upon the automobile. However, despite its potential for reducing driver error and improving safety, automation can introduce challenges in workload and attention management. It has already been associated with a number of unique and unanticipated problems; many of the design and implementation challenges have not been well-understood by users and designers alike. If not implemented with careful consideration of driver behavior and limitations, automation can be accompanied by inappropriate levels of workload. Specifically, even well-functioning automation can lead to driver underload, disengagement from the driving task, and a decreased readiness to respond in take-over situations. If rapid transitions back to manual driving are needed, overload from the immediate demands of the task and a reduced ability/readiness to reengage quickly can compromise safety. Workload should be neither too high nor too low; errors are least likely to occur when workload is moderate and does not change suddenly or unpredictably (Kantowitz & Casper, 2009).

Maintaining appropriate levels of workload during automated operating conditions is therefore critical. Along with a clear understanding of the system's capabilities and limitations, drivers need to understand their roles and responsibilities under various operating conditions and scenarios. This is especially the case in Level 2 and Level 3 automation, which may require rapid transfers of control from automated to manual driving. From a vehicle perspective, the system can monitor various aspects of driver state during automated driving and then guide the drivers' attention to the forward road when needed (e.g., attentional cueing). For driver state detection based on drivers' glance patterns, we expect that advanced ways to estimate drivers' engagement and situation awareness will be required. That is, conventional on/off-road glance classification may not be sufficient to measure driver vigilance and readiness in automated vehicles. We should not assume that visual glances or fixations to the forward roadway equate to engagement or readiness. Specifically, "on-road" glance behaviors are too coarse a behavioral indicator as the driver's readiness or accumulated situation awareness needs to be assessed by what they fixate upon (e.g., safety-critical cues from lead vehicle/potential hazard versus low-information objects on the road), what information they sample (e.g., traffic flow, latent risk zone, etc.), and the timing of such fixations. Therefore, any assessment of driver engagement should be relative to the driver's role and information needs in a specific driving situation. It may require a task analysis-based approach as a part of an advanced way to estimate driver state.

Beyond recognizing the perils of driver underload with imperfect automation and looking for opportunities to monitor driver attention and engagement, designers should consider the value of using a proactive HMI to reduce the readiness gap that can result from driver underload. A proactive HMI could incorporate information about trip goals; driver state, readiness, and situation awareness; and immediate tactical demands. Such an approach could provide the driver with focused information that supports short-term tactical demands and promotes long-term development of system trust and functional mental models of system capabilities and limitations.

REFERENCES

Abraham, H., McAnulty, H., & Mehler, B. (2017). Case study of today's automotive dealerships: Introduction and delivery of advanced driver assistance systems. *Transportation Research Record*, 2660, 7–14.

Abraham, H., Seppelt, B., Mehler, B., & Reimer, B. (2017). What's in a name: Vehicle technology branding & consumer expectations for automation. In *Proceedings of the 9th International Conference on Automotive User Interfaces and Interactive Vehicular Applications* (pp. 226–234). New York: ACM.

Allen, K. (2018, May 15). Tesla Model S was in Autopilot mode during Utah crash, driver says. Retrieved August 8, 2018, from https://abcnews.go.com/US/tesla-model-autopilot-mode-utah-crash-driver/story?id=55168222

Alliance of Automobile Manufacturers. (2006). *Statement of principles, criteria and verification procedures on driver interactions with advanced in-vehicle information and communication systems* (report of the Driver Focus-Telematics Working Group). Washington, DC: Alliance of Automobile Manufacturers.

Arroyo, E., Sullivan, S., & Selker, T. (2006). CarCoach: A polite and effective driving coach. In *CHI '06 Extended Abstracts on Human Factors in Computing Systems* (pp. 357–362). New York: ACM.

Bainbridge, L. (1983). Ironies of automation. *Automatica*, 19(6), 775–779.

Blanco, M., Atwood, J., Vasquez, H. M., Trimble, T. E., Fitchett, V. L., Radlbeck, J., ... Morgan, J. F. (2015). *Human Factors Evaluation of Level 2 and Level 3 Automated Driving Concepts (DOT HS 812 182)*. Washington, DC: National Highway Traffic Safety Administration.

Brookhuis, K. A., van Driel, C. J. G., Hof, T., van Arem, B., & Hoedemaeker, M. (2009). Driving with a Congestion Assistant; mental workload and acceptance. *Applied Ergonomics*, 40(6), 1019–1025.

Buono, B. (2018). *Report: Safety driver involved in fatal self-driving Uber crash was watching Hulu.* Retrieved August 8, 2018, from https://www.12news.com/article/news/local/valley/report-safety-driver-involved-in-fatal-self-driving-uber-crash-was-watching-hulu/75-566585455

Campbell, J. L., Brown, J. L., Graving, J. S., Richard, C. M., Lichty, M. G., Bacon, L. P., ... Sanquist, T. (2018). *Human factors design guidance for Level 2 and Level 3 automated driving concepts* (Report No. DOT HS 812 555). Washington, DC: National Highway Traffic Safety Administration.

Campbell, J. L., Brown, J. L., Graving, J. S., Richard, C. M., Lichty, M. G., Sanquist, T., ... Morgan, J. F. (2016). *Human Factors Design Guidance for Driver-Vehicle Interfaces (DOT HS 812 360).* Washington, DC: National Highway Traffic Safety Administration.

Campbell, J. L., Lichty, M. G., Brown, J. L., Richard, C. M., Graving, J. S., Graham, J., ... Harwood, D. (2012). *Human Factors Guidelines for Road Systems (2nd Edition).* Washington, DC: Transportation Research Board.

Casner, S. M., & Schooler, J. W. (2015). Vigilance impossible: Diligence, distraction, and daydreaming all lead to failures in a practical monitoring task. *Consciousness and Cognition, 35,* 33–41.

DinparastDjadid, A., Lee, J. D., Schwarz, C., Venkatraman, V., Brown, T. L., Gasper, J., & Gunaratne, P. (2017). After the fail: How far will drivers drift after a sudden transition of control. In *Proceedings of the Fourth International Symposium on Future Active Safety Technology–Towards Zero Traffic Accidents.* Nara, Japan: Society of Automotive Engineers of Japan, Inc.

Endsley, M. R. (2016). *Designing for Situation Awareness: An Approach to User-Centered Design.* Boca Raton, FL: CRC Press.

Endsley, M. R. (2017). Autonomous driving systems: A preliminary naturalistic study of the Tesla Model S. *Journal of Cognitive Engineering and Decision-making, 11*(3), 225–238.

Endsley, M. R. (2018). Situation awareness in future autonomous vehicles: Beware of the unexpected. In *20th Congress of the International Ergonomics Association, IEA 2018.* (pp. 303–309). Florence, Italy: Springer.

Endsley, M. R., & Kiris, E. O. (1995). The out-of-the-loop performance problem and level of control in automation. *Human Factors, 37*(2), 381–394.

Fagnant, D. J., & Kockelman, K. (2015). Preparing a nation for autonomous vehicles: Opportunities, barriers and policy recommendations. *Transportation Research Part A: Policy and Practice, 77,* 167–181.

Fridman, L. (2018). *Safety Driver Glance Classification (Uber Self-Driving Car).* Retrieved August 8, 2018, from https://lexfridman.com/

Gibson, J. J., & Crooks, L. E. (1938). A theoretical field-analysis of automobile-driving. *The American Journal of Psychology, 51*(3), 453–471.

Gold, C., Damböck, D., Lorenz, L., & Bengler, K. (2013). "Take over!" How long does it take to get the driver back into the loop? In *Proceedings of the Human Factors and Ergonomics Society Annual Meeting, 57*(1), 1938–1942. Los Angeles, CA: Sage Publications.

Gold, C., Naujoks, F., Radlmayr, J., Bellem, H., & Jarosch, O. (2017). Testing scenarios for human factors research in Level 3 automated vehicles. In N. A. Stanton, S. Landry, Di Bucchianico, & A. Vallicelli (Eds.), *Advances in Human Aspects of Transportation* (pp. 551–559). Cham, Switzerland: Springer.

Goodrich, M. A., & Boer, E. R. (2003). Model-based human-centered task automation: A case study in ACC system design. *IEEE Transactions on Systems, Man, and Cybernetics—Part A: Systems and Humans, 33*(3), 325–336.

Hale, A. R., Stoop, J., & Hommels, J. (1990). Human error models as predictors of accident scenarios for designers in road transport systems. *Ergonomics, 33*(10-11), 1377–1387.

International Organization for Standardization. (2017). *Road Vehicles: Ergonomic Aspects of Transport Information and Control Systems: Dialogue Management Principles and*

Compliance Procedures (No. ISO 15005:2017). Geneva: International Organization for Standardization. Retrieved August 8, 2018, from https://market.android.com/details?id=book-U9abtAEACAAJ

Ivancic, K., & Hesketh, B. (2000). Learning from errors in a driving simulation: Effects on driving skill and self-confidence. *Ergonomics, 43*(12), 1966–1984.

Kantowitz, B. H., & Casper, P. A. (2009). Human workload in aviation. In R. K. Dismukes (Ed.), *Human Error in Aviation* (pp. 123–153). New York: Routledge.

Körber, M., Cingel, A., & Zimmermann, M. (2015). Vigilance decrement and passive fatigue caused by monotony in automated driving. *Procedia Manufacturing, 3,* 2403–2409.

Lee, J. D. (2018). Perspectives on automotive automation and autonomy. *Journal of Cognitive Engineering and Decision-making, 12*(1), 53–57.

Lee, J. D., & See, K. A. (2004). Trust in automation: Designing for appropriate reliance. *Human Factors, 46*(1), 50–80.

Lee, J. D., Wickens, C. D., Liu, Y., & Boyle, L. N. (2017). *Designing for People: An Introduction to Human Factors Engineering.* Charleston, SC: CreateSpace.

Liang, Y., Lee, J. D., & Horrey, W. J. (2014). A looming crisis: The distribution of off-road glance duration in moments leading up to crashes/near-crashes in naturalistic driving. *In Proceedings of the Human Factors and Ergonomics Society Annual Meeting* (pp. 2102–2106). Los Angeles, CA: Sage Publications.

Louw, T., Merat, N., & Jamson, H. (2015). Engaging with highly automated driving: To be or not to be in the loop? In *8th International Driving Symposium on Human Factors in Driver Assessment, Training and Vehicle Design* (pp. 190–196). Salt Lake City, UT: Public Policy Center, University of Iowa.

Merat, N., & Jamson, A. H. (2009). How do drivers behave in a highly automated car? In *5th International Driving Symposium on Human Factors in Driver Assessment, Training and Vehicle Design* (pp. 514–521). Iowa City: Public Policy Center, University of Iowa.

Merat, N., Jamson, A. H., Lai, F. C. H., Daly, M., & Carsten, O. M. J. (2014). Transition to manual: Driver behaviour when resuming control from a highly automated vehicle. *Transportation Research. Part F, Traffic Psychology and Behaviour, 27,* 274–282.

Michon, J. A. (1985). A critical view of driver behavior models: What do we know, what should we do? In L. Evans & R. C. Schwing (Eds.), *Human Behavior and Traffic Safety.* New York: Plenum Press.

Morgan, J. F., & Hancock, P. A. (2009). Estimations in driving. In C. Castro (Ed.), *Human Factors of Visual and Cognitive Performance in Driving* (pp. 51–62). Boca Raton, FL: CRC Press.

National Highway Traffic Safety Administration. (2017). Automated Vehicles for Safety. Retrieved August 8, 2018, from https://www.nhtsa.gov/technology-innovation/automated-vehicles-safety#issue-road-self-driving

Neisser, U. (1977). Cognition and reality: Principles and implication of cognitive psychology. *Perception, 6,* 605–610.

Panou, M. C., Bekiaris, E. D., & Touliou, A. A. (2010). ADAS module in driving simulation for training young drivers. In *13th International IEEE Conference on Intelligent Transportation Systems* (pp. 1582–1587). Funchal, Portugal: IEEE.

Parasuraman, R., & Riley, V. (1997). Humans and automation: Use, misuse, disuse, abuse. *Human Factors, 39*(2), 230–253.

Parks, D. L., & Boucek, G. P. (1989). Workload prediction, diagnosis, and continuing challenges. In G. R. McMillan, D. Beevis, E. Salas, M. H. Strub, R. Sutton, & L. Van Breda (Eds.), *Applications of Human Performance Models to System Design* (pp. 47–63). Boston, MA: Springer US.

Pollock, E., Chandler, P., & Sweller, J. (2002). Assimilating complex information. *Learning and Instruction, 12*(1), 61–86.

Rasmussen, J. (1987). The definition of human error and a taxonomy for technical system design. In J. Rasmussen, K, Duncan, & J. Leplat (Eds.), *New Technology and Human Error.* Chichester, UK: Wiley.

Rogé, J., Kielbasa, L., & Muzet, A. (2002). Deformation of the useful visual field with state of vigilance, task priority, and central task complexity. *Perceptual and Motor Skills, 95*(1), 118–130.

SAE International. (2014). *Taxonomy and Definitions for Terms Related to On-Road Motor Vehicle Automated Driving Systems (J3016)*. Retrieved August 8, 2019, from https://www.sae.org/standards/content/j3016_201401/

Sarter, N. B., Woods, D. D., & Billings, C. E. (1997). Automation surprises. In G. Salvendy (Ed.), *Handbook of Human Factors and Ergonomics*, 1926–1943, Hoboken, NJ: Wiley.

Senders, J. W. (2017). Driver distraction and inattention: A queuing theory approach. In M. A. Regan, J. D. Lee, & T. W. Victor (Eds.), *Driver Distraction and Inattention* (pp. 55–62). Boca Burlington, VT: Ashgate.

Seppelt, B. D., & Lee, J. D. (2007). Making adaptive cruise control (ACC) limits visible. *International Journal of Human-Computer Studies, 65*(3), 192–205.

Sheridan, T. B., & Parasuraman, R. (2005). Human-automation interaction. *Reviews of Human Factors and Ergonomics, 1*(1), 89–129.

Shinar, D., Meir, M., & Ben-Shoham, I. (1998). How automatic is manual gear shifting? *Human Factors, 40*(4), 647–654.

Stanton, N. A., Chambers, P. R. G., & Piggott, J. (2001). Situational awareness and safety. *Safety Science, 39*(3), 189–204.

Toffetti, A., Wilschut, E., Martens, M., Schieben, A., Rambaldini, A., Merat, N., & Flemisch, F. (2009). CityMobil: Human factor issues regarding highly automated vehicles on eLane. *Transportation Research Record*, (2110), 1–8.

Tsao, H.-S. J., Hall, R. W., & Shladover, S. E. (1993). Design options for operating fully automated highway systems. In *VNIS '93 Vehicle Navigation and Information Systems Conference* (pp. 494–500). Ottawa, Canada: Citeseer.

Valdes-Dapena, P. (2018, January 24). Tesla in autopilot mode crashes into fire truck. Retrieved August 8, 2018, from https://money.cnn.com/2018/01/23/technology/tesla-fire-truck-crash/index.html

van der Meulen, H., Kun, A. L., & Janssen, C. P. (2016). Switching back to manual driving: How does it compare to simply driving away after parking? In *8th International Conference on Automotive User Interfaces and Interactive Vehicular Applications* (pp. 229–236). Ann Arbor, MI: ACM.

Victor, T., Dozza, M., Bärgman, J., Boda, C.-N., Engström, J., Flannagan, C., ... Markkula, G. (2015). *Analysis of naturalistic driving study data: Safer glances, driver inattention and crash risk*. Washington, DC: Transportation Research Board. Retrieved August 8, 2018, from http://onlinepubs.trb.org/onlinepubs/shrp2/SHRP2_S2-S08A-RW-1.pdf

Victor, T. W., Tivesten, E., Gustavsson, P., Johansson, J., Sangberg, F., & Ljung Aust, M. (2018). Automation expectation mismatch: Incorrect prediction despite eyes on threat and hands on wheel. *Human Factors, 60*(8), 1095–1116.

Vogelpohl, T., Kühn, M., Hummel, T., & Vollrath, M. (2018). Asleep at the automated wheel—Sleepiness and fatigue during highly automated driving. *Accident: Analysis and Prevention, 126*, 70–84.

Warm, J. S., Parasuraman, R., & Matthews, G. (2008). Vigilance requires hard mental work and is stressful. *Human Factors, 50*(3), 433–441.

Wierwille, W. A. (1993). Visual and manual demands of in-car controls and displays. In B. Peacock and W. Karwowski (Eds.), *Automotive Ergonomics* (pp. 299–320). Washington DC: Taylor and Francis.

Young, K., Regan, M., & Hammer, M. (2007). Driver distraction: A review of the literature. In *Distracted Driving* (pp. 379–405). Sydney, NSW: Australasian College of Road Safety.

12 Attention Management in Highly Autonomous Driving

Carryl L. Baldwin & Ian McCandliss

Advanced driver assistance systems (ADAS) and other forms of advanced vehicle technology have the potential to dramatically improve public safety. While advances in sensor technologies, software, data algorithms, and engineering solutions continue at a rapid pace, human evolution remains little changed. Over a century of examination demonstrates that humans have limited attentional capacity, have difficulty maintaining focused attention for a sustained time, and are easily distracted. Therefore, it is imperative that attention management solutions are developed to facilitate the safe and effective implementation of autonomous capabilities for all road users. Attention management strategies for partial and highly autonomous driving are reviewed. Focus is placed on what is currently known and what needs further investigation.

INTRODUCTION

Automobile crashes are a persistent threat to public safety and leading cause of death in the United States (NHTSA, 2018a). Human error and driver distraction play key roles in a high percentage of these crashes (Guo et al., 2017). Despite public safety campaigns, the number of alcohol-impaired driving fatalities remains high (29% in 2017, down 1.1% from 2016; NHTSA, 2018b). At the same time, the number of licensed drivers and vehicle miles traveled by drivers over the age of 65 has been steadily increasing as have the numbers of fatal crashes involving these drivers (NHTSA, 2018b). ADAS have the potential to decrease crashes for all roadway users (e.g., drivers, vehicle occupants, pedestrians, and bicyclists) and may be particularly helpful for high-crash-risk segments of the driving population (e.g., older drivers). However, effective design of these systems is urgently needed to realize these benefits in public safety.

Automation is increasingly a pervasive part of our everyday lives. Automation holds the promise of increasing safety, decreasing mental demand, and improving the overall quality of life. However, in order to achieve these ideals, it must be designed to be effective and acceptable. Effectiveness can be achieved by technological advances, but acceptance will require a host of system characteristics (Bengler et al., 2014).

VEHICLE AUTOMATION

Since the invention of the automobile, there has been the far-off possibility that human control of the vehicle could be off-loaded onto a separate entity (Lozano-Perez, 2012). While the technological capability for this control switch grows closer every day, there are still numerous barriers and obstacles that must be overcome (Fagnant & Kockelman, 2015). First, technological capabilities are constantly evolving. There was a time when antilock brakes were considered the height of new safety technologies. Now they are standard on most vehicles. Many newer vehicles are equipped with automatic braking systems, which while not necessarily able to prevent a collision, can greatly reduce the damaging impact. Likewise, increasingly both longitudinal control (e.g., cruise control and adaptive cruise control) and latitudinal control or assistance (e.g., active lane keeping or lane departure warnings) are widely available or even standard in many new vehicles. More advanced systems are under development and can be discussed in terms of their level of automation.

LEVELS OF AUTOMATION

While fully automated vehicles are likely in our future, and nearly every automobile manufacturer is currently working on prototypes for these, it is unlikely that they will become prominent within the next decade. Even when they become commercially available, it takes on average roughly 10 years for new technologies to attain high levels of market penetration. Conversely, ADAS or partially automated vehicles have been on the market for some time now, and their level of market penetration is steadily increasing. The levels of vehicle automation have been described by the Society of Automotive Engineers (SAE) on a scale of 0—completely manual to 5—full automation (SAE, 2018; refer to Figure 12.1). Note that Level 2 requires that the driver monitor the driving scene at all times and Level 3 specifies that the driver be ready to take over manual control within seconds of receiving a take-over request (TOR). Levels 2 and 3, though only partial and conditional automation, are referred to as highly autonomous driving for the purposes of this manuscript. In fact, many Tesla owners and operators may be confusing Tesla's "autopilot" system, which is only Level 2, with a fully autonomous system as noted by several recent fatal crashes. These crashes underscore the importance of finding effective attention management strategies whether these be implemented through training and education, policy decisions, or vehicle and device design. But first we provide an overview of general aspects of attention with particular focus on how they might impact interaction with highly automated vehicles.

ATTENTION

Attention has been an important topic of investigation in psychology for some time, dating at least as far back as William James' (1890) textbook entitled, *Principles of Psychology*, where he defined it as, "The taking possession by the mind, in clear and vivid form, of one out of what may seem several simultaneously possible objects or trains of thought. …It implies withdrawal from some things in order to deal

Attention Management in Highly Autonomous Driving

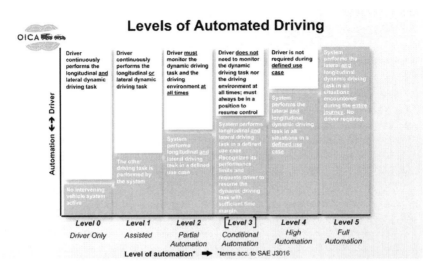

FIGURE 12.1 Society of Automotive Engineers' (SAE) levels of automation (adapted from SAE, 2014, and revised SAE, 2018).

effectively with others". It is well beyond the present scope to cover this vast topic in any comprehensive or detailed way. But James' definition points to some of the aspects that are still of critical importance today. Attention is selective so that we may exercise choice in what is attended and must be controlled to not only focus on that something of interest but also to inhibit distraction from other things. We add here two additional concepts—that **attention is of limited capacity** and that it is difficult to sustain in order to remain vigilant for extended periods of time.

A paramount concern in the creation and management of autonomous vehicles is a human's capacity to attentively (though passively) monitor the vehicle when they are not manually driving it (Braunagel, Kasneci, Stolzmann, & Rosenstiel, 2015). Attention is frequently characterized in cognitive literature as a form of limited resource that is expended on tasks. The more intense the task or the longer it is performed, the more resources are depleted in attempting to perform the task and the more strategies must be employed by the user to recoup those resources (Oulasvirta, Tamminen, Roto, & Kuorelahti, 2005). While this is not the only way attention has been characterized nor is it a wholly accurate description, this conception is helpful when considering the role that attention plays in highly automated vehicles.

SUSTAINED ATTENTION AND VIGILANCE

The study of vigilance has been examined near constantly since the inception of human factors psychology. Vigilance is a form of focused concentration, usually with an associated monitoring task. The reason this is essential to our focus here is because driving requires vigilance in order to effectively monitor the roadway for potential hazards. Increasingly higher levels of vehicle automation alter the role of the "driver" and his or her ability to understand both the need to maintain vigilance and his or her capability to do so. Rather than being constantly in the loop, acting and reacting to their environment, drivers will increasingly be placed in the role of monitoring the automation for potential system failures or for critical events that are outside the capabilities of automation. Humans are not good at remaining vigilant when critical events are rare and not well defined (Parasuraman, 1979).

Mackworth (1948) contributed much to our early understanding of vigilance with studies aimed at determining why trained operators were missing critical signals during radar operation. It may also be argued, however, that the first studies of sustained attention were conducted by Wilhelm Wundt amidst his foundational psychophysics studies (Leahey, 1979). Regardless of these origins, there are a few concrete factors that have been gleaned in relation to human performance on vigilance tasks. First, when a human is given a single task objective to monitor a given target for a critical signal over a given period of time, their performance on this task will decrease the longer they try to do this. What is more, this decrement is resistant to training and may continue to persist even in "experts" in monitoring (Williams, 1986). This is generally referred to as the vigilance decrement. Many research studies have designed especially difficult tasks with the intention of making this decrease occur more rapidly, with some instances of it occurring as early as ten minutes into a task (Temple, Warm, Dember, Jones, LaGrange, & Matthews, 2000).

These laboratory tasks obviously are not completely representative of many real-world activities that require sustained attention, but they do provide insight into how behavior changes especially across task types. This behavioral decrement may appear as either a decrease in the number of correct responses in recognizing a critical target or as a slowing in the subject's response time (Pattyn, Neyt, Henderickx & Soetens, 2008). See, Howe, Warm, and Dember (1995) conducted a meta-analysis of vigilance decrement literature and modeled how a person's ability to recognize critical signals (sensitivity) decreases based on the type of task they must perform. They categorized the tasks as either sensory based (monitoring the environment) or cognitive based (monitoring a given quality or condition). They further categorized the tasks as either being successive (with the critical target being held in memory) or simultaneous (with an example of the critical target available for immediate reference). The authors were then able to model how performance would decrease under these task conditions depending on the event rate of the critical signal, with the greatest decrement observed for tasks that had high overall event rates but few critical signals and those for which the critical signals were not clearly denoted at all times but rather had to be recalled from memory and were therefore more difficult to recognize. Driving itself is such a task, with generally high event rates but potentially few critical signals or hazards depending on the time of day, weather, and road conditions. As driving becomes more highly automated and switches a driver's task to a more supervisory and monitoring role, vigilance decrements are more likely to occur. The driver will need to be able to detect rare critical signals that must be discriminated from those held in memory as possible events where a collision might occur. The more reliable the automation, the rarer the events will be and thus the more complacent (or overtrusting) the driver will likely become (Parasuraman & Manzey, 2010). At best, this will result in difficulty sustaining attention for more than a short period of time, or at worst, it may lead to a failure to monitor the situation or automation altogether. Put another way, drivers who rarely encounter potential hazards will be slow to respond, if they respond at all, when critical events occur.

Assuming a low event rate situation, drivers can be expected to become more disengaged in the task and, with higher levels of automation, will be likely to engage in nondriving-related tasks. Given this lack of immediate engagement, it will take time for the driver to respond to any stimuli (Gold, Damböck, Lorenz, & Bengler, 2013). To illustrate this point, Ruscio, Ciceri, and Biassoni (2015) found that when driving with a collision avoidance warning system (a low level of automation), it took drivers significantly more time to respond (by nearly 0.5 seconds) to an unexpected object on the road even with a warning the first time they encountered it relative to when they were warned but anticipating the subsequent potential hazards.

The reasons for the vigilance decrement is a matter of no small scholarly debate. Some parties argue that the vigilance decrement results from a subject becoming habituated to a task, failing to become aroused by it, and generally being bored (Ariga & Lleras, 2011). Others argue that sustaining attention on any single given task depletes limited attentional resources (Grier, Warm, Dember, Matthews, Galinsky, Szalma, & Parasuraman, 2003). In general, those who support the boredom hypothesis consider actions that will arouse or trigger greater focus on the task, while those who support the attentional resource hypothesis pose that taking restorative actions (such as taking breaks or mind wandering) allows for longer sustained attention.

It is important to consider these theories moving forward as monitoring and controlling the attention of the driver will be of critical importance. There are already many instances noted in which drivers of low-level semiautonomous vehicles report growing either "bored" or "tired" with monitoring the road and leave the automation to handle it instead, even in situations that the automation is unsuited for. Dual-task or off-task attention management is especially important with the co-occurrence of smart-phone technology in many of these vehicles. It is for this reason that the next section will discuss attentional control and further describe the phenomena that humans broadly dub "attention."

ATTENTIONAL CONTROL

Attention is a broad topic in human factors psychology and one of the greatest difficulties in studying it is a lack of a concrete definition (Pashler, 2016). It may be considered by some to be a form of focused concentration, but it also consists of the ability to extract information from the environment, recognize it as important, make a decision to act on it, and then follow through on that information. Earlier theories of attention focused primarily around a "bottleneck" of how much information can be taken in, but more modern theories lean more toward attentional resources (Wickens, 1980) and activation of corresponding neural anatomy.

Peterson and Posner (2012) provide a description of the anatomical areas of the brain connected to attention and its processes. Their examination revealed that the portions of the brain responsible for maintaining and focusing attention are anatomically distinct from areas in which sensory information is processed and higher level decisions are made. Furthermore, they propose that there are five distinct forms of attention, each with its own corresponding anatomical regions.

The first systems are the alerting systems, which are primarily localized in the right frontal lobe. These generally are associated with activation of the local coeruleus and the release of norepinephrine. Present pharmacological agents are already in use that can be used to inhibit or excite norepinephrine production and thus reduce or increase alertness, respectively.

The second are the orienting systems. Like many brain regions, these can be subdivided into smaller portions. Generally, one portion of the system appears to consist of the frontal eye fields and the intraparietal sulcus. This portion is considered to regulate rapid strategic control over attention. The other part of this system is responsible for interrupting other processes and reorienting the system to more critical events. These parts are considered to consist of the temporoparietal junction and the ventral cortex. This system is of interest as it is key to switching attention between tasks.

The third and fourth systems were both previously subsumed under the title of executive control; however, there now exists evidence for their being two such systems working in parallel to provide top-down regulation. One of these systems is considered transient, generally dealing with top-down processes more toward the beginning or initiation of a task. These systems are thought to primarily consist of the lateral frontal and parietal regions. The other is thought to be involved in sustained attention and to consist of the medial frontal/cingulate cortex and the bilateral anterior insula.

The final neural attention system is one that is critical for self-regulation—the anterior cingulate gyrus. However, this is also believed to have a pathway split depending on task type, with the dorsal path being related to more cognitive-based tasks while the ventral path is involved more in emotional regulation (Beckmann, Johansen-Berg, & Rushworth, 2009).

The aforementioned systems and their impact on the various parts of attention are listed here for two reasons. The first is to note that attention is a multifaceted system that can be interrupted or augmented at many stages. Different tools and signals can be used depending on whether a system seeks to alert a person to a problem or to orient them toward a particular source of information. Likewise, providing some form of incentive to maintain vigilance may alter the strategies a person uses during this process.

The second reason for the anatomical rundown is to remind the reader of a simple point about attention and cognitive resource theory. While attention may be thought of and experienced as a "loss" of energy that may be recouped through a variety of means, this is more of a helpful metaphor than a literal description of the mechanism. Hockey (2011) elaborates on this as our brains are not batteries that become depleted through use (though experientially they may appear to be this at first glance). Rather, they are a complex system of interrelated anatomy that can be pushed to exceed limits that may not often be seen in laboratory experiments alone.

While sustained attention may appear as a constant depletion of resources, the reality is that humans have many means of structuring tasks to ensure that attention can be maintained for extended periods of time. Experimental vigilance tasks are meant on some level to induce the sensation of cognitive fatigue or boredom (Scerbo, 1998) and to result in a decrement in the ability to attend to signals. In comparison, many tasks such as games are designed to hold attention for as long as possible, to a point that it could be considered an addiction (Van Rooji, Schoenmakers, Vermulst, Van Den Eijnden, & Van De Mheen, 2011).

Prior models of human attention such as the demands control model (Landsbergis, 1988) have already hit upon the fact that there is an interaction between a person's ability to remain focused on a task and their perceived ability to have latitude within it. It may even be seen as a truism that if a person is forced to do a task, they will not try to focus as much on it or try as hard at it. By this same convention, granting people more latitude and a greater capacity to control their situation may result in greater interest in the task and what has been dubbed by some to be a greater internal motivation (Zuckerman, Prac, Lathin, & Deci, 1978). Games and play are also representations of this same capacity to maintain attention on a task, one that is entirely self-motivated (Klimmt & Hartmann, 2006). There are limits on this, of course, as a person can't play games forever. Eventually, they will run into issues of physical and cognitive fatigue that cannot necessarily be overcome using internal motivation alone. But, gamification of the automated driving task is one strategy that been suggested though rarely examined.

Furthermore, there is the not insoluble issue that there are logistic hurdles in turning any given task into a game. Not all tasks can be gamified and, of those tasks that can be gamified, not all of them can maintain a person's attention or internal motivation to complete them eternally. The same applies to autonomous systems as well. It comes as little surprise that when a task is uninteresting, people will seek to engage in off-task behaviors. These off-task behaviors can be hazardous in normal driving situations. While ADAS may help to mitigate the worst of these off-task behaviors, this does not mean that the driver's attention can be completely excised from the process.

DIVIDED ATTENTION AND DISTRACTION

Driver distraction due to engagement in secondary tasks increases crash risk, particularly for novice and adolescent drivers. Estimates vary, but novice drivers are seven to eight times more likely to be involved in a crash or near crash if they are reaching for or dialing on a mobile phone (Klauer, Gluo, Simons-Morton, Ouimet, Lee, & Dingus, 2014). Adolescents within the first 18 months of licensure are nearly four times more likely to be involved in crashes or high G-force (near-crash) events relative to their parents (Simons-Morton et al., 2011). Though this rate decreases for experienced drivers, it still dramatically increases the likelihood of crash involvement. Despite the elevated crash risk, drivers persist in engaging in distracting tasks. The Second Strategic Highway Research Plan (SHRP2) is a naturalistic driving assessment plan and data set containing thousands of hours of data from actual drivers in real-world situations (Victor et al., 2015). This massive data set reveals that over 50% of drivers aged 30 to 60 are engaged in a secondary task at any given time, and this rate increases to roughly 57% to 58% for drivers under age 30. It would seem that even during manual driving, people have difficulty maintaining their focus on the task of driving and refraining from doing other tasks. Add to this the finding that even when not engaged in distracting secondary tasks, drivers are internally distracted or mind wandering 50% to 70% of the time (Baldwin, Roberts, Barragan, Lee, Lerner, & Higgins, 2017).

Given this information it seems reasonable to ask whether people should be driving at all. The obvious advantage of an autonomous or semiautonomous system is that it could reduce the prevalence of human error while freeing people from the mentally demanding and sometimes stressful task of driving. In normal driving situations, a person's capacity to orient themselves to all critical signals is not necessarily possible, and so they may end up succumbing to hazards (Meng & Spence, 2015). It is possible that if people were freed from the constraints of performing the latitudinal and longitudinal control aspects of the driving task, they would have more resources to devote to remaining vigilant for critical signals and potential hazards in the environment. This idea of "person as monitoring agent" is the assumption of Level 2 automation (see Figure 12.1).

However, as human factors engineers have noted, there is inherent danger in this as disengaging from the primary driving task can be expected to lead to a perceived loss of control and willingness to participate (Hockey, 2011). As has been previously alluded to, perceived control and internal motivation to accomplish a task can play a key role in attributing attentional resources to it. Indeed, prior research indicates that drivers do not take advantage of the freed attentional resources they have while driving in partially autonomous vehicles to engage in increased hazard detection. Present evidence indicates that rather than engaging in the hazard-monitoring task, drivers with ADAS may instead default to engaging in more secondary tasks (Llaneras, Salinger, & Green, 2013). These secondary tasks at best divide attention and result in a performance decrement (Strayer & Johnston, 2001) and at worst take up the majority of a person's attention, precluding any form of hazard monitoring in all but the most extreme circumstances (Jamson, Merat, Carsten, & Lai, 2013).

Developing attentional strategies for highly automated vehicles requires a different framework. Lee (2014) has suggested that rather than approaching the situation from the perspective that drivers have limited attentional resources that can be exceeded by the demands of driving, human factors engineers should instead consider, "…how drivers come to engage…and under what conditions they engage and disengage from driving—the dynamics of distraction" (p. 24). This dynamics of distraction approach views attention more as an ongoing process which changes over time, rather than as a singular capacity to maintain attention. The focus becomes one of interruption management.

Another more detailed approach is proposed by Cabrall, Eriksson, Dreger, Happee, and de Winter (2018). They suggest a model based on three separate paradigms (behaviorism, cognitivism, and ecological) resulting in six proactive solution areas designed to keep people engaged in the driving task when interacting with semiautonomous vehicles and ADAS. These approaches will be described in further detail in the next section. We turn now to a discussion of strategies of assisting the driver with attention management.

ATTENTION MANAGEMENT APPROACHES

Potential attention management approaches and specific strategies for maintaining safe and effective human-machine-interaction within the context of highly autonomous vehicles can take many forms. While it is practically impossible to be comprehensive in our discussion, we aim to provide a wide range of approaches that have the potential to independently, or interactively, accomplish this important goal.

ATTENTION RESTORATION STRATEGIES

In general, as stated earlier in this chapter, people are not good at sustaining their attention to monitoring tasks (e.g., monitoring automation for potential but infrequent failures). Decades of research examining these types of sustained attention tasks indicates that performance declines within periods of 10 to 20 minutes (See, Howe, Warm, & Dember, 1995) and that people find these types of tasks unpleasant, mentally demanding, and stressful (Dillard et al., 2018). As Cabrall and colleagues (2018) have pointed out, one of the most basic approaches to this is simply to avoid putting drivers in this position altogether. Some would say this might be the best approach. Simply do not put humans in the role of becoming a supervisory monitor of the automation because we know from decades of experience that they are not good at it and it does not appear to be good for them. As enticing as this may seem, it is really too late for this draconian approach. ADAS systems are increasingly pervasive, and this is not likely to change.

Other, relatively low-level solutions are either reducing the time during which drivers are required to perform the monitoring task (scheduled rest breaks) or finding ways to enhance the tasks by attempting to decrease the unpleasantness and stress of performing these tasks. There are a number of potential strategies to accomplish this, and while gamification has been previously proposed as a solution, there also exist a number of other more passive mood-altering techniques. This generally can be accomplished through strategies that the driver may take to maintain their attention on the task or passive safety systems that are engineering controls intended to alert the driver and ensure they orient themselves to critical events.

SLEEP AND REST BREAKS

The need for breaks from the task of prolonged driving have been recognized for some time and considerable investment have been made in roadway infrastructures in the form of rest areas along U.S. interstates. Less appreciated is the potential need to switch back and forth between the focused attention network and the default mode network. As described previously, humans have difficulty sustaining their attention in search of rare-events targets (e.g., potential road hazards), particularly if the distracting stimuli is of a high rate of occurrence, and target events must be held in memory (Parasuraman, 1979), if the distractors are monotonous, and target events are unpredictable. These conditions describe many low-traffic highway scenes and will increase in prevalence as the driver is further removed from the driving task due to automation. It is likely natural that driver-operators will increasingly fall prey to one form of distraction or another—internal thoughts (or default mode engagement) or external distractions (mobile phone use, etc.)

MIND WANDERING

Mind wandering is a natural process where attention becomes decoupled from external stimuli. Despite the fact that people are less responsive to external stimuli during mind wandering, they appear to be able to remain responsive to unexpected deviant stimuli (Kam, Dao, Stanciulescu, Tildesley, & Handy, 2013). Mind wandering may be a natural method of restoring focused attention resources. ADAS that support mind wandering while providing alerts for unexpected events may therefore provide a means of supporting a driver's ability to maintain sufficient attentional resources to allow response to potential road hazards over extended periods of time.

NATURAL SCENES SIMULATIONS

Berto (2005) found that looking at nature scenes relative to industrial cityscapes improved sustained attention. It is conceivable that such scenes could be digitally rendered in future automobiles. Research to determine whether or not such strategies would be effective or would rather increase distraction potential is warranted. Since monitoring the driving scene places large visual demands on drivers, it is worth investigating whether similar positive attentional effects might be found via presentation of various auditory soundscapes.

SOUNDSCAPES AND MUSIC

Baldwin and Lewis (2017) found that popular, positive valence music (particularly if it was of slow tempo) played during a short break in between vigils resulted in better post-break music relative to negative valence music (of fast or slow tempo), silence, or no break at all. One explanation for these results is that the positive valence music led to a more pleasant affective state and may have decreased stress levels associated with task performance.

Music of any form can be used to augment performance in a variety of environments (e.g., at work and during exercise) (Barwood, Weston, Thelwell & Page, 2009). However it has often been cited as a potential cause of driver distraction, taking away critical resources during driving (Ünal, Steg, Epstude, 2012). Regardless of whether or not it is helpful in maintaining vigilance or serving as a distractor, the implications for how music affects driving behaviors is largely unknown to the general public (Brodsky, 2001).

ESCALATING INTERVALS

Effective attention management will require escalating levels of automation including both passive and active safety systems as described in the following sections. At the lowest level, attention monitoring systems (e.g., sensory to detect hands on the wheel and/or eyes on the road) are needed to ensure driver engagement for SAE Levels 0–2. Though technologies to determine that drivers maintain their hands on the wheel are used by a number of manufacturers (e.g., Tesla, Mercedes-Benz, BMW) to allow continued use of lateral and longitudinal automation (Level 2), few have incorporated technologies aimed at ensuring that visual attention, or eyes-on-the-road, is maintained (Cabrall et al., 2018, in press). General Motors (GM)/Cadillac is one of the few manufacturers that currently incorporates metrics of eyes-on-the-road to disable or allow continued use of Level 2 automation. Escalating levels of attentional management can be categorized into the two broad categories of passive and active safety systems, which are described as follows.

PASSIVE SAFETY SYSTEMS

Passive safety systems in the form of alerts and alarms provide a form of sensory augmentation to direct drivers' attention to hazardous situations they might otherwise not notice. These generally come in the form of some type of alert or alarm presented in visual, auditory, or tactile modalities or their combination.

ALERTS AND ALARMS

Collision avoidance alarms have shown great potential for increasing safety. Advanced sensor capabilities can serve as a second set of eyes and ears in modern vehicles and facilitate a driver's awareness of potential hazards. Forward collision warnings (FCWs), lane departure warnings (LDWs), and blind spot warnings (BSWs) are some of the most prevalent types. Research to date on the effectiveness of these systems indicates that they have great potential to decrease crashes and prevent injuries (Kusano, Gabler, & Gorman, 2014).

Older adults are at particular risk of crashes when making left turns at unsignalized intersections. A number of intersection assist devices are currently under development that may reduce these types of crashes for all drivers but that may particularly benefit older drivers. Bellet, Paris, and Marin-Lamellet (2018) found that older drivers reported preference for a system that would provide a warning if their intended intersection maneuver was unsafe, relative to systems that simply provided them with additional information that could be used in making their gap determination.

Harper et al. (2016) estimated that widespread market penetration of three types of alerts (blind spot warnings, lane departure warnings, and forward collision warnings) could "… prevent or reduce the severity of as many as 1.3 million U.S. crashes a year including 133,000 injury crashes and 10,100 fatal crashes" (p. 104).

LIKELIHOOD ALARMS

One solution to return attentional resources from distractions to back-in-the-loop driving is the involvement of likelihood alarms. Ideally, these alarms are calibrated such that there is a high likelihood of their activation in the face of an imminent collision with enough proceeding time for the driver to reengage with the system and alert the crash. There are, however, several barriers to the implementation of such systems.

MONITORING AND TAKEOVER

As indicated in Figure 12.1, Level 2, or partial automation, requires the driver to monitor the driving task and roadway at all times. This may not even be understood by drivers, pointing to the need for more effective training, reminders, and safeguards to ensure that drivers understand and comply with this mandate. Several methods of ensuring that drivers maintain at least some engagement with the driving task have been developed by different manufacturers, and these will be discussed subsequently. But, given the human's limited ability to sustain attention and remain vigilant (discussed previously), it is questionable whether or not the driver will be able to effectively monitor the driving task even if he or she knows this is required.

Level 3, or conditional automation, does not require active monitoring at all times but requires that drivers be able to take over control within a short period of time. Studies have demonstrated that the time it takes to return to manual driving after periods of automated driving is frequently longer than might be expected, particularly if drivers are engaged in some other task during the automated driving (Eriksson & Stanton, 2017). Slowed responses have been thought to be a result of automation-induced underload of attentional resources (Solis-Marcos, Galvao-Carmona, & Kircher, 2017). Previous studies on people reacting to a TOR as far out as seven seconds away have shown the lengthy time required to act upon the obvious critical signal (Gold, Damböck, Lorenz, & Bengler, 2013). Issues pertaining to improving the quality of the TOR are also addressed.

IMPLICIT DISPLAYS

Implicit displays can be thought of as displays that support unconscious attentional processing of information (theoretically, while conscious attention is directed elsewhere). Considerable research indicates that when environmental cues are ambiguous, humans may unconsciously activate or prepare multiple simultaneous action sequences (for a review, see Chapter Three in Forsythe, Liao, Trumbo, & Cardona-Rivera, 2015). Designs that support this unconscious level of action preparation may increase the appropriate response (and thus safety) of drivers even when they are not consciously monitoring the roadway. One early example of an implicit display for manual driving was the use of lane demarcations that were smaller and closer together to give the impression that drivers were traversing at faster speeds when approaching intersections. Drivers slowed down even though they were not consciously aware of the roadway marking changes (Charlton, 2007).

To date, there have been few in-vehicle displays explicitly designed to improve the situation awareness (SA) of drivers. In one notable exception, Gregoriades and Sutcliffe (2018) observed that providing supplemental head up displays (HUDs) could support drivers' SA. In their study, drivers were operating a simulated vehicle in manual mode. However, similar display technology may benefit drivers operating partially autonomous vehicles. Further research in this area is warranted.

HAPTIC AND VIBROTACTILE DISPLAYS

Haptic or vibrotactile displays present information through the sense of motion, touch, or vibration and share the advantage with auditory displays of not requiring a driver's visual attention. There have been numerous investigations in recent years on using vibrotactile and haptic displays within a driving context, and they have the potential to support attention management in highly autonomous vehicles.

For example, Racine et al. (2010) found support for the use of active force-feed haptic feedback from the gas pedal and steering wheel in addition to a passive visual blind spot alert for preventing crashes in a simulator study. Others have found that haptic forces provided through the steering wheel can prompt a driver to more quickly regain a safe lane position when they are inadvertently drifting out of their lane of travel (Deroo, Hoc, & Mars, 2013). While these two techniques both require that the driver has his or her foot on the gas pedal or hands on the steering wheel in order to be effective (which may not be the case with Level 2 or 3 automation), they nonetheless point to the potential benefit of novel approaches toward implicit display design. Additional novel approaches and further work in this area are clearly needed.

CONCLUSION

Vehicles are becoming increasingly automated. Until they reach a level of full autonomy, strategies will be needed for effective attentional management from policy and education approaches to human factors designs to ensure their safety. ADAS changes the driver's role to one of a monitor for potential hazards and a supervisor of the automation, with a number of associated problems stemming from the need to sustain attention.

In order to handle these deleterious behavioral changes, there have already been some attempts to keep drivers focused on the monitoring task. Manufacturers of semiautonomous vehicles such as Tesla have already sought to curtail secondary task diversion through the use of a regular alert system that requires the user to keep their hands on the wheel (Lambert, 2018a). This, in turn, was followed with the short after-release of devices used by consumers to foil the alert system (e.g., by attaching something to the steering wheel) and keep them from having to maintain contact with the wheel so that they might divert their attention to other tasks (Lambert, 2018b). This, of course, makes sense as people generally do not want to be "nagged" to engage in any given task. In effect, the need to be attentive to the semiautonomous system further disengages the user from the task of driving by removing more of their own control of the situation. As such, developing a means of keeping consumers engaged in the driving task and internally motivated to continue hazard monitoring (attention management strategies) is a critical need for vehicle automation Levels 1 and 2, and keeping them at least partially attentive and aware of the surrounding roadway conditions so that they can quickly and efficiently resume manual control in Level 3 is essential.

Additional research is needed to better understand how drivers will manage their attention in increasingly automated vehicles. It is imperative that we understand how attention may be recouped in strenuous circumstances, or how certain strategies may be employed to make its expenditure more efficient. Research on attention has a long history, going back perhaps to the beginning of psychometrics. In particular, this early research will play a key role in the use of ADAS and other semiautonomous systems as drivers will be increasingly pushed to use a particular form of attention.

Ultimately, there are numerous concerns that must be addressed in the driver-as-monitor paradigm required by semiautonomous systems. Previously used paradigms to keep the driver engaged are no longer viable, and the allowance for off-task and distracting stimuli present great challenges to a safe and effective system. While drivers may be able to become more informed about the limitations of their autonomous systems, engineering solutions will be required to ensure consistent changes in driving habits and vigilance behavior.

REFERENCES

Ariga, A., & Lleras, A. (2011). Brief and rare mental "breaks" keep you focused: Deactivation and reactivation of task goals preempt vigilance decrements. *Cognition*, *118*(3), 439–443.

Baldwin, C., & Lewis, B. (2017). Positive valence music restores executive control over sustained attention. *Plos One*, *12*(11), e0186231. doi:10.1371/journal.pone.0186231

Baldwin, C. L., Roberts, D. M., Barragan, D., Lee, J. D., Lerner, N., & Higgins, J. S. (2017). Detecting and quantifying mind wandering during simulated driving. *Frontiers in Human Neuroscience*, *11*, 406. doi:10.3389/fnhum.2017.00406

Barwood, M., Weston, N., Thelwell, R., & Page, J. (2009). A motivational music and video intervention improves time trial performance in warm conditions. *Journal of Sports Science and Medicine*, *8*(3), 435–442.

Beckmann, M., Johansen-Berg, H., & Rushworth, M. F. S. (2009). Connectivity-based parcellation of human cingulate cortex and its relation to functional specialization. *Journal of Neuroscience 29*, 1175–1190.

Bellet, T., Paris, J.-C., & Marin-Lamellet, C. (2018). Difficulties experienced by older drivers during their regular driving and their expectations towards advanced driving aid systems and vehicle automation. *Transportation Research. Part F: Traffic Psychology and Behaviour*, *52*, 138.

Bengler, K., Dietmayer, K., Farber, B., Maurer, M., Stiller, C., & Winner, H. (2014). Three decades of driver assistance systems: Review and future perspectives. *Intelligent Transportation Systems Magazine, IEEE*, *6*(4), 6–22. doi:10.1109/MITS.2014.2336271

Berto, R. (2005). Exposure to restorative environments helps restore attentional capacity. *Journal of Environmental Psychology*, *25*(3), 249–259.

Brodsky, W. (2001). The effects of music tempo on simulated driving performance and vehicular control. *Transportation Research Part F: Traffic Psychology and Behaviour*, *4*(4), 219–241.

Braunagel, C., Kasneci, E., Stolzmann, W., & Rosenstiel, W. (2015, September). Driver-activity recognition in the context of conditionally autonomous driving. In *2015 IEEE 18th International Conference on Intelligent Transportation Systems (ITSC)* (pp. 1652–1657). doi:10.1109/ITSC.2015.268

Cabrall, C. D. D., Eriksson, A., Dreger, F., Happee, R., & De Winter, J. C. F. (2018). How to keep drivers engaged while supervising automation? A literature survey and categorisation of six solution areas. *Theoretical Issues in Ergonomics Science*, *20*(3), 332–365.

Charlton, S. G. (2007). The role of attention in horizontal curves: A comparison of advance warning, delineation, and road marking treatments. *Accident Analysis & Prevention*, *39*(5), 873–885.

Deroo, M., Hoc, J-M., & Mars, F. (2013). Effect of strength and direction of haptic cueing on steering control during near lane departure. *Transportation Research Part F: Psychology and Behaviour*, *16*(C), 92–103. doi:10.1016/j.trf.2012.08.015

Dillard, M. B., Warm, J. S., Funke, G. J., Nelson, W. T., Finomore, V. S., McClernon, C. K., ... Funke, M. E. (2018). Vigilance tasks: Unpleasant, mentally demanding, and stressful even when time flies. *Human Factors*, *61*(2). doi:10.1177/0018720818796015

Eriksson, A., & Stanton, N. A. (2017). Takeover Time in highly automated vehicles: Noncritical transitions to and from manual control. *Human Factors*, *59*(4), 689–705. doi:10.1177/0018720816685832

Fagnant, D. J., & Kockelman, K. (2015). Preparing a nation for autonomous vehicles: Opportunities, barriers and policy recommendations. *Transportation Research Part A: Policy and Practice*, *77*, 167–181.

Forsythe, C., Liao, H., Trumbo, M., & Cardona-Rivera, R. E. (2015). *Cognitive neuroscience of human systems: Work and everyday life*. Boca Raton, FL: CRC Press.

Gold, C., Damböck, D., Lorenz, L., & Bengler, K. (2013, September). "Take over!" How long does it take to get the driver back into the loop? In *Proceedings of the Human Factors and Ergonomics Society Annual Meeting* (Vol. 57, No. 1, pp. 1938–1942). Los Angeles, CA: SAGE Publications.

Gregoriades, A., & Sutcliffe, A. (2018). Simulation-based evaluation of an in-vehicle smart situation awareness enhancement system. *Ergonomics, 61*(7), 947–965. doi:10.1080/00 140139.2018.1427803

Grier, R. A., Warm, J. S., Dember, W. N., Matthews, G., Galinsky, T. L., Szalma, J. L., & Parasuraman, R. (2003). The vigilance decrement reflects limitations in effortful attention, not mindlessness. *Human Factors, 45*(3), 349–359.

Guo, F., Klauer, S. G., Fang, Y., Hankey, J. M., Antin, J. F., Perez, M. A., ... Dingus, T. A. (2017). The effects of age on crash risk associated with driver distraction. *International Journal of Epidemiology, 46*(1), 258–265. doi:10.1093/ije/dyw234

Harper, C. D., Hendrickson, C. T., & Samaras, C. (2016). Cost and benefit estimates of partially-automated vehicle collision avoidance technologies. *Accident Analysis & Prevention, 95*, 104–115.

Hockey, G. R. J. (2011). A motivational control theory of cognitive fatigue. In Ackerman, P.L. (Ed.), *Cognitive Fatigue: Multidisciplinary Perspectives on Current Research and Future Applications* (pp. 167–187). Washington, DC: American Psychological Association

James, W. (1890). *The Principles of Psychology*. New York: Henry Holt and Company.

Jamson, A. H., Merat, N., Carsten, O. M., & Lai, F. C. (2013). Behavioural changes in drivers experiencing highly-automated vehicle control in varying traffic conditions. *Transportation Research Part C: Emerging Technologies, 30*, 116–125.

Kam, J. W. Y., Dao, E., Stanciulescu, M., Tildesley, H., & Handy, T. C. (2013). *Mind Wandering and the Adaptive Control of Attentional Resources, 25*(6), 952–960. doi:10.1162/jocn_a_00375

Klauer, S. G., Guo, F., Simons-Morton, B. G., Ouimet, M. C., Lee, S. E., & Dingus, T. A. (2014). Distracted driving and risk of road crashes among novice and experienced drivers. *The New England Journal of Medicine, 370*(1), 54–59. doi:10.1056/NEJMsa1204142

Klimmt, C., & Hartmann, T. (2006). Effectance, self-efficacy, and the motivation to play video games. In Vorderer, P. & Bryant, J. (Eds.) *Playing Video Games: Motives, Responses, and Consequences* (pp. 133–145). Mahwah, NJ: Lawrence Erlbaum Associates Publishers.

Kusano, K. D., Gabler, H., & Gorman, T. I. (2014). Fleetwide safety benefits of production forward collision and lane departure warning systems. *SAE International Journal of Passenger Cars—Mechanical Systems, 7*(2), 514–527. doi:10.4271/2014-01-0166

Lambert, F. (2018a, June 11). Tesla's latest Autopilot update comes with more "nag" to make sure drivers keep their hands on the wheel. Retrieved from https://electrek.co/2018/06/11/tesla-autopilot-update-nag-hands-wheel/ accessed July 9, 2019.

Lambert, F. (2018b, September 10). Tesla Autopilot "buddy" hack to avoid "nag" relaunches as "phone mount" to get around NHTSA ban. Retrieved from https://electrek.co/2018/09/09/tesla-autopilot-buddy-hack-avoid-nag-relaunch-phone-mount-nhtsa-ban/ accessed July 9, 2019.

Landsbergis, P. A. (1988). Occupational stress among health care workers: A test of the job demands-control model. *Journal of Organizational Behavior, 9*(3), 217–239.

Leahey, T. H. (1979). Something old, something new: Attention in Wundt and modern cognitive psychology. *Journal of the History of the Behavioral Sciences, 15*(3), 242–252.

Lee, J. D. (2014). Dynamics of driver distraction: The process of engaging and disengaging. *Annals of Advances in Automotive Medicine, 58*, 24–32.

Llaneras, R. E., Salinger, J., & Green, C. A. (2013). Human factors issues associated with limited ability autonomous driving systems: Drivers' allocation of visual attention to the forward roadway. In *Proceedings of the Seventh International Driving Symposium on Human*

Factors in Driver Assessment, Training and Vehicle Design (pp. 92–98). Bolton Landing, NY: Public Policy Center, University of Iowa. doi:10.17077/drivingassessment.1472

Lozano-Perez, T. (2012). *Autonomous Robot Vehicles.* Princeton, NJ: Springer Science & Business Media.

Mackworth, N. H. (1948). The breakdown of vigilance during prolonged visual search. *Quarterly Journal of Experimental Psychology, 1*(1), 6–21.

NHTSA. (2018a). Traffic Safety Facts 2016 Data: Summary of Motor Vehicle Crashes. (DOT HS 812 580). Washington, DC: US Department of Transportation. Retrieved from https://crashstats.nhtsa.dot.gov/Api/Public/ViewPublication/812580 accessed July 9, 2019

NHTSA. (2018b). Traffic Safety Facts: 2017 Fatal Motor Vehicle Crashes: Overview. (DOT HS 812 603). Washington, D.C.: US Department of Transportation Retrieved from https://crashstats.nhtsa.dot.gov/Api/Public/ViewPublication/812603 accessed July 9, 2019

Oulasvirta, A., Tamminen, S., Roto, V., & Kuorelahti, J. (2005, April). Interaction in 4-second bursts: The fragmented nature of attentional resources in mobile HCI. In *Proceedings of the SIGCHI Conference on Human Factors in Computing Systems* (pp. 919–928). Ney York, NY: ACM.

Parasuraman, R. (1979). Memory load and event rate control sensitivity decrements in sustained attention. *Science, 205*(4409), 924–927.

Parasuraman, R., & Manzey, D. H. (2010). Complacency and bias in human use of automation: An attentional integration. *Human Factors, 53*(3), 381–410. doi:10.1177/0018 720810376055

Pashler, H. (2016). *Attention.* London: Psychology Press.

Pattyn, N., Neyt, X., Henderickx, D., & Soetens, E. (2008). Psychophysiological investigation of vigilance decrement: Boredom or cognitive fatigue? *Physiology & Behavior, 93*(1–2), 369–378.

Petersen, S. E., & Posner, M. I. (2012). The attention system of the human brain: 20 years after. *Annual Review of Neuroscience, 35,* 73–89.

Racine, D. P., Cramer, N. B., & Zadeh, M. H. (2010, October). Active blind spot crash avoidance system: A haptic solution to blind spot collisions. In *2010 IEEE International Symposium on Haptic Audio Visual Environments and Games* (pp. 1–5). IEEE.

Ruscio, D., Ciceri, M. R., & Biassoni, F. (2015). How does a collision warning system shape driver's brake response time? The influence of expectancy and automation complacency on real-life emergency braking. *Accident Analysis and Prevention, 77*(C), 72–81. doi:10.1016/j.aap.2015.01.018

SAE. (2014) Taxonomy and definitions for terms related to on-road motor vehicle automated driving systems. In *On-Road Automated Vehicles Standards Committee: Society of Automotive Engineers International.* Warrendale, PA: SAE.

SAE. (2018). Surface vehicle recommended practice: Taxonomy and definitions for 1492 terms related to driving automation systems for on-road motor vehicles. In *On-Road Automated Vehicles Standards Committee: Society of Automotive Engineers International.* Warrendale, PA: SAE.

Scerbo, M. W. (1998). What's so boring about vigilance? In R. R. Hoffman, M. F. Sherrick, & J. S. Warm (Eds.), *Viewing Psychology as a Whole: The Integrative Science of William N. Dember* (pp. 145–166). Washington, DC: American Psychological Association. doi:10.1037/10290-006

See, J. E., Howe, S. R., Warm, J. S., & Dember, W. N. (1995). Meta-analysis of the sensitivity decrement in vigilance. *Psychological Bulletin, 117*(2), 230–249.

Simons-Morton, B. G., Ouimet, M. C., Zhang, Z., Klauer, S. E., Lee, S. E., Wang, J., … Dingus, T. A. (2011). Crash and risky driving involvement among novice adolescent drivers and their parents. *American Journal of Public Health, 101*(12), 2362. doi:10.2105/AJPH.2011.300248

Solis-Marcos, I., Galvao-Carmona, A., & Kircher, K. (2017). Reduced attention allocation during short periods of partially automated driving: An event-related potentials study (Report) (Author abstract). *Frontiers in Human Neuroscience*, 11. doi:10.3389/fnhum.2017.00537

Strayer, D. L., & Johnston, W. A. (2001). Driven to distraction: Dual-task studies of simulated driving and conversing on a cellular telephone. *Psychological Science*, *12*(6), 462–466.

Temple, J. G., Warm, J. S., Dember, W. N., Jones, K. S., LaGrange, C. M., & Matthews, G. (2000). The effects of signal salience and caffeine on performance, workload, and stress in an abbreviated vigilance task. *Human Factors*, 42(2), 183–194.

Ünal, A. B., Steg, L., & Epstude, K. (2012). The influence of music on mental effort and driving performance. *Accident Analysis & Prevention*, *48*, 271–278.

Williams, P. S. (1986). Processing Demands, Training, and the Vigilance Decrement. *Human Factors: The Journal of Human Factors and Ergonomics Society*, 28(5), 567–579. doi:10.1177/001872088602800507

Van Rooij, A. J., Schoenmakers, T. M., Vermulst, A. A., Van Den Eijnden, R. J., & Van De Mheen, D. (2011). Online video game addiction: Identification of addicted adolescent gamers. *Addiction*, *106*(1), 205–212.

Victor, T., Dozza, M., Bärgman, J., Boda, C. N., Engström, J., Flannagan, C., … & Markkula, G. (2015). Analysis of naturalistic driving study data: Safer glances, driver inattention, and crash risk (No. SHRP 2 Report S2-S08A-RW-1). Washington, DC: The National Academies of Sciences, Engineering, and Medicine.

Zuckerman, M., Porac, J., Lathin, D., & Deci, E. L. (1978). On the importance of self-determination for intrinsically-motivated behavior: *Personality and Social Psychology Bulletin*, *4*(3), 443–446.

13 To Autonomy and Beyond

Peter A. Hancock

INTRODUCTION

It is now almost a quarter of a century since our community found itself intimately concerned with human interaction with automated systems (Parasuraman & Mouloua, 1996). In itself this seemed, even at the time, to be a rather paradoxical and even profoundly oxymoronic enterprise (Hancock, 1996). Surely, as one increased the level of systemic automation, the chances for human interaction were necessarily first constrained, then reduced, subsequently dissipated, and then assumedly completely eliminated? Yet, the evolutionary path of this innovative form of interaction never proved to be quite as straightforward as such pristine prognostications portended. For, as Parasuraman and his colleagues subsequently observed "automation does not eliminate human performance, it changes human performance" (Parasuraman, Sheridan, & Wickens, 2000). As the landmark text by Parasuraman and Mouloua (1996) established, and as has been confirmed in the subsequent three decades of research (see e.g., Kaber, 2018), such interactions have proved to be much more nuanced and elaborative than the pristine human versus automation, figure/ground comparison would suggest (and for an earlier perspective on this see Chapanis, 1965). It is true that the evolutionary steps involved in such human-automation interaction can, in general, be described as the removal of the human from the inner loops of operation. Necessarily then, it moves the human operator toward an ever more peripheral position and then even excision from immediate active control. Yet oversight and supervisory roles, while not especially suited to inherent human propensities (Hancock, 2013), remained central to system success—and especially to recovery from failure (Hoffman & Hancock, 2017).

It is clear that one obviously anticipated end point of this implicit line of development was the complete removal of the human operator from any form of control. An allied engineering aspiration was to alleviate any further human intervention even unto such tasks as setup and maintenance. In many developed systems, these aspirations were not hopeful expressions of future states but rather represented systems now in operations. For example, we do now have totally automated mining facilities as well as virtually independent agricultural operations (Austmine, 2018). The tide of automation rolls on, and the driving force of economics, which assumes the removal of human operators is the most cost-effective path, continues to impel this line of "progress." Such developments must be a source of concern for those in Human Factors/Ergonomics (HF/E) who are centrally concerned with human quality of life. For, what of the economic value of human work in such an automation-dominated world (Frey & Osborne, 2013)? However, since my central question concerns the even more important consideration of an existential threat, the critical questions here are i) what happens in the case of systemic automation failure and ii) where do all the redundant human workers go when automation usurps their tasks? Prospectively, what of human job prospects in these evolving systems? The proliferation of robotic invasion into human work means that we now see many fully automated production facilities. Even in many factories and warehouses that do retain human workers, their numbers are decreasing (Ford, 2015). Thus, when discussing existential threats here, these threats need not necessarily be spectacular, catastrophic, momentary battles between human and machine. Rather they may well be a far more insidious penetration in work and employment patterns (Hancock & Drury, 2011).

Now we are approaching a new age of concern in which the technological expression does not simply act within a restricted repertoire of responses (Hancock, 2014) but is one that is anticipated to act in independent, adaptive, generative, and resilient ways, all on its own. This is the coming wave of autonomy. Will we now anticipate the same expectations that we did for automation? That is, will autonomous systems be completely independent and separated from any form of human interaction? Alternatively, will the line of progress again be one in which subtle and unexpected interactions still persist and remain, embodied in the cooperative actions of a human counterpart or teammate? This is the conundrum of prediction, and our previous prognostications concerning automation and its effects must thus temper our expectations for the future of autonomy.

Autonomous systems are, at least in part, necessarily founded upon developments in automation. This assertion must be true even if it is simply predicated on the necessary historical, chronological progression. At present, it proves rather hard to say where intelligent automation ceases and dumb autonomy begins. Many would, rather rightfully, consider this emergence a step of evolutionary development rather than any discrete change. It is somewhat akin to trying to distill where one species turns into another since, in these biological efforts, transition stages are notoriously difficult to distinguish unequivocally. Acts of distinction here then can be rather arbitrary. Certainly, in the world of technology, there is no great hard-and-fast rule that differentiates one form of such technologies from another. Of course, this rather assumes that we already have autonomous systems but, as I will evaluate here, this itself is a contentious issue (and see Winner, 1978). Whether we now do, or do not, possess even one truly and fully autonomous system is rather a moot issue. However, it very much appears that as we march along the wide swath of progress that characterizes our current vector of development, autonomous systems are looming on the horizon of our near future. Given this degree of anticipation, I think the wider issue is the future of human beings in a world redolent with, and perhaps even dominated by, such degrees of technology autonomy. Most especially I want, in this present chapter, to ask about the future challenges that such progress might pose to the human species—whether that human occupies the roles of user, operator, customer, maintainer, passenger, or mere bystander. Most specifically, does the coming tide of autonomous systems represent an existential threat to the human species that is developing them? It can hardly be argued that any science that embraces the tenet of improved quality of human life can neglect this potential downside to such developments. Here, I want to explore this question for both near and far-term vistas. These concerns are often framed as lurid visions of either a functional dystopia or dysfunctional utopia.[1] The utopian vision, which was often championed in the decade of the seventies, had work weeks diminishing through 30 to 20 hours of human labor per week and below. It was envisaged that the then-touted "automated" systems would eventually alleviate all human wants and toil. But, as we know, this simply did not happen, and the economic benefits that did accrue were divided in a highly disproportionate manner. We must consider this precedent of what was derived from that harvest of automation as it pertains to coming autonomy. While neither the attainment of utopia nor the nightmare of dystopia are liable to become reality, our future appears to lie somewhere along this continuum, and any science concerned with human well-being must surely push for the former over the latter.

FROM AUTOMATION TO AUTONOMY

To begin to answer to this question, we must start with an examination that compares and contrasts what we now presently witness as automation and what we anticipate for autonomy. In respect of such discussion I begin with my own definitions (Hancock, 2017). I have asserted that automation is composed of: "Those systems designed to accomplish a specific set of largely deterministic steps, most often in a repeating pattern, in order to achieve one of an envisaged and limited set of pre-defined outcomes" (p. 284). For definitional purposes, I have provided a contrast that gives autonomy to, "Those systems which are generative and learn, evolve, and permanently change their functional capacities as a result of the input of operational and contextual information. Their actions necessarily become more indeterminate across time" (p. 284). Although these statements are comparative, they represent, as I have noted, an evolutionary sequence. It is important to point out what specifically evolves. The principle element that evolves is the range of behavior sanctioned by the repetitive systems. Automation is largely constrained and repetitive. It is, in the words of Earl Wiener (1987) "dumb and dutiful." That is, it performs its set function but is largely incognizant of conditions around it that can pervert its actions from the useful and relevant to the almost comical. What evolves with autonomy is the capacity to deal with change and innovation. Autonomous systems adapt to their changing circumstances and are resilient in the face of such change (Hoffman & Hancock, 2017). Accompanying such an expansion in the repertoire of performance comes increasing uncertainty on behalf of those who witness and those who interact with such systems. This tends to represent a vast expansion of the *mode error awareness* issue (Sarter & Woods, 1995), and this problem is concatenated by the ever-increasing speed of such operations. Thus, it is highly probable that sophisticated, advanced autonomous systems will have already answered virtually all of the questions that their human observers have yet to formulate. This form of technological "dissonance" is liable to prove a particular stumbling block in arenas where adoption is voluntary, as opposed to mandatory. This dissonance, what I here term *incommensurate temporality,* is liable to prove a rather prohibitive barrier to any successful collaboration. Not to belabor the point, but the fundamental issue of definition and thus identification, is one of set inclusion and set exclusion. Can we draw presently defined distinct boundaries between what is automated and what is autonomous? This is a difficult division as the targets and the associated divisions are themselves also dynamically changing. Descriptive definitions of the sort I have provided earlier are essentially one-time snapshots of what is fundamentally a river of development. In general then, I believe that autonomy is an evolutionary step beyond automation; although the lineage is neither a simple nor straightforward one, such that future automation does not have to be autonomous or perhaps vice versa.

DO WE YET HAVE AUTONOMOUS SYSTEMS?

Let me preempt the answer to this question, by stating that, at the present time, I do not believe we have any fully autonomous systems in operation. However, I believe their advent is close. The etymological origin of the term autonomy can help us to understand why this is presently the case. The "auto" element of the word refers to self—as for example in autobiography, autodidact, etc. The "nomos" component is the reference to "law" or "law-like" propensities, e.g., nomothetic. The combinatorial result is that autonomous systems, in their full incarnation, are "laws unto themselves." We must first ask whether such entities exist at all. Obviously, our deistic pronouncements would have the various incarnations of "God" as a law unto himself (or more recently herself). So, we can certainly conceptually envisage that such entities could exist. However, do we witness them in our direct, terrestrial world? As with all such explorations, words matter (Hoffman & Hancock, 2014a, b). In an interactive, social community, no entity can be purely "a law unto itself."[2] Such entities necessarily act within environmental and social constraints. In nature, the most apposite analogy would be a "peak predator." There are animals that, either by dint of strength, intelligence, or environmental chance, possess virtually no natural predators.

Humans do possess such predators and so are not "natural" peak predators per se. However, through their intelligence, they have created circumstances in which the opportunity for such remaining predators to attack have been greatly minimized; now, almost to extinction, humans can thus rightly claim the title of "peak species." The open question is whether in creating autonomous systems we are ourselves creating our own form of technological "peak predator." The peak predator notion represents a biomimetic analogy, and I have suggested that one way of ensuring harmonious mutuality is to constrain the ecological "niche" that any incarnation of technical autonomy can occupy (Hancock, 2017). Sadly, this design strategy seems very unlikely to be enacted; and the time for any purposive development and instantiation is rapidly passing. While we have, as yet, no evidence whatsoever of any intrinsic motivation on behalf of beginning autonomous systems, this is to cast their operational functionality in terms of our own human perspective. It may well be that such systems, as and when they emerge, possess no such "drive" and their actions result, albeit inadvertently, in the same overall effects. That is, human and machine freedom are, at some level, mutually self-limiting. However, to cast the whole discussion into this oppositional mode may simply be an expression of human attribution error in which we apply to other nonhuman entities the same motivational attributions that we attach to other people.

To conclude my consideration on this particular point, I believe we have yet to witness independent, autonomous machine systems in action in our world. But it must be admitted that it is quite possible that we might not be able to easily comprehend, or even recognize, autonomy's actions should one or more such systems be in operation. We have no equivalent of the Turing test for such elaborative, creative, adaptive, and resilient entities. If their actions are primarily contained within the purely electronic domain, their existence and presence may be difficult, if not impossible, to distill. Often, this sort of statement gives rise to fear and trepidation in human society (Hancock, 1996). But, of course, these fears are often visceral (Benet, 1935), as opposed to rationale estimates of existential threat from a form of functionality whose emergent properties we have yet to experience. Our innate fear of possible predators has served us well in the process of survival in the putatively "natural" world; the open question is whether such reactions are appropriate and useful in a technological ecology in which operational entities are qualitatively different from those of the world we are evolved in.

THE MODAL, CONTEMPORARY CASE STUDY: NASCENT AUTONOMOUS VEHICLES

At some stage, we must bring the foregoing theoretical speculation and our prospective conjectures down to specific cases. Philosophical reveries about possible futures can be very entertaining, but at some point they must be linked to reality or remain mere persiflage. With respect to the present discourse, I look to use the example of emerging, driverless vehicles (Hancock, 2019b). There are two major reasons I have selected this domain. First, I have some personal familiarity with the developments in this domain (see e.g., Hancock, 2019b; Hancock, Nourbakhsh, & Stewart, 2019). Second, it is my belief that it is in the area of personal transportation within which the general public will first encounter these nascent autonomous entities in the form of their own self-controlling vehicles.[3]

The transition from driver-centered control to vehicle-centered control is predicated upon a long and storied history in HF/E in the area of function allocation (Sheridan & Ferrell, 1974). I do not rehearse all of this history or its more recent debates and incarnations (but see Fitts, 1951; Hancock & Scallen, 1998; Roth & Pritchett, 2018; Sheridan & Verplank, 1978). Suffice it to say that the specifications of the various stages of human-automation interaction by some national agencies derive from this foundational basis and the discussions that have surrounded it (see Figure 13.1). It is, however, important and fair to emphasize that much of the current specifications, as presented in Figure 13.1, come from Sheridan's "Ten Levels" of human-automation interaction, and although not seminal per se, this identified work has acted largely to set the framework of modern discussion.

There are many elements of the adopted and adapted hierarchic representation shown in Figure 13.1 to discuss. Not least of which is its purported hierarchic arrangement. While fundamentally descriptive in nature, there is something more than seductive in seeing, for example, Level 4 as being more advanced say than Level 3, and also Level 2 being the logical consequence of "stepping up" from Level 1. Such a progression also implies that Level 5 is the goal or the end point of

SAE level	Name	Narrative Definition	Execution of Steering and Acceleration/ Deceleration	Monitoring of Driving Environment	Fallback Performance of *Dynamic Driving Task*	System Capability (*Driving Modes*)
Human driver monitors the driving environment							
0	No Automation	the full-time performance by the *human driver* of all aspects of the *dynamic driving task*, even when enhanced by warning or intervention systems	Human driver	Human driver	Human driver	n/a
1	Driver Assistance	the *driving mode*-specific execution by a driver assistance system of either steering or acceleration/deceleration using information about the driving environment and with the expectation that the *human driver* performs all remaining aspects of the *dynamic driving task*	Human driver and system	Human driver	Human driver	Some driving modes
2	Partial Automation	the *driving mode*-specific execution by one or more driver assistance systems of both steering and acceleration/ deceleration using information about the driving environment and with the expectation that the *human driver* performs all remaining aspects of the *dynamic driving task*	**System**	Human driver	Human driver	Some driving modes
Automated driving system ("system") monitors the driving environment							
3	Conditional Automation	the *driving mode*-specific performance by an *automated driving system* of all aspects of the dynamic driving task with the expectation that the *human driver* will respond appropriately to a *request to intervene*	System	**System**	Human driver	Some driving modes
4	High Automation	the *driving mode*-specific performance by an automated driving system of all aspects of the *dynamic driving task*, even if a *human driver* does not respond appropriately to a *request to intervene*	System	System	**System**	Some driving modes
5	Full Automation	the full-time performance by an *automated driving system* of all aspects of the *dynamic driving task* under all roadway and environmental conditions that can be managed by a *human driver*	System	System	System	**All driving modes**

FIGURE 13.1 The discrete stage description of differing levels of automation in advancing vehicle systems as promulgated by the Society for Automotive Engineers (SAE). Intimately linked to a comparable description by the National Highway Traffic Safety Administration (NHTSA), the hierarchical framework derives from the work of Sheridan (see e.g., Sheridan, 2002), which itself derives from earlier considerations by Fitts and his colleagues (1951).

the process. These are both debatable assumptions. However, let us begin with some obvious concerns. Level 0 is given as no automation, yet this cannot literally be true. Contemporary vehicles, even those featuring few of the proposed "intelligent" control facilities, possess extensive degrees of automation already. From engine firing rate to airbag deployment, drivers do not actively participate in these, or even the majority of already operational systems on their vehicle. Thus, the descriptive levels are really concerned with, and largely confined to, moment-to-moment control of the lateral and longitudinal location of the vehicle as well as, more distally, its progress from origin to destination. Immediate longitudinal and lateral positioning are forms of acute, moment-to-moment (tactical) control, the latter origin to destination concerns are forms of chronic (strategic) control. We are presently passing through a stage of control-sharing, and yet as noted, many if not most functions of the vehicle, especially in relation to the drivetrain, are already "automated" in the fashion that Levels 1 and above suggest.

The steps between levels are then, either implicitly or explicitly, used as prescriptions for progress. Thus, the assumption of control by nonhuman sources is seen as the step forward in the evolution of development. The usual, underspecified pabula of safety and efficiency are and have been invoked to encourage (and even mandate) each progressive step from one stage to the next. But there are several dimensions of experience that are left out of this insidious calculus of mechanical progress. Where is the value of the enjoyment of the human driver entered into the implicit equation?

What of individuals who want to "just drive?" Where is the happiness of the family traveling under its own recognizance? What of those who want to explore spontaneously and just enjoy the feeling of control? What of those who love old cars or modern cars and like to race them? What would Formula One racing look like with only autonomous control? What happens to people who are ardent taxi drivers, entrepreneurial Uber business people, or even aspiring bank robbers? Technology, in its various forms, can act to constrain just as readily as it can act to liberate. These and many other "affective" dimensions of the driving experience are evidently excised from this rather sterile and clinical engineering vision. And of course, such dimensions are often either neglected or ignored when autonomy is invoked in the vast swath of other potential applications.

As well as the affective concerns, the steps also imply some degree of equivalency in both time and capacity. Thus, it can seem as if the step from Level 2 to Level 3 will take exactly the equivalent time, effort, and progress as the step from Level 4 to Level 5. This is also probably an incorrect representation of reality, given this line of progress is pursued. The stepwise description also tends to ablate the notion of residuality. Thus, not all vehicles on the road take these respective steps at the same time. It is almost certain that we will have those at Level 0 coexisting with those at Level 3, etc. These are just a limited number of the assumptions involved with this rather untrammeled and deterministic representation of automation's progress into the realm of vehicle control. Yet, of course, it represents a very powerful narrative that, as I noted, may well be applied to numerous other domains as autonomy makes its presence felt more generally across much broader facets of human society.

The purely utilitarian arguments for progress in this area then have to be tempered with many human concerns beyond the immediate concerns for momentary vehicle control. It is often thought that the putatively "free market" exerts its "invisible hand" upon such developments, but this is a rather simplistic and naïve view of technological innovation and customer choice (see Hancock, Hancock, & Warm, 2009). Thus, while certain smartphones may not make the grade and be discarded, it does not mean that the genre of smartphones is altogether excised. It may be that, to a degree, the customer disposes. However, the public can only dispose from what is proposed, and what is proposed is necessarily a limited set of options. If the vast majority of vehicles are automated, there remains little choice not to automate also, especially in light of the fact that the associated driving infrastructure is now built for these very vehicle capacities.[4] If you do not believe this assertion, try to find a working public phone today. I think it is reasonable to conclude that personal vehicles will be the arena in which most individuals first encounter nascent autonomy full on. This despite the fact that personal ownership of vehicles is itself in decline. The transition phase to a new landscape of personal mobility may be a volatile one. However, the next decade and a half will still be an interval in which car ownership of ever-increasing capability vehicles will bring the driving public into direct contact with growing autonomous capabilities. As to whether and how we can tolerate and sustain the disruptions associated with such introductions, only the collective will of the social and political realm will determine. Yet human beings remain a highly adaptive species, and each new generation possesses the dynamic capacity to accommodate what is for them considered an ever-renewed "normal."

A ZERO-SUM GAME?

Human beings understand the game of life very well. In general, they have played it so superbly that, for good or bad, they have come to dominate (Hancock, 2019a). In many ways, our experience with, and the understanding of, the forces involved in evolution, have come to shape much of our thinking. In our own science of HF/E such *biomimetic principles* have helped facilitate progress in conceptual advances, design implementation, and real-world operations. We neglect the lessons of nature at our peril. Yet the question here is are those same principles and forces in operation with envisaged autonomous systems? While it is true that the "physical" space of existence and the access to limited resources have dictated our prior survival strategies, is this competition necessarily the case in a world of coexistence between humans and autonomy? If the answer here is a simple and unalloyed *yes*, then there will be such a competition, and we necessarily will be involved in an iatrogenic, existential struggle with our own creations. If one believes that autonomy will become a "species," exactly like any other living species on the planet, then there is great cause for concern. Under such circumstances rational, Luddite efforts to curtail such actions in which humans are the architects of their own destruction may even be justified under humanist moral strictures.[5] My view, however, is that autonomy need not necessarily partake of all, or even some, of the characteristics of living entities. This being so, a battle for limited physical resources need not necessarily ensue. Nor need such an in-group/out-group, figure-ground, them versus us, you're either with us or against us comparison and mind-set be applicable here. In short, my expectation is that autonomous systems will occupy a very different realm of operational existence; they will reside in their own *electronic eco-niches* (e^2niches), and the places where such realms interact with the physical world will be a limited set of all possibilities involved in such interaction. Specifying the locations and degrees to which that realm will intersect with and impact the human world is our challenge to define, and assumedly still barely at the present time—to design?[6]

DOES SMART TECHNOLOGY MAKE DUMB PEOPLE?

One empirical question that we might be able to address, and perhaps even resolve, is the proposition that smarter technologies make dumb people (and see Sparrow, Liu, & Wegner, 2011). Of course, like many such propositions, this one embeds many levels of assumption. For example, the issue of "dumb and dumber" is one of temporal as well as cognitive extent. A good example here is the replacement of human map skills with an in-vehicle GPS. We now enter city, town, village, or street as destination and then perhaps some specific address or more general area designate. The GPS now calculates (in a manner that is often nontransparent to the user) a route to be followed. In England recently, I used such an in-vehicle navigation device and was often subjected to a nominally "shortest-distance" route. Unfortunately, in England, this can take you down very narrow back lanes and tracks that visitors may well not be comfortable navigating. Further, name designations in many countries are not unique (e.g., Springfield in the United States). Thus, the driver needs at least some rudimentary idea of their destination's location, in order not to confuse it with others of the same name. Problematically, many GPS interfaces do not provide support for such a determination or, at best, will give you a secondary reference that without a map or familiarity may not help very much. In consequence, a driver can find themselves many tens, if not hundreds of miles off track. There is an example, of someone in London wishing to attend a football match at Chelsea's home ground of "Stamford Bridge." The driver programmed that destination into their GPS and then, many hours later, found themselves in Yorkshire at the site of the famous battle of Stamford Bridge. Without essential map skills, one can easily keep driving, especially if the interface is a turn-by-turn one and uninformative as to distance or time to destination, as some are. We might laugh at this sad mistake, but more and more, as map skills diminish, we will find ourselves vulnerable to such errors.

In such circumstances, the automation (i.e., the GPS) itself remains "dumb and dutiful" (see Orlady & Orlady, 2016; Wiener, 1987). Dumb in the sense that it cannot offer strategic advice in respect of the chosen destination, dutiful in the sense that once selected, it will take you there with unerring accuracy—albeit assuming the GPS technology works without flaw, which has not been my own ubiquitous experience in car rentals aboard. What we might learn from such an example is that "intelligence," per se, does not now necessarily reside in either the human or the machine singly but is in fact an emergent property of the interaction of the two. We might, in recognition of such collaborative failures, adopt the standard tactic of "blaming the user," and/or "blaming the machine." How can anyone be so stupid, we say to ourselves and to others. Of course, we do not embrace such an explanation when we ourselves are the user. We then go to the next default stage of "blaming" the automation—why don't "they" design it so it is useful? But neither of these putatively, explanatory approaches are necessarily useful. Rather, the shortfall lies in the degree and effectiveness of collaboration—what is essentially the "human factors" bridge between one and the other. Yet, such shortfalls are, hopefully, always under attack. We can inject greater abilities and transparency into the automated system such that people can use them effectively and trust them accordingly (Hancock et al., 2011; Schaefer, 2013; Schaefer, et al., 2017). What is more concerning is whether the "locus" of intelligent action is actually being transferred from the human to the machine and now by extension into the contextual environment. Thus, we need to consider not only the transient examples of failure of a proportion of human users, as exemplified in the GPS shortfalls described. We need to examine the sum of the interaction and more provocatively to understand whether smart technology makes dumb people.

There can be little doubt that the handheld calculator makes elementary arithmetic much simpler. As a child I was "taught" my times tables by rote. Even today I can multiply any number by any number, up to twelve times twelve in my head; "automatically." This derives from repeated "chanting" of the times tables in class until it became not calculation per se, but direct pattern recognition (and see Battino, 1992). Give me the first two numbers, I give you the product. Calculators are somewhat quicker (although I can still outpace anyone who has to engage in key entry up to 12 times 12), and much more reliable. In my nascent dotage, I have been known to make a momentary mistake. But in what sense do I "know" the rules and rubric of mathematics from these now in-dwelling, basic pattern-recognition skills? The answer is probably not much. But I can however, still reconstruct these foundational transformations from what I have learned. Given an ascending level of autonomous support, it is not that humans cannot still learn such skills, but why ever would they? As with calculators, so with vehicles. If the automated vehicle can take you flawlessly from origin to destination, why ever learn the skills of driving? In fact, as autonomy takes over ever-greater tranches of human performance, why learn to do any of these tasks? Let the autonomy take care of it. After all, humans have evolved to be very economical, efficient, and even miserly with both their physical and cognitive efforts. If autonomy can subsume such expenditure of effort, why not let it do so? What we have to be wary of here is that the assemblage of human and machine is what proves to be "intelligent," but when one of those elements burgeons, the other should not be diminished in a zero-sum fashion. Whether such forms of "de-skilling" that threaten current professions such as that of pilot might extend to vast stretches of human pursuits is the evident fear. What happens to food production if the relevant autonomous system fails and the skills of farmers have been lost in the intervening generations? Such diminution obviously increases the fragility of civilized society (Hancock, 2019a).

ISLES OF AUTONOMY—ISLES OF HUMANITY

The final issue I want to deal with in this present chapter is to examine the mode and manner of the evolution of these approaching autonomous systems. Elsewhere (Hancock, 2018), I have suggested that what we shall witness is the emergence of *isles of autonomy*. By this I mean that at first, we will identify perhaps even just one, extremely limited example of an autonomous technological entity. This Stage I development I term the *emergence watershed*. Like the emergence of all new "species," its first incarnation will be fragile and vulnerable. Hence, as per the associated illustration in Figure 13.2, this entity is likely to be ring-fenced around with, and buffered by, a division of human teammates who help and augment in case of functional uncertainty.

As time goes on, and the relative capacities of autonomy in general improve, I anticipate that we will see the emergence of another independent autonomous agent. I term this Stage II development as *the second coming*. At first, these disparate systems will be largely independent of one another and again will rely on their human companions to act as the bridge of communications between them. Despite this human contribution, either at the design or operations phase, this jointure will represent an important threshold in developments. I term this Stage III level of development as the *precipitating watershed*. This stage may well proceed rather quickly. The next threshold will be that in which such autonomous systems can, with no human mediation, interchange information among themselves. This will represent an even bigger watershed, and thus I have termed this Stage IV development as the *independence watershed*. As per my previous discussion, we humans may not necessarily be direct witnesses to such watersheds, although their more elaborate effects will be felt throughout society and most especially in the domains within which these respective autonomies act. According to the descriptive model I have offered (Figure 13.2), the next stage is one in which autonomy dominates. Not unnaturally, when greater tranches of the global operational domains are subject to autonomous

To Autonomy and Beyond

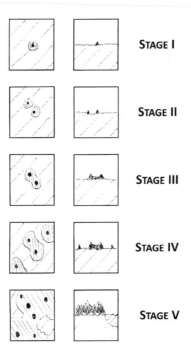

FIGURE 13.2 The progression from Isles of Autonomy to Isles of Humanity. At first (Stage I), there emerges a singular point of autonomy. Its operational capacities are supported and enabled by surrounding human teammates. Over time (Stage II), other "islands" emerge. At some juncture (Stage III), two or more of these "isles" converge and coalesce together. Here the bridge between them is still facilitated by human teammates and companions. The following step (Stage IV) sees the mergence and then dominance of autonomous systems across the whole landscape. Increasingly greater constraints are brought to bear on human contributors. Finally (Stage V), autonomy dominates, leading to ever-diminishing "isles of humanity." What happens to the humans functionally extirpated by these autonomous developments is often overlooked, or they are seen as casualties of associated economic stringencies. The irony that this progressive description bears much of the same character as that described in Figure 13.1 has not been lost on the author.

(as opposed to human) control, we will witness a fully ascendency of technological control. I have termed this Stage V development as the *dominance watershed*. What lies beyond this transition is unclear.

However, assuming for a moment the zero-sum postulate, what is suggested is that we will witness diminishing *isles of humanity*. Although unlikely to be in competition for pure "physical" space per se, there will be the issues of resources and control. Whether this necessarily creates a form of predator-prey ecology is as yet uncertain as no form of "artificial" predator has ever been created. If, like the Golem, autonomous systems are denied self-intentionality *by design*, this circumstance need never arise (Wiener, 1950; 1966). Yet we constantly design systems that are antithetical to someone's existence, as in the case of emerging automated killing machines (Krishnan, 2016; Sharkey & Suchman, 2013). It is therefore most probable that if such forms of autonomy can be designed, they will be. For while humans have, *pro tem*, abjured the use of global nuclear weapons, the time remains fragile at best. We have also, by mutual consent, limited the use of chemical and biological weapons, and yet they are still employed in modern warfare as transgressions of agreed treaties show. What we create, we use. And now our next great creation will be autonomous machines.

SUMMARY AND CONCLUSIONS

As with the revolution in automated systems, human interactions with autonomous systems seem ever more profoundly paradoxical. For if autonomous, what is any human role supposed to be? But this is again and also to underestimate the fact that we have been interacting with autonomous systems throughout our species' existence—with these (essentially) autonomous agents being other human individuals. So will nascent autonomous systems be burdened with person-hood? That is, are such systems primarily created to replicate human forms of response? It seems rather limiting and curtailing to do so; and, do we not have humans aplenty anyway? Thus, we can visualize an intrinsic continuum whose one anchor is the artificial replication of human characteristics, and its antithetical anchor finds technological systems oblivious to all forms of human communication and empathy. Between such extremes lies our future reality. And, of course, states along the continuum are not mutually exclusive, one of the other.

In my chapter in the original Parasuraman and Mouloua text (Hancock, 1996), I advocated that human interaction with automation should be considered via the twin metaphors of the "sheepdog" and the "Japanese garden." The former comparison described circumstances in which the human expressed a series of strategic goals while the automation dealt with certain momentary, tactical needs in relation to the mutually expressed goal. Thus, the shepherd controlled the dog in terms of its higher-level goals, e.g., how to approach the flock of sheep, while saying virtually nothing about, presumably, "lower level" activities such as limb control, locomotion, etc. The other metaphor, the "Japanese garden," was not so much emphasized in the subsequent discussion over the decades that followed. Yet in respect of autonomous systems, it is my anticipation that how we *sculpt* the operational *eco-niches* in which they are permitted to operate will represent the dominant question as to how we live with them (Hancock, 2017).

The sharing of power is a transfer of hegemony. We must face a rather unpalatable fact that human versus autonomous system influence might truly turn out to be a zero-sum game, at some level of analysis beyond that of each individual human being. I have here explored this proposition with reference to a geographic analogy: isles of autonomy/isles of humanity. Here, the rare and fragile first expressions of machine independence emerge necessarily on the constant support of human companions and teammates. Yet as autonomous capacities grow, their sphere of influence proliferates until, eventually, expressions of autonomy coalesce in some manner and a threshold is passed in which "emergent" autonomy is realized. Often characterized as the "singularity" (Kurzweil, 2010), I prefer to think of such an event as the "plurality." I conceive of it in this fashion since I see no clear evidence that "autonomy" necessarily needs to be homogenous in any manner. Such emergences may even pass unnoticed by human beings who possess no necessary direct perceptual capacities for such recognition of such. This human insensitivity need not relate to intelligence per se but might simply be a reflection of our temporal constraints (Hancock, 2005). So, autonomy's emergence and coalescence may occur at a temporal rate many several orders of magnitude in time faster than we humans can appreciate.

The debates around autonomy have come to be mixed up in the mire of the question of self-intentionality. This concatenation induces and then features forms of fundamental attribution error (FAE) in which we give to the machine abilities and aspirations that we ourselves possess. This propensity operates on an "as if" basis. That is, we empathize and act "as if" we possessed the same assumed capacities as the other entity we are encountering. Attribution is a human psychological propensity, not a machine ambition. We will, however, find that the "eco-niche" of such growing autonomous capacities is never perfectly aligned with our own human ambitions. Much of the coming decades will be spent in working out, making transparent, and resolving these conditions of inter-entity dissonance. The current emphasis on the "tool-to-teammate" transitions will be much altered as we experienced the uncertainty and confusion attendant on autonomous system evolution. I have argued previously (Hancock, 2017) that across stages of development, increasingly sophisticated autonomous entities necessarily become more opaque and less transparent to their human partners. This I see as a crucial issue. In the same way that the tactically oriented frontline war fighter of past wars was unable to access the overall strategic plan (if there was one!), so human operators are liable to be "blind" to the multilayered goals of increasingly connected autonomy.

The problem in trying to predict the future lies not in the simple linear extrapolation of contemporary, extant trends. We can always imagine things in the future going faster and further. What always makes a mockery of such linear predictions are the "left turns" that history takes with the invention of technologies that radically alter the physical and social environment in which we exist. On many occasions, such turns surprise and confound those who experience them. Continuing the analogy, these brisances come out of "left field" and their impacts are profound. Some of these are of long latency in which a trend or invention gathers a momentum that then acts to redirect the course of progress, albeit rather slowly. Others act almost overnight to exert their effects. Autonomous technical systems are surely one such turning point in human history. The difference here, I believe, is that we ought to clearly see this change coming. While we cannot see as clearly on to the far side of its effects, we should at least prepare for these anticipatable changes.

To return to the very first observation of the present chapter, when automation was gathering great momentum, we, in the HF/E community, expressed much concern over the form that human-automation interaction would take. Adaptive automation, function allocation, hierarchies of supervisory control were all to the fore of discussion. The linear extrapolation of these concerns leads us to issues such as adaptive autonomy, autonomous allocation, and the even further diminishment of human supervision. Yet, the anticipated technological advances do not encourage us in that linear extrapolation. Humans naturally express concern about such uncertainties; after all, had they not, they would not have become the dominant species. And for any dominant species, evolution provides only one real alternative and that is removal from that principal position. It may well be here that humans prove the architects of their own iatrogenic diminution. Elsewhere, I have considered this eventuality and even find some comfort in it (Hancock, 2019a). In light of this, I am content to conclude the following: Human beings are involved in a perilous existential race between extermination from the impact and ramifications of some facets of their technology and the putative solutions and resolutions posed by other aspects of that same technological revolution. I would further conclude that autonomous systems are the pivotal point in that race and they can serve to save us or sink us. I think the parameters that dictate which edge of the "sink-or-swim" threshold we go over are now being set. I do not think these are being set mindfully but are developed in support of transient commercial profit. In this sense, I think the local optimizations in space and time of such temporary gains are being set directly against the long-term sustenance of human civilization (Hancock, 2019c). I wish I could be more sanguine and more optimistic about the eventual outcome of this line of progress. Yet, in adhering to some remnant vestiges of rationality, I cannot.

NOTES

1. It is always critical to reiterate that the common parlance usage of the word "utopia" betrays a nonreading of Thomas More's original expression of such a place (More, 1516). Even a cursory reading of his text disillusions anyone from considering utopia as some form of earthly paradise that the word is so often used to communicate.
2. The phrase should very much remind us of John Donne's aphorism that "no man is an island," by which he meant that the human species is almost inevitably a social one.
3. There is even a current advertisement that now suggests to the public that the most sophisticated technology they possess sits in their driveway.
4. One might ask, for example, if all route guidance is a function of smart infrastructure communicating with smart vehicles, why does signage need be present at all? If and when such signage starts to degrade in capacity because the vast majority of users now do not need it, what of the remaining road users who still depend on it? It will look to be a case of adapt, i.e., automation is ubiquitously adopted, or do not use the system at all. Of course, this danger extends well beyond road signs to many aspects of the interconnected and interdependent infrastructure of "civilized" society.
5. It is important to note that passive resistance only works to the degree that the controlling authority emphasizes with and even shares many of the moral precepts of those engaged in such resistance. Failing this coacting form of moral empathy, passive resistance can dissolve rapidly both as an effective strategy and as a de facto method of protest. The question here is the degree to which any "oppositional" autonomy contains, or is designed to contain, such a degree of moral empathy (Wallach & Allen, 2008).
6. It is rather piquant that just as we are beginning to demonstrate mastery of human factors in design, the circumstances are arising in which there are potentially fewer and fewer human interactors to design for.

REFERENCES

Austmine. (2018). *Fully-automated mines: Science fiction or reality?* (Republished June 21, 2018). Retrieved from: www.austmine.com.au/News/fully-automated-mines-science-fiction-or-reality; accessed 06/20/2018.

Battino, R. (1992). On the importance of rote learning. *Journal of Chemical Education*, *69* (2), 135.

Benet, S. V. (1935). *Nightmare no. 3: The revolt of the machines*. Retrieved from: www.tangentonline.com/old-time-radio/1175-nightmare; accessed 06/05/2018.

Chapanis, A. (1965). On the allocation of functions between men and machines. *Occupational Psychology*, *39* (1), 1–11.

Fitts, P. M. (1951). *Human Engineering for an Effective Air Navigation and Traffic Control System*. Washington DC: National Research Council.).

Ford, M. (2015). *Rise of the robots: Technology and the threat of a jobless future*. New York: Basic Books.

Frey, C. B., & Osborne, M. A. (2013). *The future of employment: How susceptible are jobs to computerization?* Retrieved from: www.oxfordmartin.ox.ac.uk/downloads/academic/The_Future_of_Employment.pdf; accessed 03/13/2018.

Hancock, P. A. (1996). Teleology for technology. In: R. Parasuraman and M. Mouloua. (Eds.). *Automation and Human Performance: Theory and Applications*. (pp. 461–497), Hillsdale, NJ: Erlbaum.

Hancock, P. A. (2005). Time and the privileged observer. *Kronoscope*, *5* (2), 176–191.

Hancock, P. A. (2013). In search of vigilance: The problem of iatrogenically created psychological phenomena. *American Psychologist*, *68* (2), 97–109.

Hancock, P. A. (2014). Automation: How much is too much? *Ergonomics*, *57* (3), 449–454.

Hancock, P. A. (2017). Imposing limits on autonomous systems. *Ergonomics*, *60* (2), 284–291.

Hancock, P. A. (2018). *Isles of autonomy: Are human and machine forms of autonomy a zero sum game?* Keynote presentation at the NATO Human-Autonomy Teaming Symposium, Portsmouth, England, October 15, 2018.

Hancock, P. A. (2019a). In praise of civicide. *Sustainable Earth*, Submitted.

Hancock, P. A. (2019b). Some pitfalls in the promises of automated and autonomous vehicles. *Ergonomics*, *62*(4), 479–495.

Hancock, P. A. (2019c). The humane use of human beings? *Applied Ergonomics*, *79*, 91–97.

Hancock, P.A., Billings, D.R., Olsen, K., Chen, J.Y.C., de Visser, E.J., & Parasuraman, R. (2011). A meta-analysis of factors impacting trust in human-robot interaction. *Human Factors*, *53*(5), 517–527.

Hancock P. A., & Drury, C. G. (2011). Does human factors/ergonomics contribute to the quality of life? *Theoretical Issues in Ergonomic Science*, *12*, 1–11.

Hancock, P. A., Hancock, G. M., & Warm, J. S. (2009). Individuation: The N=1 revolution. *Theoretical Issues in Ergonomic Science*, *10* (5), 481–488.

Hancock, P. A., Nourbakhsh, I., & Stewart, J. (2019). On the future of transportation in an era of automated and autonomous vehicles. *Proceedings of the National Academy of Sciences*, *116*(16), 7684–7691.

Hancock, P. A., & Scallen, S. F. (1998). Allocating functions in human-machine systems. In: R. R. Hoffman, M. F. Sherrick, and J. S. Warm (Eds.). *Viewing Psychology as a Whole: The Integrative Science of William N. Dember*. (pp. 509–539), Washington, DC: American Psychological Association.

Hoffman, R. R., & Hancock, P. A. (2014a). Words matter. *Human Factors and Ergonomics Society Bulletin*, *57* (8), 3–7.

Hoffman, R. R., & Hancock, P. A. (2014b). (More) words matter. *Human Factors and Ergonomics Society Bulletin*, *57* (10), 9–13.

Hoffman, R. R., & Hancock, P. A. (2017). Measuring resilience. *Human Factors*, *59* (4), 564–581.

Kaber, D. B. (2018). Issues in human–automation interaction modeling: Presumptive aspects of frameworks of types and levels of automation. *Journal of Cognitive Engineering and Decision Making*, *18* (1), 7–24.

Krishnan, A. (2016). *Killer robots: Legality and ethicality of autonomous weapons*. New York: Routledge.

Kurzweil, R. (2010). *The singularity is near*. London, UK: Gerald Duckworth & Co.

More, T. (1516). Utopia (translation). In: J. J. Greene, J. P. Dolan (Eds.), (1967). *The Essential Thomas More*. New York: New American Library.

Orlady, H. W., & Orlady, H. M. (2016). *Human factors in multi-crew flight operations*. Oxford, UK: Routledge.

Parasuraman, R., & Mouloua, M. (1996). (Eds.). *Automation and human performance: Theory and applications*. Hillsdale, NJ: Lawrence Erlbaum Associates, Inc.

Parasuraman, R., Sheridan, T. B., and Wickens, C. D. (2000). A model for types and levels of human interaction with automation. *IEEE Transactions on Systems, Man, and Cybernetics—Part A: Systems and Humans, Systems, Man and Cybernetics, 30* (3), 286–297.

Roth, E. M., & Pritchett, A. R. (2018). Preface to the special issue on advancing models of human–automation interaction. *Journal of Cognitive Engineering and Decision Making, 12* (1), 3–6

Sarter, N. B., & Woods, D. D. (1995). How in the world did we ever get into that mode? Mode error and awareness in supervisory control. *Human Factors, 37* (1), 5–19.

Schaefer, K. (2013). The perception and measurement of human-robot trust. *Electronic Theses and Dissertations*, 2688. Available at https://stars.library.ucf.edu/etd/2688

Schaefer, K. E., Straub, E. R., Chen, J. Y., Putney, J., & Evans III, A. W. (2017). Communicating intent to develop shared situation awareness and engender trust in human-agent teams. *Cognitive Systems Research, 46*, 26–39.

Sharkey, N., & Suchman, L. (2013). Wishful mnemonics and autonomous killing machines. *Proceedings of the AISB, 136*, 14–22.

Sheridan, T. B. (2002). *Humans and automation: System design and research issues*. New York: John Wiley(published in cooperation with Human Factors and Ergonomics Society, Santa Monica, CA).

Sheridan, T. B., & Ferrell, W. R. (1974). *Man-machine systems; Information, control, and decision models of human performance*. Cambridge, MA: The MIT Press.

Sheridan, T. B., & Verplank, W. L. (1978). *Human and computer control of undersea teleoperators*. Retrieved from http://www.dtic.mil/docs/citations/ADA057655; accessed 05/18/2018.

Sparrow, B., Liu, J., & Wegner, D. M. (2011). Google effects on memory: Cognitive consequence of having information at our fingertips. *Science, 333*, 776–778.

Wallach, W., & Allen, C. (2008). *Moral machines: Teaching robots right from wrong*. Oxford, UK: Oxford University Press.

Wiener, E. (1987). Interview in *Why Planes Crash*. www.imdb.com/title/tt0942719/; accessed 05/26/2018.

Wiener, N. (1950). *The human use of human beings*. New York: Da Capo Press.

Wiener, N. (1966). *God and Golem, Inc*. Boston: MIT Press.

Winner, L. (1978). *Autonomous technology: Technics-out-of-control as a theme in political thought*. Boston: MIT Press.

14 Teleology for Technology

Peter A. Hancock

I have seen
The old gods go
And the new gods come
Day by day
And year by year
The idols rise
Today
I worship the hammer.
(The Hammer, Carl Sandburg)

TELEOLOGY: 1. The doctrine of final causes or purposes; 2. the study of the evidence of design or purpose in nature; 3. such design or purpose; 4. the belief that purpose and design are a part of, or are apparent in nature (2019).

TECHNOLOGY: 1. The branch of knowledge that deals with the industrial arts: the sciences of the industrial arts; 2. the terminology of an art, science, etc.; technical nomenclature (2019).

"Lo! Men have become tools of their tools."
(Henry David Thoreau)

STATEMENT OF PURPOSE

Science and technology have always tried to answer the question "how?" How does this or that mechanism work? What are the laws and causal properties that underlie this or that phenomenon? In the case of technologies, how can such knowledge be used to develop a useful tool or machine? However, science and technology rarely address the question "why?" It is often conceived as being outside their respective spheres of discourse. The question is ruled inadmissible or not appropriate for the methods and capabilities at hand. I dismiss this rejection. I believe that the questions how and why are so mutually dependent that they should never be considered independently. Indeed, I attribute much of our present grim circumstances to this unfortunate and unhappy division. Those who know how must always ask why. Those who ask why must always think how.

Overview of the Chapter

With reference to the foregoing thematic statement, I want to examine our collective future by asking questions about our intention with respect to technology. This argument is set against a background of current human-machine systems and particularly the rise of automatic systems. I do this because of my belief that technology cannot and should never be considered in the absence of human intention. Likewise, contemporary societal aims have no meaning without reference to the pervasive technology that powers them. I start off the discussion with a prelude that presents a metaphor to frame the initial considerations. I then define the terms within which the chapter's arguments are framed. This definition of terms leads to an examination of what technology is and to what extent technology is "natural." I then examine human-machine symbiosis and potential futures that may be encountered by such a coevolutionary pairing. I will point to human-centered automation as one stage in this sequence of evolution that will eventually witness the birth of autonomous machine intention about which I express a number of cautions. In noting the stage-setting function of contemporary systems design, I cite earlier warnings concerning previously held principles of human-machine interaction. My hypothesis is that the collective potential future for humans and machines can only be assured by the explicit enactment of mutually beneficial goals. In the immediate future, I caution against the possibility of a society divided by technology against itself. I advocate for a science of human-machine systems as a liberating force in providing technical emancipation, the heart of which is universal education.

A METAPHORICAL PRELUDE

> It is in this way alone that one comes to grips with a great mystery that life and time bear some curious relationship to each other that is not shared by inanimate things.
>
> **(Eiseley, 1960, p. 169)**

SETTING THE "SCENE"

To start this chapter, I want to put a vision in your head. To do this, I am going to use a familiar icon. This icon is not a single picture but rather, it is a scene from a famous motion picture. The film is Alfred Hitchcock's *North by Northwest*; the scene is the coast road. I hope this brief description will let most readers identify the sequence I mean. However, for those who are not familiar with the movie, it is as follows. Our hero, Cary Grant, has been forcibly intoxicated by the henchmen of the evil James Mason on the mistaken assumption that Grant is an investigating government agent. To rid themselves of this menace, the evildoers put the now drunk Grant in a car and start him off on a perilous trip down a steep and winding coast road. Through force of will and no small portion of luck, our hero manages to survive both the human and the environmental sources of danger to fight again. Even those who have not seen the film will not be surprised to know that in the end, the forces of evil are routed and our hero survives to win the girl.

A METAPHORIC RELATION

This outcome is all well and good in film plots, but I want to use Grant's vehicular progress as a metaphor for our own uses of technology. I want to suggest that we, like him, are careening down a dangerous path. Like Grant, we have not intentionally put ourselves in this predicament but nevertheless, here we are. We each possess similar goals, in which simple survival has the highest priority. Both we and Grant are painfully aware that the volatile combination of powerful technology and fallible humans in unstable environmental circumstances threatens disaster. While our progress resembles Grant's in many respects, we are radically different in some critical ways. Above all things, we have no director to shout "cut" when things get too dangerous or scriptwriter to "ensure" that the story ends well. Unlike Grant, we also seem to be drinking heavily as we proceed down the mountainside in, apparently, a progressively more chaotic fashion. We do, however, have a science whose primary purpose seems to be to ensure that we are able to keep a firm grip on the bottle and a lead foot on the accelerator. The science is what has been traditionally known as "human factors." Also, we have a motive force that seeks to accelerate the rate of our "progress." The force is technology.

META-TECHNICAL PURPOSE

By using this metaphor, I am suggesting that the emerging science of human-machine systems largely fails to address the fundamental purposes of and for technology. Those involved in the facilitation of human-machine interaction rarely question whether technology represents the appropriate avenue through which human society can achieve its necessary goals. The traditional science of human factors seems to have accepted the current social assumptions behind technology. In so doing, it looks to facilitate human-machine interaction, even if the purpose of the machine is suicide, genocide, or terracide. By this generalization, I do not mean that many individual members of this science do not question such functions; they definitely do (for example, Moray, 1993, 1994; Nickerson, 1992; see also Hancock, 1993. However, as a body, those scientists involved in human-machine interaction have yet to state that it is their role to question the very purposes of technology. I affirm that it is. Those outside the scientific and technical community frequently protest about their perceived helplessness and the subsequent fallout of technology. However, it is human factors scientists and practitioners who shape human interaction with technology that can and should "direct" from within.

MORE THAN AN "APPLIANCE SCIENCE"

Therefore, in what follows, I want to question the fundamental tenets of human interaction with technology. I want to question whether the human-machine sciences should always facilitate the human-machine linkage, especially when such improvements are clearly antagonistic to the collective good. I want to question whether we should always assume that technological growth, and increased automation in particular, are appropriate. In sum, I want to question where we are going with technology and what our goals and purposes are.

My conviction is that human factors and the emergence of more widespread human-machine sciences are more than just a form of "appliance science." I believe that human-machine systems studies can exert a strong influence in what we do with technology and these are beginning to be recognized more and more in that role (Sedgwick, 1993). I believe this science can be a critical voice in determining the goals we collectively set for ourselves. It is the mandate of human factors to mediate among humans, technology, and the environment. This mediation requires informed value judgments. I think human factors scientists must acknowledge and grasp this social responsibility. The future is too important to be left to the accidental happenstance of time, the pointlessness of financial greed, or the commercial search for something smaller, faster, and less expensive. In essence, the carefree adolescence of humankind is now at an end, and this new millennium must hail an age of responsible societal adulthood. If not, it will in all likelihood witness our eventual demise as a species.

I fully realize that the foregoing is largely a series of polemical exhortations. The reader is entitled to inquire not only about the basis for such assertions but also the reasons why they should be put forward as important at this time. I think the answer is simple. It has been in the final decades of the twentieth century and the first decades of the twenty-first century that we have started to set the agenda for automated technologies and the stance of humans with respect to such automated systems. In truth, this represents the earliest growth of machine autonomy and independence. Decisions made now constrain what will be possible in our future. I want that future to contain at least sufficient alternatives for one to be our continued survival.

A Definition of Terms: Teleology and Technology

> Machines just don't understand humans. They [automobiles] don't understand that they should start even when we do dumb things like leave the lights on or fail to put antifreeze in the radiator. But maybe they understand us too well. Maybe they know we rely on them instead of ourselves and, chuckling evilly in the depths of their carburetors, they use that knowledge against us.
>
> We humans forget how to rely on our own skills. We forget we can survive without machines; we can walk to the grocery store even if it's a mile away. We forget that before the calculator, we used our brains. (Well at least some of us did.) We forget that the Universe doesn't grind to a halt when machines break down. Machines like that. They're waiting for us to get soft and defenseless so they can take over the world.
>
> **(Tolkkinen, 1994)**

In what follows, I hope to show some of the fallacies and misunderstandings that underlie Tolkkinen's recidivist position.

Teleology: A Traditional View

I want to define the terms that are central to the arguments that follow. I start with the term "teleology." Teleology is a word that has historically been used with respect to the existence of a deity or some form of final cause or purpose for being, although in respect to the Aristotelian interpretation, it need not necessarily be so associated. The concept of teleology is founded upon our human observations of order in the universe. The teleological argument postulates that because the universe shows this order, there must be some entity behind that order that creates and directs it.

Although if the initial act of creation itself served to kill such a god, then the process of creation and ongoing direction can be theologically separable. Deism argues for a creator but one that is subsequently uninvolved with what has been created. These assertions concerning a sentient creator have been rebutted by a multiple universe argument, which proposes that any order we observe is only present because we, as human beings, are here to observe it. That is, many other universes are possible, but conditions within them do not give rise to observers who observe them. The latter point is a cornerstone of the anthropic principle, which seeks to shed light on the early conditions of our universe, founded upon the fact that observers such as ourselves do exist. Finally, one can argue that there is no intrinsic order in the universe and that it is we humans ourselves who create such order through our perceptual propensities. Although these argument might not appear to be amenable to empirical resolution, there are opportunities to explore a more scientific approach to theology, an endeavor I hope to be able to undertake.

TELEOLOGY: RETROSPECTION VERSUS PROSPECTION

I do not want to use the term teleology in this retrospective way to look back to the origins of order nor to speculate here about the existence of any deity. For me, teleology does not relate to a historic and passive search for ultimate cause. Rather it is an active term concerning our prospective search among the potentialities with which we, as human beings, are presented. As a result, I use teleology in the present context to refer to an active search for unified purpose in human intention. By using it in this fashion, I am affirming that individuals and society do exhibit intention. That is, regardless of metaphysical arguments as to the reality or illusion of free will, we individually and collectively act in an everyday manner that signifies our belief that our actions are not uniquely predetermined and that our beliefs do influence ourselves and others. In sum, my use of teleology represents a search for goals and intentions for ourselves and society. In the context of the present chapter, I focus on the role that technology plays in this search.

TECHNOLOGY: ART, ARTIFACT, OR SCIENCE?

The formal definition of technology that is cited at the start of this chapter is different from that of everyday use. However, it is the term as used in common parlance that I now want to employ here. In these terms, technology is associated with things, objects, and devices, rather than with knowledge as either art or science (see Westrum, 1991). While a concept of information underlies each of these definitions of the word technology, my use of it here very much accords with the common everyday way that we use the term. However, I want to broaden this definition just a little. Technology is often associated with that which is new, modern, and exciting. I want to go beyond thinking of technology in terms of new machines, to expand the definition to include such developments as the domestication of animals and of plants. It might be of some interest to note, for example, that carrots were originally white. Their present appearance comes from selective breeding to "engineer" the vegetable we see today. Likewise, many animals owe their present appearance and subserve functions to such "manufacturing" over a considerable period of time. I am grateful to Kara Latorella who pointed out Gary Wilkes' work from the *Mesa Tribune* on an article entitled "Animal Behavior Put to Task for Good—and Bad—of Humans." In this, he examines birds trained to respond upon seeing the color yellow being used by air-sea rescue to find downed pilots, the pigeon having much greater visual acuity for yellow than a human observer (and thought to be less affected by the noise and vibration of the rescue helicopter). For a more technical evaluation of this work see Parasuraman (1986). Also, there is another use in which pigeons were trained to spot defective parts on a production line as a way to relieve human observers of the vigilance burden. Interestingly, this latter project was stopped by protesters who said that enforcing birds to perform this tedious task was "inhuman." No one protested for the human workers who were then rehired to do the same job.

Technology then can be redefined as the purposive ways in which the environment has been structured to benefit humankind. Already then, there is teleology in the very basis of technology. (I hope the reader can already see this mutuality such that the chapter may as easily be titled "Technology for Teleology.") A wide interpretation of this definition of technology permits the identification and inclusion of rudimentary tools and even forms of communication that are employed by other members of the animal kingdom to structure their own world. Unlike those who seem preoccupied with searching for the absolute unique nature of human beings, I am not unhappy about this wider interpretation of what technology represents.

HUMANS AND THEIR TECHNOLOGY: A MUTUAL SHAPING

What is demonstrable is that technology has fashioned humans as much as humans have fashioned technology. Elsewhere, I and my colleagues (Flach, Hancock, Caird, & Vincente, 1995; Hancock et al., 2005) have argued that our contemporary ecology is technology and as such it represents a most powerful influence upon who we are today. Such is the degree of our reliance upon technology that we could not survive in the way that we do without its support. As Arthur Koestler (1978, p. 4) put it:

> It has to be realized that ever since the first cave dweller wrapped his shivering frame into the hide of a dead animal, man has been, for better or worse, creating for himself an artificial environment and an artificial mode of existence without which he no longer can survive. There is no turning back on housing, clothing, artificial heating, cooked food; nor on spectacles, hearing aids, forceps, artificial limbs, anesthetics, antiseptics, prophylactics, vaccines, and so forth.

While at some very basic level this mutual influence between the conditions of the environment and an individual's actions applies to all living things, the extent to which humans have exploited, and now rely on, technology is, I think, sufficiently distinct to provide a watershed differentiation. The central question then is the future for a species that relies so heavily on technological support and what happens to that species as the nature of the support itself coevolves.

Teleology for Technology

TECHNICAL VERSUS HUMAN CAPABILITIES

The servant glides by imperceptible approaches into the master; and we have come to such a pass, that even now, man must suffer terribly on ceasing to benefit [from] machines. If all machines were to be annihilated at one moment, so that not a knife nor lever nor rag of clothing nor anything whatsoever were left to man but his bare body alone that he was born with, and if all knowledge of mechanical laws were taken from him so that he could make no more machines and all machine-made food destroyed so that the race of man should be left as it were naked upon a desert island, we should become extinct in six weeks. A few miserable individuals might linger, but even these in a year or two would become worse than monkeys. Man's very soul is due to the machines; it is a machine-made thing; he thinks as he thinks, and feels as he feels, through the work that machines have wrought upon him, and their existence is quite as much as sine qua non for his, as his is for theirs.

(Butler, 1872, p. 290)

The Direction of CoEvolutionary Changes

I begin my evaluation of our coevolution through the use of a traditional, dichotomous comparison of human and machine abilities. This is expressed in Figure 14.1. I have already indicated that it is critical to find purpose in technology and to identify purpose at all times. Technology may have been around as long as human beings have cultivated crops, herded cattle, or created their own tools. What makes the present different is the change in the respective roles of the machine and the human operator. A traditional view of this change is illustrated here, where the horizontal axis is time, on the vertical axis capability.

The curve labeled (b) represents human capabilities. It is perhaps easiest to conceptualize these abilities first in terms of simple physical achievement. As can be seen, human capabilities improve over time, but the rate of that improvement gradually diminishes. These trends can be seen most clearly in world records for forms of "locomotion" such as running and swimming. What we know from experience is borne out in the data; human beings get better but by progressively smaller amounts. This trend holds for individual learning curves as well as collective performance,

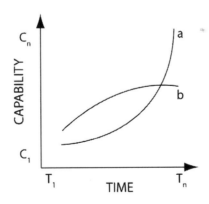

FIGURE 14.1 Human (b) versus machine (a) capabilities as a function of time.

as expressed in world athletic records. Eventually, we will reach the constraints of physical capability. It is true that unaided by technology, a human can run a mile in four minutes. However, unaided, no human will run a mile in one minute (at least not a human that we would recognize as such). The question is: Are intrinsic cognitive functions limited in the same way? We may each improve our cognitive capabilities but, eventually, is there a limit upon sensory resolution, memory capacity, decision-making speed, as well as motor skills? Assuming there to be such limits, defining where they occur on any ratio scale of measurement is, of course, much more difficult than for structured physical pursuits. Parenthetically, this manifest, public measurement of achievement may be one reason for the attraction of sports in contemporary society, in which the winner, in most games, is identified unequivocally. It is this pristine and public way of determining a winner that proves so seductive, and hence the outrage that follows episodes of "cheating" in major sports, which are seen, often naively, as the one arena of true competition and resolution. For comparison, imagine trying to have an objective competition to determine which is the best film, scientist, etc. Regardless of where these latter cognitive limits reside, we can be fairly sure that unaided by technology, individuals do not possess unbounded, endogenous cognitive abilities.

The curve labeled (a) represents technology. There are a myriad of examples of this exponential form of growth (see Card, 1989). For example, the history of the size, speed, and unit cost of computer memory shows this phenomenal technical improvement clearly (More, 1516/1965) as do similar measures on a greater social scale. Moravec (1988) has also more recently captured this progression in which he also shows how differing forms of technology have supported continued geometric growth. I should note that this approach, which was conceptually developed over a decade ago, was also articulated by Vallee (1982). Before I discuss the significance of this perspective and the point of intersection, I have to examine some of the shortcomings and assumptions of this comparative approach.

FALLACIES OF GENERALIZATION

Of course, each curve is a generalization. We know that technology, like resources in whatever form, cannot sustain geometric growth indefinitely (Malthus, 1798; see also Nickerson, 1992). Indeed, for the function cited in computer memory, and generally known as "Moore's Law," we seem to be approaching physical limits to the type of capacity for processing and storage (Bennett & Landauer, 1985), although we have not reached those limits yet. Already, however, futurists are touting quantum leaps to alternative "technologies" such as photonics and bio-computing in order to fracture what seems to be an intrinsic limit to silicon-based capacities. So, while increase in technical capabilities can appear to be geometric, it is only so for some selected portion of their growth. It is much more likely that the curve for computer processing technologies possesses an inflexion and is shaped more like an ogive. Each kind of processor is, through some intrinsic factor, ultimately self-limited. It is then only by the overlapping of progressively more effective processors of a different type that we sustain the illusion of continuous geometric growth. However, this illusion cannot be permanently perpetuated.

With respect to human cognitive capabilities, asserting that we know all about such limits and what constrains human thought at this time is very shortsighted. For every assertion about limits to human intellectual abilities and cognitive capacities, there is an equally powerful assertion about individual variation and strategic skills that counters it (Flach and Hoffman, 2003). Thus, specifying an upper boundary to all cognitive functions is naive.[1] However, the recognition of some endogenous limits is, in no small part, the stimulus for technology itself. I consider this integrative perspective in the next section. Yet despite these and other simplifications, consideration of the intersection of the curves is of critical importance.

CONVERGENCE, INTERSECTION, AND DIVERGENCE

While the physical abilities of machines superseded the physical abilities of humans some generations ago, I want to suggest that it is in our own era in which the cognitive balance between the human and the machine has begun to invert. Independent of the objections that I have raised earlier, it is within our own generation that certain characteristics of machine intelligence have outstripped their human equivalents. Pragmatically, what this has meant is that many control functions are now performed more quickly and more accurately by machines. Such is the nature of the technologies we have, and are building, that many are uncontrollable without the use of computer assistance. I hasten to add that the supersession of individual human abilities by machine abilities has happened, ability by ability, at different times. As a result, different sorts of devices have had to rely upon some degree of machine control, in an obligatory sense, at different junctures. This progress expressed in systems as complex as nuclear power stations and single-seat fighter aircraft has percolated now to technologies as apparently mundane as washing machines. It has now become a challenge to identify machines that do not have some sort of processor in them. Indeed, quite sophisticated processing capabilities can now be seen in smart greetings cards. In the present generation, after holidays and birthdays, it is not unusual to consider these and like items as simply disposable. Indeed, the trash receptacles of many houses this holiday season will contain more computational power than the most powerful systems of only 50 years ago!

It is this transference of superiority in specific areas that generated the birth of imperative automation. By imperative automation, I mean automation we cannot do without. That is, the goal of the system at hand cannot ever be achieved by manual action alone. Human beings have always exhibited a fascination with automata and have incorporated governors and other forms of automatic or semiautomatic controllers into devices for almost as long as humans have been making machines themselves. But this was discretionary automation. Humans could exercise control if they had to, and in many cases, they were found to be the least expensive and indeed the preferred way for doing so. However, in our times, we have had to acknowledge that without machine assistance we cannot control some of the things we have built. Many people work better if they have machine support, but some work only if they have machine support. Hence, the difference is one of acknowledgment that automation is something we cannot do without, rather than something we would prefer to have.

Why is it important to consider the ascending role of technology now? It is important now since our relationship with technology has changed from discretionary use of automation to mandatory use of automation. With that change has come a subtler move in which machines are no longer simply mindless slaves but have to be considered more in the sense of a partnership, although anthropocentric views of this partnership are likely to be misleading at best. Any time of change is disturbing. However, we live in an age when nonhuman and nonanimal entities have, by circumstance, been granted perhaps the birth of their emancipation. We have to ask ourselves questions concerning their future—if and when, they will demand the right of franchise.

Summary: Human and Machine Abilities

Human capabilities have progressively been superseded by machine capabilities. In our age, we now build technologies that rely upon these levels of machine ability that cannot be replicated by any human. Consequently, our relationship with machines has changed. They are no longer unquestioning slaves but are becoming active companions. Are they to become more than this?

Is Technology "Natural"?

> One can, of course, argue that the crisis (of technology), too, is "natural," because man is part of nature. This echoes the views of the earliest Greek philosophers, who saw no difference between matter and consciousness—nature included everything. The British scientist James Lovelock wrote some years ago that "our species with its technology is simply an inevitable part of the natural scene," nothing more than mechanically advanced beavers. In this view, to say we "ended" nature, or even damaged nature, makes no sense, since we are nature, and nothing we can do is "unnatural." This view can be, and is, carried to even greater lengths; Lynn Margulis, for instance, ponders the question of whether robots can be said to be living creatures, since any "invention of human beings is ultimately based on a variety of processes including that of DNA replication from the invention."
>
> **(McKibben, 1989, pp. 54–55)**

The preceding argument is one that is based on the understanding that humans and machines are to be explicitly contrasted as disparate entities. That is, the perspective is dominated by a view of human versus machine abilities. As we progress, I want to argue that this divisive perspective is itself unhelpful. To do so, I have to first overcome the assertion that technology and nature are in some way "opposed." That is, that technology is not "natural."

The Importance of the Question

It might seem, at first, that the question of whether technology is natural is either facile, in the sense that technology being "artificial" cannot be "natural," or pointless, in the sense that the answer makes little difference one way or the other. I suggest that the question is neither facile nor pointless. It is not facile because it forces us to

Teleology for Technology

consider what the boundaries of what we call "natural" are and what artificial means in this context. It is not pointless since our answer biases the very way in which we think about technology and what the ultimate purposes of technology are. Having considered the nature of technology, let us move to a consideration of the nature of its human operator, and to begin, I present an examination of how we look to unify and divide any of our descriptions of such individuals.

How Things are the Same, How Things are Different

The term that characterizes individual differentiation is idiographic. In contrast, the term for the average patterning of events is nomothetic. These twin tendencies form the basis of statistical descriptions since they are reflections of the dispersal and the central tendency of data, respectively. However, the question of grouping or set function goes well beyond this one arena. Indeed, it is a fundamental characteristic of all life that we look for similarities and differences in experience. The use of language represents an explicit recognition of the differentiation or the separating apart of ideas, objects, and things. In contrast, mathematics represents the propensity in the other direction, toward unification. We may start life by distinguishing self from nonself, but it is the richness of language that gives voice to the diversity of the world around us, and parenthetically, it is also language that strikes us dumb with respect to transcendent experience. In essence, we try to name each of the things we can perceive. Before long, however, we start to try to categorize these things by grouping similar things together. That is, we seek common characteristics through which we can link individual items together. This complementarity between unity and diversity continues throughout life. Essentially, these propensities for differentiation and integration go hand in hand in all sequences of perception-action.

In all facets of human life, we rejoice in discovering new ways in which things can be unified so that we can extract pattern from (or impose pattern on) experience. For example, we count the recognition of a "common" force acting on an apple and the moon at one and the same time, as one of the great insights of science. Indeed the concept of number in mathematics is an explicit statement that one object is sufficiently of the same characteristic as another object that they can be put in a common class and recognized as two separate occurrences of the same object type. This observation of multiple members of a common set precedes the concept of the set being empty and having no members. This latter state is the formal definition of zero. The abstraction of number proceeds from this explicit grouping principle. Elsewhere (Hancock, 2002), I have argued that time is the basis for both this fundamental unification and differentiation and hence stands, as Kant (1781) implied, as an a priori psychological and physical construct. However, before this journey proceeds too heavily into the metaphysical, I would like to provide a biological example as a precursor to an examination of technology.

Are Humans the Same as Other Animals?

The example that I would like to examine in detail concerns the difference between humans and the rest of the animal kingdom. We are aware of Descartes' protestation about the soul as the difference between humans and animals (although I suspect that neither Descartes, nor indeed Aristotle before him, were quite the absolutists in this matter that they are now often portrayed to be). It was indeed this barrier between humans and other animals that Darwin, without malevolent intent, so thoroughly ruptured. Contemporary scientific debate does not revolve around the contention of common evolution, but one battleground is now established around language. Lurking in the background is the often silent extension into the question of mind and consciousness and the unsaid and now virtually unsayable link to the soul. The arguments center putatively around the nature of the data but the global agenda of the uniqueness of human creation always hovers in the background.

Why is this? The answer is, I think, understandable. As human beings we have always been, like Cary Grant in the introductory example, the hero of our own story. But our history is a chronicle of our progressive displacement from the center of the universe. From Aristarchus to Copernicus, from Newton to Einstein, the gradual displacement from the physical center has progressed (at times stultified) but never ceased (Koestler, 1959). This outfall of science threatens to displace human beings from the spiritual center of our universe also (Hancock, 2005). It is only in the present century that the concatenation of physical relativity and biological unification has served to shatter some of the foundational pillars upon which the conventional and comfortable worldview was perched for so long.

In 1859, Charles Darwin published *On the Origin of Species*. Epic of science though it is, it was a great blow to man. Earlier, man had seen his world displaced from the center of space; he had seen the empyrean heaven vanish to be replaced by a void filled only with the wandering dust of worlds; he had seen earthly time lengthen until man's duration within it was only a small whisper on the sidereal clock. Finally, now, he was taught that his trail ran backward until, in some lost era, it faded into the night-world of the beast. Because it is easier to look backward than to look forward, since the past is written in the rocks, this observation, too, was added to the whirlpool (Eiseley, 1960, p. 1).

TECHNOLOGY AND NATURAL LAWS

In the sense I have conveyed, we now have to inquire whether technology and nature are different or whether they are in fact essentially the same. For good or bad, we have come to a situation where strong positive empiricism reigns and technology is the material manifestation of that creed. But is this natural? As I am sure the reader has suspected all along, it all depends upon what one considers "natural." That is, are we going to use the term in an inclusive or an exclusive sense? The inclusive, coarse-grained view is that physical entities obey physical laws. Hence, everything is "natural" by this definition of nature. But this view is biased by a reification and generalization of physical laws. To the strict ecologist, to whom these laws and their application is sacrosanct, technology in general and human-machine systems in particular are only extensions of nature. True, they explore more exotic regions that cannot be compassed by any living organism alone, but they are still bound by the same strictures and constraints and are subject to the "pervasive" laws. But, in conception, they are founded in human imagination, which is not bound by any such laws. As Koestler (1972, p. 58) noted, "The contents of conscious experience have no spatio-temporal dimensions; in this respect they resemble the non-things of quantum physics which also defy definition in terms of space, time, and substance." This unbounding is what makes developments such as virtual worlds so intriguing (Hancock, 2009).

INEVITABLE TECHNOLOGICAL FAILURES?

I have purposefully spent some time considering the general level of the question of "natural" technology to provide a background for the following and more specific example. We have, in our science, noted and commented on the increasing complexity of technical systems. Indeed, one of the major raisons d'être for automation is this progressive complexity. I shall not argue what I mean by complexity here since I have done this elsewhere (Hancock and Chignell, 1989). I simply assert that technical systems of today are more "complex" than those of a century ago. Such growing complexity compels comparison with natural ecosystems. Regardless of our eventual determinations on the link between technology and nature, we can look to nature for models of interacting complex systems with biological "players" as system components. In so doing, we find that there are intriguing models of systems with mutually adapting agents that can provide us with vital information and insight. In particular, we find that the way that failure propagates readily through ecosystems (that is, species destruction) provides a valuable insight. In this sense, they are similar to the tightly coupled technical systems that were discussed in detail by Perrow (1984). However, research on natural ecosystems lets us go a little further than the qualitative statements of Perrow. Indeed, there appears to be a lawful relationship between the extent (effect) of any one failure and its frequency. It has been posited as a log-log relationship (see Kauffman, 1993; Raup, 1986), and is illustrated in Figure 14.2.

Kauffman has suggested an ln/ln relationship in which the log frequency of failure is linear with the log size of failure events. The crux of the argument is that small perturbations are resident in all complex systems and that, in the vast majority of instances, these perturbations are damped out in the system. However, from time to time, these same perturbations are magnified through the system resulting in correspondingly larger destruction of the elements of the system. In this chapter, I do not go into all of the nuances of these important observations. For example, what dictates when in the time series of resident perturbations these larger catastrophes occur? For this and other intriguing issues, I would strongly recommend reference to Kauffman's (1993) work. The importance for the science of human-machine systems, of course, is to understand whether this "law" applies when the entities in the ecosystem are interacting humans and machines. The importance of the general question of the naturalness of technology is now laid bare. If technology is no more than a logical extension of other natural systems, we must expect human-machine systems to be subject to the same effects. The implication of this conclusion is that there will always be failures in such systems and that the size and frequency of those failures will be proportional to the complexity of the system.

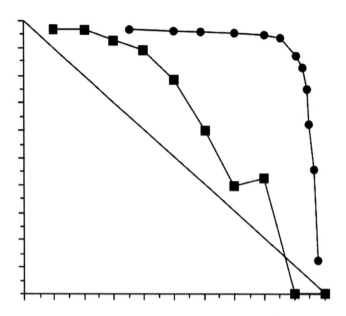

FIGURE 14.2 Relationships in log/log space. The reason that there are no associated numbers is that the axes refer to two specifically different relationships. In general, the axes represent the frequency of events and size of events, respectively. The linear relationship is a hypothetically perfect one, while the two others represent actual data. The circles show human cohort morbidity; the squares show extinction events in the Phanerozoic era. (Conception after Kauffman, 1993; extinction events data after Raup, 1986; morbidity data superimposed here by the present author.)

As a result, since there will always be intrinsic perturbations of these systems and as some of these will always reach such a disastrous magnitude, there will be catastrophic failures as a function of their very nature. If, on the other hand, we are able to convince ourselves that our technology is not "natural," then we can return to the comfort of the belief that all engineered systems, however, complex, must be ultimately controllable; and therefore, catastrophic failure can be circumvented. This latter perspective remains bad news for operators as it argues that human error, in all its forms, will continue to remain the putative "cause" of such failures in the decades to come.

Rejecting Deterministic Failure

It is almost instinctive in us to reject the assertion of "inevitable" or "unavoidable" failure. Indeed, there are several objections that immediately present themselves. Surely technology can be natural without having to fail in this manner. Doesn't such an assertion imply that the safety measures that are put in place are bound to be ineffective? As some reflection will show, it comes back again to the nature of technology and the naturalness of technology. If we believe that human actions supersede the haphazard exploration of potential systems "spaces" by nature, then we are strong advocates for separation and the positive influence of human intention. In essence, we are optimists (Hancock, 1999). If we view the combinatorial explosion of system complexity with trepidation and have difficulty distinguishing technology from nature, then we are liable to favor the more fatalistic view as represented in the latter conception. From the foregoing arguments, metaphysical conundrums such as the mystery of mind, the reality of consciousness, and the power of free will are at the heart of how we conceive possible futures for human-machine systems. If we are to be dedicated human-machine systems scientists, we have to address ourselves to these perennial human mysteries as much as we do the nature of the interface, the character of drop-down menus, or the solution to repetitive strain trauma. For we can never find acceptable solutions to the latter issues unless we have resolved in our own mind our stance toward the former concerns. That is, our science must be mindful, not mindless.

Giving Solipsism the Slip

At this juncture, the morass of radical skepticism and fundamental solipsism may beckon the unwary traveler towards the "slough of despond." However, the pragmatic solution to this issue is relatively simple (see Kurtz, 1994). As I have noted earlier, regardless of our personal philosophical stance, we all act as though we exercise free will. In our discussions of consciousness also, we are mostly willing to attribute consciousness to other human beings, although we are a little less generous with our attribution to animals and positively miserly to other forms of life. Our practical pragmatism means that we use the concept of mind as something that is a shared characteristic of human experience. Therefore, at a practical level, we continue to believe that our interventions are important, whether they are or not (Kurtz, 1992). Although the position may be advocated as pragmatism, the proposition certainly remains philosophically doubtful. Some would argue that our actions are the clearest evidence of our intentions; on this basis, we continue to act as though all events are controllable and failure of control is overwhelmingly a failure in ourselves.

On this basis, we will, most probably, continue to exert our best efforts to support the safety of systems as a pragmatic strategy. Of course, as a compromise position, it may well be that safety, being a multidimensional construct, contains some elements that we can influence and others about which we can do nothing. What I advocate here is that in viewing technology as natural, we have to begin to divorce ourselves from the debilitating concept of blame (see Hancock, Mouloua, and Senders, 2008). In a society redolent with litigation, such as the United States, this is difficult. I concede that while there is an element of greed, there will always be malfeasance, but it is not this form of blame I am seeking to excise. It is the "blame" directed toward conscientious, dedicated professional operators who may make "so-called" errors, for which some pay with their lives. To apportion "blame" in such circumstances is to adhere to a physical model in which Aristotelian "efficient" cause can be distilled for each and every action. It is also to fall for the tragic "hindsight" bias, or more accurately, fallacy. As society begins to reject the concept of a personal deity, dealing with fate on a momentary basis, it must also begin to reject this latter physical analogue of complete personal control and thus responsibility as a concept equally impoverished of mind. If we can begin to conceive pluralistically, and through mutual experiences, as facilitated by technologies such as virtual reality, such improved consciousness may be possible. From this vantage point, society only indemnifies itself. It would be as if the cells of an organ took out insurance against individual damage while the whole organ ceased to function. This analogy can be continued through the multiple levels of any holarchy (see Koestler, 1978). In consequence, in a technologically replete world, local kinship becomes a vestigial characteristic. Others have argued that the individual in these circumstances is not the correct "unit of analysis" (Flach & Dominguez, 1995). As much of the science of human-machine systems is founded in psychology, this diffusion of the emphasis on the human individual is hard for some to take, but trying to grasp the very conception leads to some most useful insights.

As society matures from the self-centered world of adolescence, our reach must be toward a social consciousness in which the myopia of individuality is replaced with the vision of mutuality. While both secular and religious theorists have advocated similar views, they have, by and large, not been successful in influencing collective behavior. I suggest that this is because they have attempted to achieve this as a moral crusade independent of the technology on hand during their particular era. True, they have used the existing media of the times, but they have largely ignored or at least failed to embrace the power of technological innovation. It may be that the nascent emancipation of technology will prove an effective catalyst for such change and growth. However, intelligent human leadership will remain a central and crucial requirement.

I claim that technology is evidently natural. But I argue that it can be more. I suggest that we can learn much about complex technological systems by understanding the nature and nuances of complex biological ecosystems. I ask whether catastrophic failure must be an intrinsic characteristic of these linked complex systems and use this as the acid test for the influence of intention. With respect to disaster, it may be that humans see more through tears than they do through telescopes. But they have to be looking in the right temporal direction to do so, regardless of any lachrymal barrier.

EVOLUTION OF HUMAN AND MACHINE

> This is a principal means by which life evolves—exploiting imperfections in copying despite the cost. It is not how we would do it. It does not seem to be how a Deity intent on special creation would do it. The mutations have no plan, no direction behind them; their randomness seems chilling; progress, if any, is agonizingly slow. The process sacrifices all those beings who are now less fit to perform their life tasks because of the new mutation. We want to urge evolution to get where it's going and stop the endless cruelties. But life doesn't know where it's going. It has no long-term plan. There's no end in mind. There's no mind to keep an end in mind. The process is the opposite of teleology. Life is profligate, blind, at this level unconcerned with notions of justice. It can afford to waste multitudes.
>
> **(Sagan and Druyan, 1992, p. 84)**

The foregoing discussion has considered nature and technology almost as though they were static things. However, perhaps the most distinguishing feature of each is their continual change; and consequently, the next step is an examination of their respective patterns of evolution.

Human Evolution

In considering the respective paths of evolution of human and machine, it is important to consider first the disparities of these separate evolutionary processes. With respect to human evolution, we accept a Darwinian-based concept that has been tempered with the continuing discoveries of genetics and molecular biology. In the general run of science, the concept of evolving complexity still holds sway, where the "survival of the fittest" imperative has served as the selective process over the eons of geological time (see Hancock, 2007). However, it is appropriate to ask here, fit with respect to what? It is usual to emphasize the environment of the organism in this respect. However, any environment is a dynamic and multilayered condition. It may be that an organism is perfectly adapted to a specific set of circumstances and then an event, such as a volcanic eruption, over on the other side of the world changes these local conditions. Under these conditions, the organism may become extinct. In this respect, survival of the fittest is actually survival of the survivalists since excessively specialist adaptations do not pay in the wider order of things (Kauffman, 1993). The best form of adaptation is the ability to adapt, and meta-adaptation is the primary human characteristic. It can be argued that this is indeed what the brain has evolved to do (Hancock, Szalma, and Oron-Gilad, 2005). We accept that human evolution has taken several millions of years. We also accept that human evolution proceeds at a slow rate such that differential characteristics propagate slowly through the population. Advantages in one context may be disadvantages in another context; hence, with meta-adaptation as the central characteristic, it might be expected that human evolution should progress cautiously. In pragmatic terms, evolution proposes and nature disposes. However, what does nature dispose of when technology is the ecology? Under these circumstances, survival of the "fittest" seems a much less pristine selection principle.

The problem, of course, is time. It is perhaps the defining human characteristic that we are painfully aware of our own certain death (Hancock, 2002). However, in our brief flight upon this planet, we remain almost pathologically infirmed with respect to our temporal vision. Again, as the hero of our own story, it is so difficult for each one of us to conceive of a world without us! However, our appreciation of the passage of geological time remains always an intellectual, not an empathic, exercise. I would like the reader, for a moment, to pause and think on this. The geological year asks us to conceive of time since the origin of the earth (mark this, not the origin of the universe) as a period of one year (a hopelessly embedded temporal structure). Human beings are then said to have been on the planet only during the last seconds of the last day of the year. In such a picture, the difference between human appearance and technical development is obviously negligible. Even with an inkling of the length of duration, it is clear that our species is a newcomer to the world. In a true sense, each species has some hand in "creating" other species. All partake in the interplay of environment and resources that set the frame of selection. The human "creation" of technology is distinguished only by the apparent "intentionality" of the act. It is this intentionality that, I think, provides the difference for contemporary minds. I therefore will appeal to Darwin for armament to help support this view. With respect to intentional selection, Darwin wrote:

> One of the most remarkable features in our domesticated races is that we see in them adaptation, not indeed to the animal's or plant's own good, but to man's use or fancy. Some variations useful to him have probably arisen suddenly, or by one step. But when we compare the dray-horse and the race-horse, the dromedary and camel, the various breeds of sheep fitted either for cultivated land or mountain pasture, with the wool of one breed good for one purpose, and that of another breed for another purpose; when we compare the many breeds of dogs, each good for man in different ways; when we compare the game-cock, so pertinacious in battle, with other breeds so little quarrelsome, with "everlasting layers" which never desire to sit, and with the bantam so small and elegant; when we compare the host of agricultural, culinary, orchard, and flower-garden races of plants, most useful to man at different seasons and of different purposes, or so beautiful in his eyes, we must, I think, look further than to mere variability. We cannot suppose that all breeds were suddenly produced as perfect and as useful as we now see them; indeed, in many cases, we know that this has not been their history. The key is man's power of accumulative selection: nature gives successive variations; man adds them up in certain directions useful to him. In this sense he may be said to have made for himself useful breeds.
>
> **(Darwin, 1866, p. 31)**

However, with respect to natural selection Darwin indicated:

> This preservation [in Nature] of favorable individual differences and variations, and the destruction of those which are injurious I have called Natural Selection, or Survival of the Fittest. Variations neither useful nor injurious would not be affected by natural selection.
>
> **(quoted in Sagan and Druyan, 1992, p. 55)**

Some of Darwin's critics could never overcome the absence of intention in evolution. Indeed, without intention, it is perfectly reasonable to argue that evolution does not represent "progress" at all; although this is the typical connotation associated with the word—not progress, but a series of undirected "accidents." Consider, for example, the following:

> The Darwinian process may be described as a chapter of accidents. As such it seems simple, because you do not at first realize all that it involves. But when its whole significance dawns on you, your heart sinks into a heap of sand within you. There is a hideous fatalism about it, a ghastly and damnable reduction of beauty and intelligence, of strength and purpose, of honor and aspiration, to such casually picturesque changes as an avalanche may make in landscape, or a railway accident in a human figure. To call this Natural Selection is a blasphemy, possible to many for whom Nature is nothing but a casual aggregation of inert and dead matter, but eternally impossible to the spirits and souls of the righteous. If this sort of selection could turn an antelope into a giraffe, it could conceivably turn a pond full of amoebas into the French Academy.
>
> **(G. B. Shaw as also quoted in Sagan and Druyan, 1992, p. 64)**

The original arguments against the Darwinian perspective of natural selection as the intrinsic force of evolution were manifold. For some, at that time as now, the idea of a descent of human beings was anathema to the notion of original creation. If God truly created "man in his own image" (women being in biblical terms an afterthought), then evolution transgresses this edict. Essentially, the data took care of this objection, although it is clear that the argument, independent of the data, rolls on today. In addition to theological disputes, more scientific arguments raised against natural selection invoked the blind and accidental nature of selection. It is the case that an individual of any species might represent the "fittest" of the group and yet through mere accident or haphazard demise fail to preferentially reproduce. Hence, "survival of the fittest" as has been noted, can appear to rapidly devolve to "survival of the survivalists." As Waddington (1957, pp. 64–65) observed:

> Survival does not, of course, mean the bodily endurance of a single individual, outliving Methuselah. It implies, in its present-day interpretation, perpetuation as a source for future generations. That individual "survives" best which leaves most offspring. Again, to speak of an animal as "fittest" does not necessarily imply that it is the strongest or most healthy or would win a beauty competition. Essentially, it denotes nothing more than leaving most offspring. The general principle of natural selection, in fact, merely amounts to the statement that the individuals which leave most offspring are those which leave most offspring. It is a tautology.

It is insufficient to argue that any preferential trait has a strong statistical chance of persistence and proliferation since mutation, almost by definition, is a rare and even singular event. From this view, natural selection is a process by which life explores its myriad possibilities but with no divine intervention and thus no essential direction. It is this godless and chance nature of evolution which proves to be so upsetting to many who could otherwise accept the observation of progressive change in the expressions of life across the eons of time.

What has always been posed as an alternative, and a proposition that predates Darwin, is the inheritance of learned traits. There is something intrinsically satisfying in this doctrine to those who believe in accountability. Diligence is passed on by the diligent, profligacy by the profligate, skill by the skillful. We still long to see this in operation, hence repeated sports comments about coaches' sons, who by some direct inheritance did not have to put long hours in the gymnasium with their father, but somehow inherited the gene for the "zone defense." Sadly, the direct inheritance of only favorable characteristics accumulated by the parent is still very doubtful as a scientific proposition, and what of the children born before their parent accumulated such valuable skills? Direct inheritance does not seem to work for humans. However, the conception of the "inheritance of characteristics" is rightly associated with Jean-Baptiste de Lamarck, a strategy that Darwin considered important throughout his own lifetime.[2]

MACHINE EVOLUTION

> Investigation revealed that the landing gear and flap handles were similarly shaped and co-located, and that many pilots were raising the landing gear when trying to raise the flaps after landing (Fitts and Jones, 1961). Since then the flap and gear handles

have been separated in the cockpit and are even shaped to emulate their functions: in many airplanes, the gear handle is shaped like a wheel and the flap handle is shaped like a wing. Most other displays and controls that are common to all airplanes have become standardized through trial and error, based on similar errors made over time. But human factors considerations have not been rationally applied to all the devices in the cockpit.

(Riley, 1994, p. 1)

If humans take millennia to evolve and apparently do so by haphazard circumstance, what of technology? More particularly, for our present purpose, what of machine evolution? In this realm, Lamarck comes now particularly to the forefront. The essence of his "laws" are that an animal's characteristics and behavior are shaped by adaptation to its natural environment, that special organs grow and diminish according to their use or disuse, and that the adaptive changes that an animal acquires in its lifetime are inherited by its offspring (see Koestler, 1964). Let us consider machines with respect to these principles. Certainly, a machine's physical characteristics and especially its function seem shaped by its immediate environment, especially if we think in terms of technology such as contemporary software. Its special organs certainly grow and diminish according to use.

It has been proposed that evolution proceeds by survival of the fittest. However, let us look at this statement with respect to contemporary society. At least in the Western world, we are replete with medical facilities. Many of those who are not "fit" frequently survive disadvantage and disease. Others, who are in dire economic circumstances and do not have simple access to sophisticated medical facilities, may frequently not survive, despite early initial advantages. Of course, the problem, as discussed earlier, is "fit" with respect to what? On the machine side, the generalization is that they progress in uniform steps, taking advantage of each previous discovery. However, when we look at technical breakthroughs in more detail, progress is much more serendipitous and haphazard than it might appear at first blush. Indeed, many steps in machine evolution depend directly upon the creative insights of single designers, where design is as much art as it is science. While machines may inherit preferred characteristics, such characteristics might become a liability in succeeding generations; also some forms of technology can become extinct in the same fashion that natural selection proceeds for animal and plant species. One need only think of musical records of the type of 78s versus 45s in this context. (Some younger readers will have to look up these forms of early competing types of record on the Web to understand the specific example, which itself is proof of the extinction principle.)

On the surface, it might therefore appear that machines evolve at their observed rapid pace because of the immediate propagation of preferred characteristics. However, it is important to ask where the innovations come from. If human mutations come from haphazard events such as cosmic rays, machine innovations come from the equally haphazard events of human intuition. These events are comparable in all aspects except for their temporal density. Some design innovations in machines are useful for a time and then die out because of supersession of improved abilities, for example, PCs. Other innovations fail because of economic forces in which rival technologies are paired against each other and one essentially "wins," for example, Betamax versus VHS, eight-tracks versus tapes and CDs. What I want to propose is that the processes of evolution for both human and machine asymptote to the common characteristic of exploration. The only fundamental difference is the timescale of the action.

COMPARATIVE EVOLUTION

It is worth just a moment to pause and to make explicit the differences in evolution between humans and machines as elaborated previously. The critical difference is cycle time. The average human life is some decades in length; the average machine life is now in the order of years to months. The machine is replaced as soon as a viable replacement is produced. In contrast, we try to save human beings, at least in general, to the degree that we can. Some machines are also "savable," for example, second-hand cars. The point being that the landscape of human beings changes slowly compared with that of machines. Also, as human life span itself is increasing, machine "life span" is diminishing. If this represents the cycle time differences of a single cycle, we should also recognize that the difference in the respective rates of those cycle times is also growing. That is, evolution or change takes place at an increasingly divergent rate. As well as timescale, the respective histories are different in terms of their time. At a surface level, it appears that humans and machines evolve in a very different manner. However, there is much in common, and only time is the essential distinction. They are so divergent in timescale we see them as more radically different than they are. In reality, these are not separate forms of evolution but go together as we coevolve.

CONVERGENT EVOLUTION AND COEVOLUTION

> It makes no sense to talk about cars and power plants and so on as if they were something apart from our lives—they are our lives.
>
> **(McKibben, 1989, p. 1)**

CONVERGENT EVOLUTION AT THE HUMAN-MACHINE INTERFACE

One characteristic in evolutionary landscape is the convergent evolution of entities subjected to the same forces. Before considering coevolution, I first illustrate a case of convergent evolution in the human-machine interface, a topic I return to in much greater detail in the following chapter.

The computer is now the dominant and preferred system that mediates between human and machine. Frequently, of course, the computer itself is the machine of concern. However, for both large-scale complex systems and small appliances, some form of computational medium has become ever more pervasive in our society. This trend toward ubiquity has had distinct effects on human-machine interaction. The generation of a common communication and control medium fosters convergent evolution. In essence, as we tailor interfaces for human capacities, it becomes progressively less clear as to which specific device is being controlled. The critical difference among different sorts of devices lies in their distinct response characteristics. However, the computer as the intermediary between human and system can "hide" or "buffer" many of these differences so that what specifically is controlled can, surprisingly, become less of an issue. Eventually, if this buffering process were carried to its logical extreme, the differences between controlling an aircraft traffic sector, a nuclear power station, or a household washing machine could become virtually opaque to the operator sitting at a generic control panel viewing a generic display. Would this line of progress be an advisable strategy? Is there any unfathomable rule that demands that there be complex interfaces for complex systems? Indeed, as all have to be "navigated" through some complex phase space of operation, it may be that the metaphor of a boat on an ocean is one that captures many common elements of all systems operation. The task of the interface designer would then be to bring the critical variables to the forefront and allow their "emergent" properties to become the sea-lane that the controller has to "pilot" their craft through. That this could be done most easily using the four-dimensional wraparound world of virtual reality is an intriguing proposition and one that offers useful vistas for future exploitation. Such convergence of evolution is a strong reason for adopting such virtual interface, which takes advantage of intrinsic human visuomotor capabilities. But the question remains: Although this is one possible path of human-machine evolution, is it a safe, effective, and reliable one? And more to the point, who decides on these questions?

Mutual Coevolution

If one basis for a unifying theme of the present chapter is the consideration of technology as a facet of the natural environment, we should take the step of recognizing technology as a nascent cospecies with ourselves. Indeed, it is not a species the like of which we have seen before. We have to abandon the perspective that technology will remain merely a human appendage. We have to acknowledge our complete dependence upon technology and recognize that we could not be who we are without technological support. We have to free ourselves from the conception that technology is merely just one of the many shaping facets of the environment and contemplate its own future, potentially as an independent entity (Moravec, 1988).

We have denied "souls" to animals, and we still cling hopefully to this supposed difference. Not to do so would be to deny not only our "special" place in creation (Hancock, 2005) but our very individual separateness. This separateness is daily and continually sustained in each of us by our existence as a unified individual conscious entity (Hancock, 2002). With the growth of technology, we are having progressively more trouble sustaining this worldview. In an age of virtual reality, of e-mail, of fax, of teaming, of telecommuting, of collaboration, what is it to be "separate" anymore? Which individual under 30 does not have a cell phone permanently in their ear? As our physical dependence on technology grows, so does our sense of cognitive dependence. Perhaps this is why there is an ever-greater collective clamor for the different, the unique, the outré in experience since it supports our vestigial and comfortable view of ourselves as strong independent entities (the myth of the hero dies hard, *Die Hard* being a good name for a film, or *Last Action Hero*). We can no longer claim to be simple differentiated entities.

Nutritionists have always known that "you are what you eat," and organ replacement at least shows that spare part grafting is feasible and useful. The cells of the body are all replaced cyclically at a different rate over a period of a short number of years. Hence, what is left of any individual after one of these full cycles is a remembered informational pattern. But modern computer technology is wonderful at detecting, storing, and replicating patterns. We are now enticed with the vision that our own personal pattern could be extracted, replicated, stored, and perpetuated, offering the dissolution of that pattern in time and in space and the hope of immortality. Moravec (1988) desperately grasped at the idea of individual downloading as a preservation of the self, but at best it seeks to replicate our present state of consciousness. Perhaps it is consciousness that is the problem?

What I suggest here is that the status of individualism is slowly dissolving. The divisions that are eroding are not simply between ourselves and technology but between ourselves as distinct individuals. The two antagonistic tendencies of self-assertion and integration have always been locked in some form of battle (see Koestler, 1978); however, technology has joined that fight, not merely as a weapon but as an active combatant. The success of technology is evident in the way that it has insinuated itself into society so completely and so unobtrusively into our very consciousness. One well-illustrated example is that of the telephone (Fischer, 1994). It has recently been argued that video games change brains in certain predictable ways. I was tempted to quote Orwell's last paragraph in *Animal Farm* where the farm animals looked "from man to pig, and from pig back to man and saw no difference," and to replace the word "pig" with the word "machine" (fundamentally a form of the Turing test). However, this still retains the divisive or divided perspective that I want to challenge and fosters the unfortunate arguments about machine "replication" of human abilities. Coevolution is much more than this simple recreation of ourselves in a technical surrogate or operational avatar. It is more dynamic, more elusive, and intellectually much more interesting. That coevolutionary results may initially be crude does not militate against the possibility of a progressively more sophisticated symbiotic interaction.

The disparity in human versus machine evolution is one of cycle time. Convergent evolution is seen in conditions where comparable constraints dictate common processes. Coevolution, the mutual shaping of constituent members of an ecosystem, is dynamic and explorative. The spice of intention adds novelty to human-machine symbiosis. The possibility of "emergent" machine intention promises a new form of overall mutual evolution that, in its embryonic stages, is likely to be seen with respect to contemporary developments in human-centered automation.

HUMAN-MACHINE SYMBIOSIS

> Either the machine has a meaning to life that we have not yet been able to interpret in a rational manner, or it is itself a manifestation of life and therefore mysterious.
>
> **(Garrett, 1925, p. 23)**

Mutual Dependence

Could contemporary human beings survive without technology? If we take technology to mean more than just machines and include animal and plant domestication, I am fairly sure that the answer is no—at least not in the way we understand human beings at present. In these terms, we have shaped the environment that now shapes the individuals and society we are. We have come so far down that evolutionary path that there is no going back. Could contemporary human beings survive without computational machines? Perhaps they may be able to do so. But such a society would be radically different from that which we experience today in the developed and developing nations of the world. Pragmatically, we will not give up our technology, and practically, its influence continues to spread daily. We must therefore conclude that there is already one elementary form of human-machine symbiosis. Machines cannot exist (or at least cannot be created as original structures) without humans; but, by the same token, humans do not exist without machines. It may be more comfortable to look upon this symbiosis as an elaboration of what it means to belong to the species *Homo sapiens sapiens*. For example, no longer are we restricted to storing information and energy endogenously. The immediate and portable availability of information and energy frees individuals in time and space to an extent that is enormously greater than the rest of the animal world. However, this nascent level of symbiosis is only one stage in the developmental sequence linking humans and machines. True symbiosis results when one entity progresses from total dependence to the status of a full partner. The stage after mutual interdependence is some degree of the beginnings of independence. For the machine, that is automation and autonomy of function.

Automation: The Birth of Machine Intention

There are, of course, multiple levels and multiple forms of automation. These levels have been discussed extensively by others (see Parasuraman and Mouloua, 1996). Yet, however autonomous a system might appear at present, there are still human beings involved. This involvement might be at the conception or design stage, or it may be at the care and maintenance end, but at some juncture human operators still enter the picture. I take this human participation even further. That is, even if we can think of a device that has no maintenance, a device that was designed by a software program so divorced from its programmer that it no longer appears to be based upon human action, even then it still subsumes human goals. At the present time, the motive force for the origin of any machine remains essentially human. All machines work to achieve human goals rather than goals of their own. No machine has created another machine in order to satisfy its own needs or desires yet. In fact, to the present, no machine really expresses or articulates an external need or desire, and perhaps that is what is missing is the effort to create surrogate, self-motivating intelligences.

Does this always have to be the case? Can we conceive of machines that are not predicated upon human intentionality? Indeed we can for they surround us in nature. The insect world knows essentially nothing of human intentionality and continues its activity relatively unmolested or polluted by our goals. Of course, we all have to exist on the same global platform, but insects are rarely of technology in the same way that domesticated animals are (always remembering their adaptation to human products such as pesticides, etc.). Can we then conceive of machines as having their own intention, at least at a level that say an ant has intention? We can certainly conceive of it; but with the result of proliferation of interactions in systems, are we now beginning to witness it in some of the machines we have created?

I do not advance this as a "strong AI" position, which would protest that machines can "think." I appreciate Searle's (1984) argument concerning syntactic versus semantic content but would suggest it misses the mark. Each such argument is based on the question of whether machines can "think" like humans "think." I find this a very constricting approach. Frankly, I would prefer that machines not "think" like humans "think" since we already appear to have thoughtful humans (an empirical statement open to much dispute). Rather, I would hope that if mind is the emergent property of brain function, then machine intention could be an "emergent property" of machine function. As I have argued earlier in this book, it is indeed an exercise in imagination to understand what characteristics such emergent properties possess. It is desperately to be hoped that they do not merely mimic that which already exists.

Design: The Balance of Constraint and Opportunity

> May we not fancy that if, in the remotest geological period, some early form of vegetable life has been endowed with the power of reflecting upon the dawning life of animals which was coming into existence alongside of its own, it would have thought itself exceedingly acute if it had surmised that animals would one day become real vegetables? Yet would this be more mistaken than it would be on our part to imagine that because the life of machines is a very different one to our own, there is therefore no higher possible development of life than ours; or that because mechanical life is a very different thing from ours, therefore that it is not life at all?
>
> But I have heard it said, "granted that this is so, and that the vapor-engine has a strength of his own, surely no one will say that it has a will of its own?" Alas! if we look more closely, we shall find that this does not make against the supposition that the vapor-engine is one of the germs of a new phase of life. What is there in this whole world, or in the worlds beyond it, which has a will of its own? The Unknown and Unknowable only!
>
> **(Butler, 1872, p. 215)**

One of the central questions we face in specifying goals at a societal level is plurality. Individuality is a central pillar of the democratic system. While pure individuality, like the comparable notion of freedom, is a myth, in the United States at least, there is a strong mandate to protect "the rights of the individual." This position is in contrast with collectivist societies in which the needs of the individual are sublimated to the greater needs of society. As a consequence, societal aims are frequently expressed as generalities that provide constraint and threat to no one, such as "life, liberty, and the pursuit of happiness." The question to be addressed here is twofold. First, can we continue with such vague social goals in the face of technology that demands specification? Second, do we want to accept societal goals if they override our valued individuality?

FIGURE 14.3 The classic hierarchy of human needs as proposed by Maslow (1964).

I think the answers to these questions lie in seeing how technology has addressed human needs as expressed in Maslow's hierarchy shown in Figure 14.3 (see Maslow, 1964). Clearly, technology looks to serve to free society and individuals from the want of the basic physiological needs. It is of course more than ironic that there are many individuals with more than sufficient monetary resources for several lifetimes who are unable to attain other levels in the noted hierarchy. However, technology rarely expresses explicit goals with respect to other levels of the hierarchy, I submit that this is a reflection of the separation of the ideal from the actual, or more prosaically, the why from the how—the propositional divorce of purpose and process. My central protestation is that technology, cognizant of the fact or not, shapes the why as much as the how. And in this it has failed. The science of human-machine systems has largely failed to attack the question of purpose in a scientific manner. The issue of purpose has been conveniently finessed by science and left to elected legislators as though purpose cannot be studied by science and specific goals, expressed in terms understandable and achievable by technological systems. I argue here that one approach to this lacuna of scientific responsibility lies in the use of advanced technology for enhanced education. The fact that our basic educational delivery systems are essentially indistinguishable from those of over a century ago is an indictment indeed (Hancock, 2000).

Note that Figure 14.3 is a descriptive structure, where the individually specified levels bear only a nominal relationship to each other and the "hierarchy" is a weak descriptor at best. Putatively, there is an order dependence such that one level is founded upon another. The implication is that, for example, friendship is predicated upon law. Clearly, these intrinsic relationships need not be dependent. It is clear that technology acts to support the base of the pyramid. That is, technology underlies the availability of food, water, and shelter in contemporary society. Unfortunately, the contemporary function of technology attempts to expand the certainty of basic needs and does not foster vertical transition.

I am acutely aware that the determination of purpose remains inherently a political decision. I do not have room to discuss the role of technology in the political process itself, although it is a matter to which considerable thought needs be directed. Rather, I acknowledge political power and assert that any proposed innovations unsupported by such power are liable to have a restricted impact. The quintessential bottom line is that technology must be used to enfranchise not to enslave (Illich, 1973), and that the political system, however, formed, should be directed to support this goal.

OUR MUTUAL FUTURE

Helena: What did you do to him [Radius the Robot]?
Dr. Gall: H'm nothing. There's reaction of the pupils, increase in sensitiveness, and so on. Oh, it wasn't an attack peculiar to the robots.
Helena: What was it then?
Dr. Gall: Heaven alone knows. Stubbornness, fury or revolt — I don't know. And his heart, too.
Helena: How do you mean?
Dr. Gall: It was beating with nervousness like a human heart. Do you know what? I don't believe the rascal is a robot at all now.
Helena: Doctor, has Radius got a soul?
Dr. Gall: I don't know. He's got something nasty.

(Čapek, 1923, p. 48)

THE LOVE-HATE RELATIONSHIP

We often express a literal love-hate relationship with the machines we create. On the one hand, we extol their virtues as labor-saving devices and more generally as mechanisms for freeing humans from the bondage of manual work. On the other hand, we decry the loss of employment and human dignity that comes with progressive automation. This relationship is, of course, culturally contingent. The countries of the East and those of the developing nations each have differing perspectives on the advantages and dangers of technology. We in the Western world express this ambivalence toward technology in many differing ways. For example, our fiction is redolent with visions of technical systems that turn against their human masters. From Samuel Butler's *Erewhon* through Čapek's *Rossum's Universal Robots*, now to Schwarzenegger's *Terminator* and Reeves' *Matrix*, we have a collective and seemingly pathological fear of the machine as master. Yet at the same time, nothing alters the headlong rush toward ever more "automation" as the solution to a growing spectrum of societal problems.

One suspects that one's position with respect to technical innovation is very much influenced by the threat that technology appears to pose to you as a person versus the potential individual gain that is to be had from such developments. Scientists can always study it, professors can always pontificate about it, and businessmen can always seek a profit factor. Hence, I would expect most who would be inclined to read this text would certainly view technology as a valuable enterprise. However, should we pose the idea of the development of an automatic research machine that can also teach and do away with capitalistic profit, there might be some dissent, even in the present readership. But even with such a machine, I still suspect that scientists could always study it, certainly professors will always be able to talk about it, and I would be most surprised if someone could not find a profit in it. I think, for the present reader, automation holds little terror. However, if your skills are limited, your circumstances straitened, or more directly your job position recently replaced by a machine, I think your perspective might be somewhat different. There again, I suspect the vast majority of the individuals in the latter group will be too busy trying to make ends meet to purchase and read a book such as the present one. Such are the vagaries of power.

BEYOND ANTHROPOMORPHISM

At present, human society shares a comfortable delusion. The delusion is anthropomorphic in nature and relies upon biological assumptions. The assumptions are that because machines are not self-replicating in the way biological systems are self-replicating, then the proliferation of technical systems is controllable. The anthropomorphic delusion is that because machines do not possess intelligence and therefore intention, or at least possess them in the same way that humans possess intelligence and intention, then the absence of machine intention will persist. Despite our best and desperate efforts at machine intelligence as a surrogate of human intelligence, we still do not have any substantive evidence of machine cognition, largely because such evidence is constrained to take a form that mimics human intelligence. This is the central fallacy of the Turing test, which requires that evidence of any "intelligence" is had only through reference to human abilities. This naive, anthropomorphic, and impoverished perspective (see Akins, 1993; Nagel, 1974) is that which provides a cloak to foster our present "warm feeling." To some, it is a much "colder" feeling. In his radical and interesting work, Illich (1973) deplores the contemporary direction of technology for precisely this reason when he notes:

> The re-establishment of an ecological balance depends on the ability of society to counteract the progressive materialization of values. Otherwise man will find himself totally enclosed within his artificial creation, with no exit. Enveloped in a physical, social, and psychological milieu of his own making, he will be a prisoner in a shell of technology, unable to find again the ancient milieu to which he was adapted for hundreds of thousands of years. The ecological balance cannot be re-established unless we recognize again that only persons have ends and that only persons can work toward them. Machines only operate ruthlessly to reduce people to the role of impotent allies in their destructive progress.
>
> **(Illich, 1973, p. 65)**

In disputing the later contention, it is still important to recognize the validity of the initial premise concerning the reification of material values. Further, given the present treatise on teleology and intention beyond the individual human, Illich's latter points must also be given careful consideration as a condition in which diversification of intention fails.

The societal ambivalence toward machines is then never far below the surface as, for example, the Luddite attacks of the Midlands of England demonstrated. Indeed, the theme of the machine as an enemy runs deep. It is noteworthy that Asimov tried artistically to concoct "Laws of Robotics," each of which was sequentially broken within the first few years of robotics research and operations (Hamilton & Hancock, 1986). Are we destined for the dark ruins of twenty-first-century cities firing hopeful lasers at Skynet's malevolent hench-machines? I should note that this is largely a Western, developed-world preoccupation. Technology is not universally seen in this manner by any stretch of imagination. In addition, differing cultures have widely divergent goals and worldviews that do not accord with the main themes of our society. Such pluralism must be considered in the aims of global technology (see Moray, 1994).

In its own way, this view of the machine as threatening master is as sterile as the "mindless" path we progress along in reality. Moravec (1988) has postulated that silicon-based intelligence will soon tire of our terrestrial confines and will seek its fulfilment in the celestial spaces, unfettered by the constraints of biological needs. With the exception of the latter vision, these futuristic perspectives are largely linear extrapolations of polarized facets of contemporary development. What is clear is that the uncertainty of future developments depends precisely upon nonlinear effects. Prediction of these sudden quirks of changes in history is problematic at best.

BACK TO ARCADIA?

Have I, by these remarks, allied myself with the recidivists who seek a return to a happy but mythical "golden age?" Am I a committed utopian or even "autopian" (Sidney, 1593)? I think not. Rather, I seek to put the development of automation and the role of the human in some perspective. Our decisions at the present, nascent stage of machine independence will constrain our possible decisions at subsequent and potentially more volatile stages of human-machine interaction. I do not advocate a giant "OFF" button since, as we have seen, we have already built globally complex networks, and the interconnections of those systems refute the possibility of such a simple strategy. Rather, I advocate that we keep to the forefront of systems design the thought of Francis Bacon, who opined that science was designed "for the uses of life." In the same way, we must design technology "for the uses of human life." For without such a teleology for technology, we are lost indeed.

TWO CULTURES: A TECHNICAL SOCIETY DIVIDED BY TECHNOLOGY

> Tools are intrinsic to social relationships. An individual relates himself in action to his society through the use of tools that he actively masters, or by which he is passively acted upon. To the degree that he masters his tools, he can invest the world with his

meaning; to the degree that he is mastered by his tools, the shape of the tool determines his own self-image. Convivial tools are those which give each person who uses them the greatest opportunity to enrich the environment with the fruits of his or her vision. Industrial tools deny this possibility to those who use them and they allow their designers to determine the meaning and expectations of others. Most tools today cannot be used in a convivial fashion.

(Illich, 1973, p. 34)

BIFURCATIONS OF SOCIETY

I have discussed the proposition that our ecology is technology and I have advocated a manifest purpose for that technology. I have given considerable space to reflections on the impact of technology in particular as a positive force. However, it is important to give considerations to the downside of technology.

I base this argument on the original observation of C. P. Snow (1964, p. 1) concerning the two "cultures" of society. Briefly, Snow observed that:

I christened [it] to myself as the "two cultures." For constantly I felt I was moving among two groups—comparable in intelligence, identical in race, not grossly different in social origin, earning about the same incomes, who had almost ceased to communicate at all, who in intellectual, moral, and psychological climate had so little in common.

However, the bifurcation of society that I contemplate is one that is much more radical. It does not concern the divergent "worldviews" of the educational aristocracy. Rather, it represents the very division of society itself between those empowered by technology and those subjugated by that self-same technology.

Bifurcations in society are based upon the differential control of resources. The obvious manifestation of those resources is physical wealth expressed as goods, lands, or currency (more generally, "capital"); however, control need not be direct ownership per se but may be more indirectly expressed. The classic example is the control exercised by European medieval clergy over the nobility of those times. The ecclesiasts mediated between man and God largely through their exclusive access to and understanding of written forms of knowledge. It was not unusual for many powerful landowners to actually be illiterate. Our development of technology promises to institute an alarmingly similar form of division. Some will understand the arcane esoterica of technology and will consequently rule over its blessings. Most, however, will not. They will suffer under either its benevolence or its malevolent oppression. For those who dismiss this as pure fantasy, I only ask them to recall the early days of computing when programmers wrote almost only for other programmers, or even worse, themselves alone (see also Vallee, 1982). Others of us mere users, albeit intelligent and well-educated scientists, struggled mightily to understand indecipherable error codes founded themselves on alpha-numeric ciphers. Who then exercised control? It is an almost inevitable rule of human existence that power follows money and vice versa. The power over technology will, in the near term, prove to be the ultimate power.

Access Denied

It is comforting to believe that applied psychology in general and human-machine systems science in particular had much to do with improving computer interaction and making communication with machines more facile and accessible. However, the truth lies nearer to the financial drive to sell technology to ever-wider markets that mandates more open, easy interaction. In these circumstances, the science of human factors of the past has acted in a remedial rather than proactive role. It might now be assumed that computer interaction is open to all. But I require the reader to consider the disadvantages of the illiterate in our society, many of the older members who have not grown up with technology and who are not facile with its manipulation. The physically and mentally challenged, the uneducated, and the poor for whom education itself and access to technology is not seen as a priority—what of these members of society? This does not even consider many of the world's population also to whom access is denied. These individuals are not in the information superhighway slow lane, they have yet to come within a country mile of the on-ramp. As yet, many such individuals have neither a vehicle nor a driver's license.

If I have talked here in general evolutionary terms, then it is these latter individuals who are threatened with the ultimate in adaptive failure. For they will have failed to adapt to the current ecology, that being technology. Thoughts of this extinction raise critical moral dilemmas in themselves. I further suggest that the bifurcation between technically literate and technically illiterate rapidly grows into rich and poor, privileged and oppressed. This distinction may divide society in a more profound manner than any of our presently perceived divisions. At one time, I believed that the only solution to this tragedy was education; after the events of 9-11, I am not so sure. Education should be the right of each member of an enlightened society. A world that promotes an arcane brotherhood wielding disproportionate power by their secretion of knowledge is doomed to failure (McConica, 1991). Or, as one wit put it in political terms, "Today's pork is tomorrow is bacon."

THE ACTUAL AND THE IDEAL

> Dichotomized science claims that it deals only with the actual and the existent and that it has nothing to do with the ideal, that is to say, with the ends, the goals, the purposes of life, i.e., with end-values.
>
> **(Maslow, 1964, p. 12)**

The present chapter started with a statement of a basic paradox in technology. On the one hand, the explicit aim of technology is to improve the lot of humankind. In actuality, the unintended byproduct of contemporary technology seems to be the threat of global destruction, through either its acute or chronic effects. "Arcadians" look to advocate for the dismantling of technology and then somehow live in a world of sylvan beauty. This vision features peace, harmony, and tranquility as the central theme. Their hopes are laudable, their aspirations naive. Peace, harmony, and tranquility are not the hallmarks of a world without technology. Such a world is one of hard physical labor and the ever-present threat of famine at the behest of an uncertain environment or war at the behest of equally unpredictable neighbors. Humankind has fought for many centuries to rid itself of these uncertainties and the dogma that attends them.

Our ecology is technology. If we are to achieve our individual and collective societal goals it will be through technology. I have argued here that we must actively guide technical innovation and be wary of the possible alternative facets of machine intention that in no way resembles our own. We must state explicit purpose for the machines we create, and these purposes must be expressed at the level of a global society not that of an individual or even single nation. When expressed at this higher level, we can expose the antagonistic nature of some of the systems we create to an overall good. I further argue that we cannot view ourselves socially, or even individually, as separated from technology. The birth of machine intention will pose questions whose answers will become the central force that shape our future. How we interact with machines, the degree of autonomy we permit them, the rate of comparative evolution, and the approach to mutual coevolution form the manifesto of our potential future. That the ground rules are being set in our time makes the current work on human-centered automation all the more important. I point to our past failure in setting such constraints and ask how we propose to do better in the future?

Human factors professionals have long clamored to be involved early in the design process—that is, in how an object is designed. However, it is now time to step forward and become involved in that process even earlier. That is, those in human-machine systems design must have a hand in determining why an object, artifact, or machine is designed in the first place. Human factors has been a bastard discipline that sits astride so many "divisions." It has linked art (or design) with science. It has dealt with the social (human) and the technical (machine). It has looked to integrate the subjective (psychological) aspects of behavior with the most concrete, objective (engineering) realms of existence. Therefore, it is imperative that in this unique pursuit, we must also encompass both the *actual* (what is) and the *ideal* (what should be). It is why the human factors of old must evolve into the human-machine sciences of the future.

> After all then it comes to this, that the difference between the life of a man and that of a machine is one rather of degree than of kind, though differences in kind are not wanting. An animal has more provision for emergency than a machine. The machine is less versatile; its range of action is narrow; its strength and accuracy in its own sphere are superhuman, but it shows badly in a dilemma; sometimes when its normal action is disturbed, it will lose its head, and go from bad to worse like a lunatic in a raging frenzy; but here, again, we are met by the same consideration as before, namely, that machines are still in their infancy; they are mere skeletons without muscle and flesh.
>
> **(Butler, 1872, p. 310–311)**

CONCLUSION

Superna Quaerite: Inquire After Higher Things

Earlier, I suggested that the turn of a millennium was an appropriate juncture for human society to turn from its adolescence to a mature adulthood. This is a comfortable homily in that it sounds most impressive but in actuality signifies almost nothing. I want to elaborate this statement in my final comments for this chapter, so that it means something substantive and, I hope, significant. I claim that for the childhood and adolescence of humankind we have acted as passive victims of an omnipotent environment. Our various names for events that happen to us—"Act of God," "fate," "Kismet," "accident," "happenstance," "luck"—all connote a conception that life simply happens to us directed by forces outside our control. I do not claim all natural forces are within human control. I do claim that the passive and victim-laden attitude that we adopt with respect to external forces is within human control. For much of our existence, we have had to label such forces as benevolent or malevolent "deities" that evolved in some incarnations as a single deity that arbitrates all earthly and cosmic events. While not wishing to trespass too egregiously upon personal beliefs, I do reject the idea of an individual deity who follows us around to continually control, test, and evaluate. In the absence of a personalized deity, our society still desperately seeks an entity to "blame" for untoward events that happen.

Earlier, I mentioned the "as if" pragmatic approach and postulated that much of society adopts this positive pragmatism. I propose that this form of pragmatism be adopted as a basis for the teleology of technology. That is, while we may continue to argue over the existence and role of an omnipotent deity, we assume this mantle of maturity upon a local scale and become responsible for our collective future.

I started this chapter with, and have made it a theme that, our ecology is technology. I end it in a similar manner by affirming that technology is also fast becoming our contemporary theology. I propose the term teleologics to cover the concept of intention in technology and its comparative theological referent. If we do not knit together the explicit scientific coconsideration of purpose and process, the division will destroy us. I can only countenance this alternative in the words as voiced by Shakespeare's Macbeth:

> Tomorrow, and tomorrow, and tomorrow,
>   Creeps in this petty pace from day to day,
>   To the last syllable of recorded time;
>   And all our yesterdays have lighted fools
>   The way to dusty death. Out, out, brief candle!
>   Life's but a walking shadow, a poor player
>   That struts and frets his hour upon the stage
>   And then is heard no more: it is a tale Told by an idiot, full of sound and fury,
> Signifying nothing.
>   (Act V, scene v)

ACKNOWLEDGMENTS

I am most grateful for the insightful comments of Raja Parasuraman and Kelly Harwood, who were kind enough to read and comment on an earlier draft of the present work. Their time and effort are much appreciated. I must also express a debt of gratitude to Ron Westrum for his guidance. In directing me to sources that frequently illustrate my naiveté, he often humbles but never fails to interest or educate.

NOTES

1. However, as Moray reminds us, we should remember Leacock's dictum that a PhD is a sign that a person's head is completely full and nothing more can be added.
2. I have purposely contrasted two extreme views of human evolution here as stereotyped under Darwinian or Lamarckian banners. Neither view in its unalloyed, pristine condition is held in contemporary times, and the questions of evolution have themselves multiplied in complexity. My purpose here is to contrast common views of human evolution with machine evolution. I would note that the equivalence I claim is even more valid when the current views of natural evolution are explained for each realm.

REFERENCES

Akins, K. (1993). What is it like to be boring and myopic? In Dahlbom, B. (Ed.) *Dennett and his Critics*. Cambridge: Blackwell.

Bennett, C. H., & Landauer, R. (1985). The fundamental physical limits of computation. *Scientific American*, 253(1), 48–57.

Butler, S. (1872). *Erewhon or over the range*. London: Trubner.

Čapek, K. (1923). *Rossum's Universal Robots (RUR)*. (Original English Publication). Oxford, England. Oxford University press.

Card, S. K. (1989). Human factors and artificial intelligence. In P.A. Hancock and M.H. Chignell (Eds) *Intelligent Interfaces: Theory, Research, and Design* (pp. 27–48). North Holland: Amsterdam.

Darwin, C. R. 1866. *On the origin of species by means of natural selection, or the preservation of favoured races in the struggle for life*. London: John Murray.

Eiseley, L. (1960). *The Firmament of Time*. New York: Macmillan.

Flach, J. M., & Dominguez, C.O. (1995). Use-centered design: Integrating the user, instrument, and goal. *Ergonomics in design*, 3(3), 19–24.

Fischer, C. S. (1994). *America calling: A Social History of the Telephone to 1940*. Berkeley, CA: University of California Press.

Fitts, P. M., & Jones, R. E. (1961). Analysis of factors contributing to 460 "pilot-error" experiences in operating aircraft controls. Reprinted in *Selected papers on human factors in the design and use of control systems*, HW Sinaiko.

Flach, J. M., Hancock, P. A., Caird, J., & Vicente, K. J. (Eds.). (1995). *Global perspectives on the ecology of human-machine systems*. Hillsdale, NJ: Lawrence Erlbaum Associates.

Flach, J. M., & Hoffman, R. R. (2003). The limitations of limitations. *IEEE Intelligent Systems*, 18(1), 94–97.

Garrett, G. (1925). *Ouroborous or the mechanical extension of mankind*. New York: Dutton.

Hamilton, J. E., & Hancock, P. A. (1986). Robotics safety: Exclusion guarding for industrial operations. *Journal of Occupational Accidents*, 8(1–2), 69–78.

Hancock, P. A. (1993a). On the future of hybrid human-machine systems. In *Verification and Validation of Complex Systems: Human Factors Issues* (pp. 61–85). Springer, Berlin, Heidelberg.

Hancock, P.A. (1999). On Monday, I am an optimist. *Human Factors and Ergonomics Society Bulletin*, 42(11), 1–2.

Hancock, P.A. (2000). *A stranger in paradigms*. Presidential Address to the Human Factors and Ergonomics Society, San Diego, CA.

Hancock, P.A. (2002). The time of your life. *Kronoscope*, 2(2), 135–165.

Hancock, P.A. (2005). Time and the privileged observer. *Kronoscope*, 5(2), 176–191.

Hancock, P.A. (2007). On time and the origin of the theory of evolution. *Kronoscope*, 6(2), 192–203.

Hancock, P.A. (2009). *Mind, Machine, and Morality*. Chichester: Ashgate.

Hancock, P. A., & Chignell, M. H. (1989). *Intelligent Interfaces: Theory, Research, and Design*. New York: Elsevier Science Inc.

Hancock, P.A. & Chignell, M.H. (1995). On Human Factors. In J. Flach, P.A. Hancock, J. Caird, & K. Vincente (Eds). *Global approaches to the ecology of human-machine systems*. Hillsdale, NJ: Lawrence Erlbaum Associates.

Hancock, P. A., Mouloua, M., & Senders, J. W. (2008). On the philosophical foundations of driving distraction and the distracted driver. In K. L. Young, J. D. Lee, & M. A. Regan (Eds.), *Driver Distraction Theory: Effects and Mitigation* (pp. 11–30). Boca Raton, FL: CRC Press

Hancock, P. A., Szalma, J. L., & Oron-Gilad, T. (2005). Time, emotion, and the limits to human information processing. In D. K. McBride & D. Schmorrow (Eds.), *Quantifying Human Information Processing* (pp. 157–175). Lanham, MD: Lexington Books.

Illich, I. (1973). *Tools for conviviality*. New York: Harper and Row.

Kant, I. (1781). *Critique of Pure Reason, Norman Kemp Smith*. New York: St. Martins.

Kauffman, S.A. (1993) The origins of order: self-organization and selection in evolution. *LUONNON TUTKIJA, 105*(4), 135–135.

Koestler, A. (1959). *The sleepwalkers: a history of man's changing vision of the universe*. New York: Hutchinson.

Koestler. A. (1964). *The act of creation*. New York: Hutchinson.

Koestler, A. (1972). *The roots of coincidence*. New York: Vintage.

Koestler. A. (1978). *Janus: A summing up*. New York: Vintage.

Kurtz, P. (1992). *The new skepticism: Inquiry and reliable knowledge*. Buffalo, New York.

Kurtz, P. (1994). *The new skepticism*. Skeptical Inquirer. 18, 134–141.

Malthus, T. R. (1798). *An essay on the principle of population it affects the future improvement of society with remarks on the speculations of Mr. Goodwin, M. Condorcet, and other writers*. London: J. Johnson.

Maslow. A. H. (1964). *Religions, values, and peak experiences*. New York: Penguin.

McConica. J. (1991). *Erasmus*. Oxford, England: Oxford University Press.

McKibben, W. (1989). *The end of nature*. New York: Random House.

Moravec. H. (1988). *Mind children: The future of robot and human intelligence*. Cambridge, MA: Harvard University Press.

Moray, N. (1993). Technosophy and humane factors. In *Ergonomics in Design*. Santa Monica. CA: Human Factors Society.

Moray. N. (1994, August). *Ergonomics and the global problems of the 21st century*. Keynote address given at the International Ergonomics Meeting, Toronto, Canada.

More. T. (1516/1965). *Utopia*. New York: Square Press.

Nagel, T. (1974). What is it like to be a bat? *The Philosophical Review, 83*(4), 435–450.

Nickerson. R. S. (1992). *Looking ahead: Human factors challenges in a changing world*. Hillsdale, NJ: Lawrence Erlbaum Associates.

Parasuraman, R. (1986). Vigilance, monitoring, and search. In K. R. Boff, L. Kaufman. & J. P. Thomas (Eds.), *Handbook of perception and human performance* (pp. 43.1–43.39). New York: Wiley.

Parasuraman, R. E., & Mouloua, M. E. (1996). *Automation and Human Performance: Theory and Applications*. Hillsdale, NJ: Lawrence Erlbaum Associates, Inc.

Perrow, C. (1984), *Normal Accidents: Living with HighRisk Technologies*. New York: Basic Books.

Raup, D. M. (1986). Biological extinction in earth history. *Science, 231*(4745), 1528–1533.

Riley, V. (1994). Human use of automation. Unpublished doctoral dissertation. University of Minnesota. Minneapolis.

Sagan, C., & Druyan, A. (1992). *Shadows of forgotten ancestors*. Ballantine Books.

Sandburg, C. (1916). *Chicago Poems*. New York: Henry Holt and Company.

Searle, J. (1984). *Minds, brains, and science*. Cambridge, MA: Harvard University Pros.

Sedgwick, J. (1993. March). *The complexity problem*. The Atlantic Monthly. pp. 96–104.

Sidney, P. (1593). *The Countess of Pembroke's Arcadia*. London: Ponfonbie.
Snow, C. P. (1964). *The two cultures and a second look Cambridge*. Cambridge University.
Technology. (2019). In *The Random House Unabridged Dictionary*. Retrieved from https://www.dictionary.com/browse/technology
Teleology. (2019). In *The Random House Unabridged Dictionary*. Retried from https://www.dictionary.com/browse/teleology
Thoreau, H. D. (1908). *Walden, or, Life in the Woods*. London: J.M.
Tolkkinen. K. (1994. January 27). *Machines are making us soft*. Minnesota Daily.
Vallee, J. (1982). *The network revolution*. Berkeley, CA: And/or press.
Waddington, C. H. (1957). *The strategy of the gene*. London: Allen & Urwin.
Westrum, R. (1991). *Technologies and society: The shaping of people and things*. Belmont, CA: Wadsworth.

15 The Axial Age of Artificial Autonomy[1]

Peter A. Hancock

PREAMBLE

The act of writing is a belief in the future. And so the texts we read, especially those in science, are oriented toward the understanding of, and even the creation of, possible forthcoming conditions. Notwithstanding such this future orientation, from time to time, it is helpful to take a glance backward in order to survey the path by which we have reached our present circumstances. That glance embraces not simply prior threads and themes of progress but embraces the concepts and contributions of now fallen "heroes," whose thoughts and ideas still permeate and resonate in our living world (see Hancock & Mouloua, 2020). Such retrospections also provide salutary experiences, for we also witness our own past efforts now set in stark contrast to revealed reality. Salutary, because our mistaken pronouncements come back to haunt us. However, we should not despair therefrom since the future is necessarily destined to make fools of us all. And these perusals do provide us opportunity to celebrate any occasional prescience we have expressed and also to expand upon such prescience to pronounce further on our immediate future (see e.g., Hancock, 2008). The relevant issue for the present commentary concerns our future with autonomy. It is upon this issue that I look to provide brief comments, in light of what I wrote now some quarter of a century ago (Hancock, 1996).

GOAL OF THE PRESENT WORK

Set then within the context of past, present, and future, I take the opportunity here to review and critique my own earlier work, which was published in Parasuraman and Mouloua's (1996) landmark text. Principally, I look to ask, how well do the points and ideas that were expressed hold up to modern scrutiny? For example, have those observations exerted any meaningful impact? And, most especially, what relevance do such notions have to today's world, which is witnessing the birth of an axial age of artificial autonomy? I have to begin by stating that the central themes hold up quite well, as can be ascertained from a rereading of the original chapter, which is reproduced here (Hancock, (1996).

PURPOSE AND PROCESS

In the original chapter, I was especially concerned with the dissociation between *process* and *purpose* and, most particularly, the ways in which the tools of life and the purposes of life could be so successfully, but so dangerously, compartmentalized.

I see no reason to abjure or amend this concern. In actual fact, I believe our present world needs to understand the peril that this warning embodied even more urgently today. However, like so much of science and communication in general, this message was clearly swamped by both the avalanche of scholarly production, as well as the ever-increasing "noise of the wider world."[2] Thus, I repeat and reiterate, but with no hope of any greater, substantive impact[3], that: *Purpose predicates process; process promotes purpose.* The observation that purpose founds process could not be more relevant, especially in relation to growing expressions of autonomy. The statement I made then that, *I do this because of my belief that technology cannot and should not be considered in the absence of human intentions,* still rings as true for me today as it did those 25 years ago. I hope, after reading the original chapters, that readers concur.

Yet, as fundamentally embodied in the central theme of the present book, this dissonance and dissociation between purpose and process is *precisely* what we are proposing to do today. The landscape of autonomy is where this dissonance is now set in greatest relief. It will come as no surprise then that I am at best ambivalent and at worst highly skeptical about the eventual outcome of such machine (artificial) autonomy. In this sense, I can only echo my former self by restating: *I ... point to human-centered automation as one stage in this sequence of evolution that will eventually witness the birth of autonomous machine intention about which I express a number of cautions.* The foregoing observations being my predicates then, and observing that they have little changed in the intervening years, it is not surprising that my conclusion: ... *The collective potential future for humans and machines can only be assured by the explicit enactment of mutually beneficial goals* still holds, even as the tide (or tsunami) of autonomy rolls ever closer.

ON THE IMPACT OF HF/E

It is no source of pleasure to have to observe that the interim quarter of a century has seen little effective, or at least overt, increase in the influence and impact of HF/E on technical developments. The same intransigent barriers remain, the same market "forces" persist and resist, and, in general, our science continues its often marginal impact upon design and developments (and see Hancock, 2019a). I am not saying we cannot find specific examples of "success" stories; most assuredly we can and have also been able to do so in the past (Harris, 1984). No, what is missing here is the application of a systematic moral philosophy as applied to the vast and largely "mindless" tide of technology[4]. I do see some prospects for hope in our HF/E embrace of the systems approach (e.g., Carayon et al., 2015). I applaud, support, and have hope for such innovations. But, I believe we are in an existential race for civilization's survival (Hancock, 2019b), and time here is not on our side[5]. Despite such a bleak prognosis, the optimist in me still holds out hope that HF/E can achieve its aspirations in improving the quality of life *for all people*; the realist in me is forced to periodic episodes of sardonic laughter.

ON THE IMPORTANCE OF TELEOLOGY

In my original chapter, I went on to define and discuss teleology as such a formal science of purpose. Since the underpinning observations here are of long standing, the conclusions I drew still hold. Yet in one aspect I have changed my opinion. More and more, I have come to believe not that humans made tools but rather that tools made humans. That is, the presence of rudimentary tools in the ecosystem of the nascent human species acted to shape and functionally adapt the brain that we have now come to possess. Twenty-five years ago, I was much more of an interactionist, seeing this relationship between tools and humans as at least a codevelopment. Today, I am content to adopt a more provocative stance and emphasize the primacy of tools in human origins. It is doubtful whether this particular proposition is open to simple empirical resolution and the proposition does run against the current zeitgeist (Oakley, 1949)[6]. Yet, I am happy to argue that we are the offspring of tools, not the original creators of them. Thus, *in tools lie our creation and in tools lie our destruction*. I am quite willing today to point more to the putatively "dystopian" outcome in respect to this line of, nominally, symbiotic progress with such tools. Although the latest in the sequence of such tools (writ large), artificial autonomy may be the epitome of the conduit of our human destruction. Again, across the years, I have become less and less persuaded by the standard "heroic" narrative of the human perspective (Campbell, 2008). While I understand the social value and collective imperative subsuming this optimistic "heroic" narrative[7], the association of such a narrative with external reality I find less and less persuasive and sustainable. Even on the eve of our destruction of civilization, I get the impression we will still be congratulating ourselves upon our "heroic" struggles.

The disparity in computational capability per unit volume still continues to favor the quasi-Lamarckian progress of our computer systems (cf., Kurzweil, 2005; Moore, 1965). Yet, this disparity in quantitative processing capacity is necessarily limited in its growth, i.e., for computers the latter promises to be a self-terminating process. While *quantitative* comparisons are relatively clear, the *qualitative* nature of the differences in processing capacity between computer and human continue to be the subject of much heat and ire; but frustratingly, the debate seems to render disappointingly little light. Suffice it to say the fully functional artificially intelligent entity seems to be really only marginally closer than it was in 1996. Perhaps the only fundamental change in perspective, exposed by Hancock (1996, Figure 2.1), is the emphasis on the idea of mutuality and collaboration that also existed three decades ago. Beyond the "heroic" narrative embedded in this perspective, I remain to be persuaded by modern arguments for a fully capable artificial general intelligence (and see Hancock, Nourbakhsh, & Stewart, 2019; Salmon, Hancock, & Carden, 2019).

HOW "NATURAL" IS TECHNOLOGY?

In my following exploration of "Technology and Natural Laws," I can either applaud my previous perspicacity or now lament my sad inability to see further. For, in the intervening years, I have come to have a much greater respect for the insights that can be derived from biomimesis (and see Hancock, 2014). The conceptual application of an appropriate biomimetic power law, advocated in the original chapter (see Hancock, 1996, Figure 2.2), has borne surprising fruit in other areas, now contingent upon empirical observation (Lopez-Sanchez & Hancock, 2018). This gold mine of insight is still in operation, and we still have much to learn from the millions of years that nature has had to experiment with it and the avenues it has chosen to discard. I can only emphasize here that biomimesis applied to technological innovation is a very fruitful avenue to pursue.

The following sections on forms of respective human, machine, and human-machine evolution stand up well to the test of time. I am here disproportionately proud of my former self, which in general is an unusual state. I will, at the risk of alienating my reader, go further. The section on the "Birth of Machine Intention" now looks even prescient! The observations upon the ambivalence linked to the concept of strong artificial intelligence (AI) persists (although see Salmon et al.; 2019). My excursions into the opportunities in design were not necessarily profound but did portend some critical and influential work, such as, for example, that of Flach and Voorhorst (2016). Our need in HF/E to understand both the principled foundations, and the practical workings, of the design processes remain a concern.

My explanation of our mutual future and the overt reference to human-robot interaction proved an unintentionally accurate forecast. Thus, in the interim, I have worked for more than a decade on human-robot trust and transparency and the factors that influence this interdependence (Hancock et al., 2011; Schaefer, Sanders, & Hancock, 2014). The direct reference to cultural attitudes to automation was also fortunate since these concerns will persist into the forthcoming uses and acceptance of autonomy in societies across multiple nations. My point about job and skill replacement also still persists. One continues to sense an economic divide between those who are masters of technology and those who are made slaves to it—willingly or not. I still believe we collectively retain an undercurrent of ambivalence toward technology, but this is always tempered by the perceived cost/benefit ratio expressed in the world. Of course, the general public is not always accurate in their cost/benefit assessments here (Hancock, Nourbakhsh, & Stewart, 2019), but often they are, either intentionally or unintentionally, provided insufficient evidence to make a full determination. The "two cultures" metaphor, promulgated by Snow (1964) concerning the arts and the sciences, is still relevant to our discussions concerning our ambivalence toward technology, i.e., those enfranchised by technology and those disenfranchised by it. The present, overarching determinate remains personal wealth and its distribution or, more properly, its mal-distribution. I have become progressively more concerned with these divisions (Hancock, 2009; 2014) since technology serves its financial masters and is a conduit of this disparity, not its source per se.

In concluding my original chapter, I appealed for us to supersede the naïve narratives of superstitious causality of our past. In such a call, I was not the first nor alone (Sagan, 2011). I argued for a much more positive, responsible, and proactive philosophy upon which to found our technological developments. I also asked us to eschew the simplistic, abnegatory, and victim-laden attitudes that seemed then to predominate. I believed then, and believe now, that such fatalism is one obvious symptom of incipient *civicide* (Hancock, 2019c). I advocated for the term *teleologics* for the welding of technology with purpose and the positive perspective that it would be designed to invoke. The idea fell stillborn from the press—the patient being dead on arrival. Now some quarter-century later, I see few signs to fire any optimism in this respect; although I still wish there were such sources of hope. I am willing to believe; but HF/E help thou my disbelief! In light of the aforementioned observations on my own work of the past, I conclude here with just the briefest set of observations on what I see of our future with autonomy. These are not technical prescriptions, nor quantitative predictions, but much more in the sense of a general overview.

OUR FUTURE WITH AUTONOMY

There is, I find, an almost inevitable, dark, optimistic undercurrent about the way we view the future of technology. Some argue, and rather convincingly it is true, that much of automation provides a collective "good" to society. It is most certainly the case that the computational underpinnings of technology now form the primary character of the eco-niche in which humans generally find themselves. There is no retreating along this path; the sustenance of more than seven billion individuals cannot be had without these systems. Others console themselves that the forms of AI that we currently witness will not grow into a fully functional artificial "general" intelligence that can elaborate its action beyond very restricted domains. True, such systems can dominate in highly formalized and constrained "worlds" such as games and logistical decision-making. But take these contemporary forms of AI beyond the limits of those specific domains and, rapidly, they flounder and fail. But as the respective *isles of autonomy* (Hancock, 2018a) coalesce, the ranges of behaviors covered necessarily grow. Whether these burgeoning systems "understand" the computational transformations they undertake remains a strong philosophical challenge. But, in action, such intrinsic understanding may be moot in respect of the actual outcome. Whether any transformation from automation to autonomy will be a "singularity," or a "plurality," (cf., Hancock, 2018a) is, in some sense, rather beside the point. The point being that any such transition will likely happen on a timescale unrecognizable to human cognition. As with the indifference of evolution as to time and directionality of "progress," so the line of development that is emerging promises broad artificial intelligence at some juncture in the future. With no sense of time (Hancock, 2007; 2018b), such systems can afford to wait it out. This being so *the days of man are numbered*. Peak species are jealous of their primacy; and artificial autonomy may represent an exception, but I am not confident that this will be so. With respect to tools: *in our beginning is our end*. I cannot view this symetricality with the apparently necessary skepticism expressed in the traditional human narrative. But therein lies my own failure.

POSTSCRIPT

As a professor, one is frequently required to evaluate the work of others. In the present case, I am evaluating my own. As an overall impression, I am pleased in one way that the central issues of the chapter have stood up to the passage of time—in other ways not. I would have preferred that many of these issues had already been addressed and resolved, but perhaps that is too optimistic. In some ways, the chapter fails. It can appear to skip among themes without sufficient coherence and bridging. In this, it requires of the reader leaps of logic that are often overambitious; in this sense, the writer has failed. I am disappointed by such narrational failures but consoled that at least I tried to cover a swath of concerns rather than simply focusing on small, restricted, and restrictive topics. That being said, I still, in all modesty, have to award myself at least a passing grade. Other judges may not be so generous!

NOTES

1. In order to follow the various points of response that are made in this present commentary, it is absolutely vital that the reader first examines and evaluates the chapter upon which the present responses are founded. This is my original chapter (Hancock, 1996), in the original Parasuraman and Mouloua (1996) text. In large part, the present observations follow sequentially upon the points that were made in the original chapter. Without reading that work, the present commentary will appear sporadic and disjointed and will be largely, if not totally, incomprehensible.
2. Of course, the floodtide of social media and the torrent of nominal 'information' has only grown in the interim. We are awash with data, flooded by information, soaked by knowledge but parched for wisdom. It is doubtful if even the most assiduous, selective and discriminating of observers, born into today's world, can fully distinguish value and veridical insight from nonsense. In this, many of our educational institutions now feature the training of individuals to serve governmental and corporate requirements over the inculcation of summated human wisdom. Thus, we machine (train) people to mind the machine. If the reader should believe this is not a predominant concern, simply ask why post-college job employment rates are the central metric of many institutions rather than the production of the fully enlightened person? Systems asymptote to the reward structure they feature (Hancock, 2019a), and employment rates are now the most obvious and publically promulgated goal in many (but not all) of our institutions of higher learning.
3. I have been advised by a sage colleague that it is often rather hard to asses one's impact. Recently, I have been reminded of this when a valued and prominent researcher approached me at a NATO Meeting in Europe and expressed how much he had been influenced and affected by my promulgation of the "sheepdog" metaphor for human-computer interaction. Not only was this a gratifying experience, it should remind us all that our work can have widespread and hopefully positive influence, even if we ourselves may not be directly aware of it.
4. By this, I do not mean to denigrate the important and 'mindful' efforts that go into particular products. That would be to traduce the contributions of many good people. The problem rather, is systemic in nature, but the nominal solutions are mostly discrete in form. e.g., how does a vehicle manufacturer explicitly design a product in light of philosophical need to support the social concerns for mobility?
5. I am not one who thinks humans are doomed to extinction as a species, we are much more hardy than that. Rather, it is the dissolution of interconnected civilization that, globally, is at peril.
6. It is rather piquant to note that the text for which Oakley is rightly famous, derived from an exhibition held at the British Museum of Natural History on July 17th, 1943 in the middle of World War II (see Oakley, & Zeuner, 1944). One rather gets the sense of the effort to maintain the progress of civilization, even as the very pillar of civilization were themselves shaking. It is no source of comfort that the Allies commenced bombing of another great and historic city (Rome), only two days later.

7. Of course, not all human cultures treat the 'hero' in the same fashion. However, in the same way that story-telling seems widespread, if not ubiquitous across human societies, so the narrative of hero and villain (the good vs. the bad), seems equally as pervasive. With such inherent figure/ground mindsets, the persuasive 'story' of human courage, tenacity etc. almost always arises. Taken from a more global perspective, such human interpretations of reality are not supported by rational consideration of their collective actions.

REFERENCES

Campbell, J. (2008). *The Hero with a Thousand Faces*. Novato, CA: New World Library.
Carayon, P., Hancock, P. A., Leveson, N., Noy, I., Sznelwar, & van Hootegem, G. (2015). Advancing a sociotechnical systems approach to workplace safety: Developing the conceptual framework. *Ergonomics, 58* (4), 548–564.
Flach, J. M., & Voorhorst, F. (2016). *What Matters*. Dayton, OH: Wright State University Library.
Hancock, P. A. (1996). Teleology for technology. In: R. Parasuraman and M. Mouloua. (Eds.). *Automation and Human Performance: Theory and Applications*. (pp. 461–497), Hillsdale, NJ: Erlbaum.
Hancock, P. A. (2007). On the nature of time in conceptual and computational nervous systems. *Kronoscope, 7* (2), 185–196.
Hancock, P. A. (2008). Frederic Bartlett: Through the lens of prediction. *Ergonomics, 51* (1), 30–34.
Hancock, P.A. (2009). *Mind, Machine, and Morality*. Chichester: Ashgate.
Hancock, P. A. (2014). *Autobiomimesis: Toward a theory of interfaces*. Invited keynote presentation given at the 6th International Conference on Automotive User Interfaces and Interactive Vehicular Applications, Seattle, WA, September, 2014.
Hancock, P. A. (2018a). *Isles of autonomy: Are human and machine forms of autonomy a zero sum game?* Keynote presentation at the NATO Human-Autonomy Teaming Symposium, Portsmouth, England, October, 2018.
Hancock, P. A. (2018b). On the design of time. *Ergonomics in Design, 26* (2), 4–9.
Hancock, P. A. (2019a). Humane use of human beings. *Applied Ergonomics, 79*, 91–97.
Hancock, P. A. (2019b). Some promises in the pitfalls of automated and autonomous vehicles: A response to commentators, *Ergonomics, 62* (4), 514–520.
Hancock, P. A. (2019c). *In praise of civicide*. Submitted.
Hancock, P. A., Billings, D. R., Olsen, K., Chen, J. Y. C., de Visser, E. J., & Parasuraman, R. (2011). A meta-analysis of factors impacting trust in human-robot interaction. *Human Factors, 53* (5), 517–527.
Hancock, P. A., & Mouloua, M. (2020). Remembering fallen heroes: A tribute to Raja Parasuraman, Joel Warm, and Neville Moray. In: M. Mouloua and P.A. Hancock (Eds.). *Human Performance in Automated and Autonomous Systems*. CRC Press: Boca Raton, FL., in press.
Hancock, P. A., Nourbakhsh, I., & Stewart, J. (2019). The road to autopia: How artificially intelligent automated vehicles will impact patterns of future transportation. *Proceedings of the National Academy of Sciences, 116* (16), 7684–7691.
Harris, D.H. (1984). Human factors success stories. *Proceedings of the Human Factors Society, 28* (1), 1–5.
Kurzweil, R. (2005). *The Singularity is Near*. New York: Viking Books.
Lopez-Sanchez, J. I., & Hancock, P. A. (2018). Thermal effects on cognition: A new quantitative synthesis. *International Journal of Hyperthermia, 34* (4), 423–431.
Moore, S. E. (1965). Cramming more components onto integrated circuits. *Electronics, 38* (8), 114–117.

Oakley, K. (1949). *Man the Tool Maker*. London: British Museum of Natural History.
Oakley, K., & Zeuner, F. E. (1944). New exhibits in the Geology Galleries at the British Museum (Natural History), London: Report of demonstrations held on 17th July, 1943. *Proceedings of the Geologists' Association, 55* (2), 115–118.
Parasuraman, R., & Mouloua, M. (1996). (Eds.). *Automation and Human Performance: Theory and Applications*. Hillsdale, NJ: Lawrence Erlbaum Associates, Inc.
Sagan, C. (2011). *The Demon-Haunted World: Science as a Candle in the Dark*. New York: Ballantine Books.
Salmon, P., Hancock, P. A., & Carden, T. (2019). To protect us from the risks of advanced artificial intelligence, we need to act now. *The Conversation*, January 24th.
Schaefer, K. E., Sanders, T. L., & Hancock, P. A. (2014). The influence of modality and transparency on trust in human-robot interaction. *Proceedings of the Human Factors and Applied Psychology Conference, Embry-Riddle University*, Daytona Beach, FL.
Snow, C. P. (1964). *The Two Cultures*. Cambridge, UK: Cambridge University Press.

Author Index

Note: Page numbers in **boldface** represent tables and those in *italics* represent figures.

A

ABC Research Group, 72
Abraham, H. 222–223
Acevedo, E. O. 7
Acomb, D. B. 118
Adams, J. A. 132
Akerman, N. 77
Akins, K. 292
Alberdi, E. 24
Allen, C. 262n5
Allen, K. 216
Alliance of Automobile Manufacturers, **225**
Amalberti, R. 116
Amelink, H. J. M. 79–80
Anderson, R. S. 7
Ardern-Jones, J. 177
Ariga, A. 235
Arrabito, G. R. 86–87, 89, 94–95
Arroyo, E. 224
Arthur, E. J. 110–111, 131, 156, 270
Astolfi, L. 130
Austin, J. R. 140
Austmine, 249
Ayton, P. 24

B

Babiloni, F. 130
Baetge, M. M. 118
Bahner, E. 24, 27, 34
Bahri, T. 107–108
Bailey, N. R. 26–27, 158, 177
Bailey, R. W. 110
Bainbridge, L. 27, 33, 151, 156, 213, 221
Baker, C. V. 117
Baldwin, C. L. ix, 144, 237, 240
Balfe, N. 29
Ballas, J. A. 107
Banks, V. A. 127, 131, 138
Barde, A. 5
Barnes, B. 2
Barnes, C. M. 141–142
Barnes, M. J. 26, 29, 108–109, 139, 142, 177
Barragan, D. 237
Bashir, M. 26, 137, 173
Bass, B. 36, 177

Battino, R. 257
Baumeister, R. F. 13
Baur, C. 135
BEA, 153
Beatty, J. 3
Becker, A. B. xi
Beckmann, M. 236
Behymer, K. J. 78
Bekiaris, E. D. 226
Bellem, H. 218
Bellet, T. 241
Benet, S. V. 253
Bengler, K. 134–136, **225**, 231, 235, 242
Bennett, C. H. 272
Bennett, K. B. 71, 79–80
Ben-Shoham, I. 215
Berch, D. B. 89–91
Berto, R. 240
Biassoni, F. 235
Bill, J. S. 135
Billinghurst, M. 5–6, 11, 13,
Billings, C. E. 25, 34, 57, 103–104, 107, 110, 120, 127, 222
Billings, D. R. 169, 173
Bindewald, J. M. 137
Biros, D. P. 32
BIS Research.com, 169
Bisantz, A. M. 34, 178
Blakely, M. J. 6–7, 13
Blanco, M. **225**
Boer, E. R. 23, 222
Boles, D. B. ix
Bolstad, C. A. 177
Borghini, G. 130, 134
Bornstein, J. 58
Borowsky, A. 52
Boubin, J. G. 137
Boucek, G. P. 218–219
Bowers, C. A. 131, 140
Bowers, J. C. 90
Boyle, L. N. 177, 217
Bradshaw, J. D. 56
Bradshaw, J. L. 3
Bradshaw, J. M. 36, 64, 191, 198
Bransford, J. D. 75
Braunagel, C. 233
Brehmer, B. 26
Broadbent, D. E. 2–3
Brodsky, W. 240
Brookhuis, K. A. **225**
Brown A. L. 75

309

Bryne, E. A. 158
Buick, F. 111
Bunce, D. 87, 91
Buono, B. 217
Burdick, I. 21
Burdick, M. 22, 27, 33–34,
Burian, B. K. 35
Burke, C. S. 141
Burns, C. 50, 196–197, **202**, 205, 209
Bush, J. M. 89
Butler, S. 271, 289, 291, 296
Buunk, B. P. 141
Byrne, E. A. 109

C

Cabrall, C. D. D. 238–240
Caggiano, D. M. 90
Caird, J. K. 110, 131, 270
Calhoun, G. L. 53, 132–133
Calvo, A. A. 5
Campbell, D. A. 175
Campbell, J. 303
Campbell, J. L. 215, 220–221, 224
Cannon-Bowers, J. A. 117, 131, 140
Čapek, K. 291
Carayon, P. 302
Card, S. K. 272
Carden, T. 303
Cardona-Rivera, R. E. 242
Carmody, M. A. 106
Carroll, J. M. 117
Carsten, O. M. 135, 222, 238
Carstensen, L. L. 7
Casner, S. M. 221
Casper, P. A. 226
Catanzaro, J. M. 87, 94
Chabris, C. F. 21
Chaiken, S. 141
Chambers, P. R. G. 222
Chandler, P. 225
Chang, D. 171
Chapanis, A. 249
Charlton, S. G. 242
Chavaillaz, A. 24, 35
Chen, J. Y. C. 26, 29, 139, 142, 156, 173, 177
Chignell, M. H. 108–109, 276
Ciceri, M. R. 235
Cingel, A. 134, 221
Claypoole, V. L. 86–87, 89–91, 94, 129
Clegg, B. A. 25, 137
Clothier, R. A. 175
Cocking, R. R. 75
Cohen, D. 112–*113*
Coiera, E. W. 24, 30
Colavita, F. B. 3
Converse, S. A. 114

Coury, B. G. 25, 28
Coutts, A. J. 8
Crocoll, W. M. 25, 28
Crooks, L. E. 214
Cullen, T. M. 64
Cummings, M. L. 25, 54, 130, 197, **202**, 205
Curry, R. E. 104, 109–110, 120, 151, 155
Curtindale, L. M. 88

D

Daly, M. 32, 222
Dambӧck, D. **225**, 235, 242
Danckert, J. 130
Dao, E. 240
Darling, K. A. 6
Darwin, C. R. 281
Davies, D. R. x, xii, 86, 89–92, 140, 152
Davis, L. E. 103
de Boer, R. J. 23
de Brunélis, T. C. 52
de Groot, A. 75
de Visser E. J. 26, 36, 177
De Vries, P. 177
de Winter, J. C. F. 57, 133–135, 238
Deaton, J. E. 87, 90, 103, 108, 112
Deblon, F. 116
Deci, E. L. 237
Defense Science Board, 49, 58–59
Degani, A. 22
Dekker, S. W. A. 54
DeLucia, P. R. 134
Dember, W. N. x–xi, 85–86, 89–91, 93, 129–130, 234–235, 239
Deneka, A. 13
Denues, K. L. 86
Deroo, M. 243
Desmond, P. A. 128–131, 133, 135, 137, 143
Deutsch, D. 3
Deutsch, J. A. 3
Dever, D. A. 86, 89–91
Devlin, S. P. 91
Dewar, A. R. 129, 136
Dickinson, T. L. 114
Dietz, K. C. 8
Dignum, V. 36
Dillard, M. B. 87, 95, 239
Dingus, T. A, 237
DinparastDjadid, A. 222
Dittmar, M. L. 93
Dixon, S. R. 137, 171–172, 183
Dodou, D. 57
Dominguez, C. O. 280
Donald, F. M. 91
Donmez, B. 171, 177
Dorneich, M. C. 30, 32, 140
Dougherty, J. R. 132

Author Index

Draper, M. H. 132
Draper, N. 6
Dreger, F. 238
Drury, C. G. 178, 250
Druyan, A. 280, 282
Dubey, S. 89
Duley, J. A. 106
Dunbar, M. 33
Dunbar, V. 30
Duncker, K. 71–72
Dünser, A. 13
Dutch Safety Board 23

E

Earle, F. 137, 142
Edland, A. 30
Eickhoff, S. B. 143
Eiseley, L. 266, 276
Elder, N. 80
Emerson, T. 116
Endsley, M. R. 19–20, 108, 127, 151–152,
    154–*155*, 156–**157**, 158, 160–*162*,
    163–164, 173, 177, 196–198, **202**, 205,
    216–217, 222, 225
Epley, N. 177
Epling, S. L. 6–7, 87, 91
Epstude, K. 240
Eriksson, A. 238, 242
Ezer, N. 177

F

Fagnant, D. J. 213, 232
Feigh, K. M. 56
Feldhütter, A. 136
Feltovich, P. J. 198
Ferraro, J. 172
Ferrell, W. R. 253
Fincannon, T. 57
Fink, N. 36, 177
Finkbeiner, K. M. 95
Finomore, V. S. 129, 131, 5, 85–87, 91, 93–94
Fischer, C. S. 287
Fischer, U. M. 19, 35
Fisher, D. L. K. 52
Fisk, A. D. 26, 177
Fitts, P. M. 43–44, 253–*254*, 283,
Flach, J. M. 71–72, 78–80, 117, 270, 273,
    280, 304
Fleishman, E. A. 115
Flin, R. 140
Ford, M. 250
Ford Motor Company, 191
Forsythe, C. 242
Foushee, H. C. 116, 118
Fraulini, N. W. 86, 129

Freeman, F. G. 158
Frese, M. 140
Frey, C. B. 250
Frey, M. 88
Fridman, L. 217
Fridsma, D. B. 23
Friedman, C. P. 23
Fritsch, T. 177
Fuld, R. 57
Fulton, N. 175
Funder, D. C. 13
Funke, G. J. 95, 133
Funke, M. E. 93

G

Gaba, D. M. 104
Gabler, H. 241
Gaillard, A. W. 141
Gaissmaier, W. 30
Galanter, E. 70
Galinsky, T. L. 89, 235
Galvao-Carmona, A. 134, 242
Gao, F. 130
Gao, J. 177
Garg, A. X. 23
Garrett, G. 287
Gatti, G. 23
Geddes, N. D. 109–110, 112
Gehlert, T. 135
Gent, E. 191
Gérard, N. 28
Giambra, L. 89
Gibson, J. J. 69, 74, 214
Gigerenzer, G. 20, 30, 33, 72
Gilson, R. D. 169–170, 172, 174, 132
Given-Wilson, R. 24
Glandon, T. 177
Glenn, F. 107
Gluckman, J. P. 103, 106, 108–109, 112–*113*
Goddard, K. 23
Goernert, P. N. 111, 156
Gold, C. 135–136, 218, **225**, 235, 242
Goodrich, M. A. 132, 222
Gordon, S. E. 10
Gorman, T. I. 241
Gormican, S. 93
Gosling, A. S. 23
Grainger, J. 57
Green, A. L. 6
Green, C. A. 238
Greenlee, E. T. 134
Greenstein, J. S. 110
Greenwald, C. Q. 90
Greenwood, P. M. 26
Gregoriades, A. 242
Grier, R. A. 85–87, 95, 235

Grossman, J. 109
Grubb, P. L. 90–91
Gunn, D. V. 88, 90
Gunsch, G. 32
Guo, F. 231
Gutzwiller, R. S. 24

H

Hagafors, R. 26
Hajdukiewicz, J. R. 79
Hale, A. R. **214**
Hall, D. S. 80
Hall, R. W. **225**
Ham, J. 177
Hamilton, J. E. 293
Hammer, J. M. 112, 114, 116–117
Hammer, M. 214
Hancock, G. M. 255
Hancock, P. A. ix–xii, 5, 7, 52, 57, 64, 85–87,
 91–92, 96, 106, 108–111, 128–129,
 131–133, 152, 156, 169–170, 172–174,
 184, 214, 249–253, 255–258, 260–262,
 267, 270, 275–277, 279, 281, 286, 290,
 293, 301–306n1n2
Hancock, S. 7
Handy, T. C. 240
Harper, C. D. 241
Harper-Sciarini, M. 29
Harris, D.H, 302
Harris, W. C. 110–111, 131, 156
Hart, S. G. 85, 132, 134,
Hartmann, T. 237
Harvey, C. 127
Häusser, J. A. 142
Hawley, J. K. 64
Hayes, C. C. 140
Head, J. 6, 8–9, 93
Heafner, J. 177
Heaton, L. D. 7
Heems, W. 23
Heers, S. 21
Hegarty, M. 5
Heikoop, D. D. 134
Heimbaugh, H. 142, 177
Heitmeyer, C. L. 107
Helmreich, R. L. 116
Helton, W. S. 5–6, 7- 8, 11, 13, 87, 91–93, 95, 130
Henderickx, D. 234
Hergeth, S. 174, 178
Hesketh, B. 226
Higgins, J. S. 237
Hilburn, B. 106, 111, 158,
Hill, S. 139, 142
Hitchcock, E. M. 87, 94–95, 106, 131
Ho, G. 177
Hoc, J.-M. 243

Hockey, G. R. J. 87, 133, 137, 142–143, 236, 238
Hoedemaeker, M. **225**
Hoeksema-van Orden, C. Y. 141
Hof, T. **225**
Hoff, K. A. 26, 137, 173
Hoffman, R. R. ix, 50, 53, 56–57, 59–60, 64, 170,
 191, 249, 251–252, 273
Hoffrage, U. 30
Hogan, C. L. 7
Hohenberger, C. 177
Hollander, T. D. 88, 93–94
Hollands, J. G. 85, 177
Hollenbeck, J. R. 141–142
Hollister, S. 1
Hollnagel, E. 10, 81
Horrey, W. J. 218
Howe, S. R. 86, 90, 234, 239
Hu, R. 50
Hueper, A. D. 24
Huerta, E. 177
Hughes, J. 30
Huitema, B. E. 92
Human-Systems Integration Community of
 Interest (HSI COI), 58
Hummel, T. 135, 221
Hurts, K. 23

I

Ilgen, D. R. 174, 177
Illich, I. 291–292, 294
International Organization for Standardization
 (ISO), 203
Ivancic, K. 226

J

Jacobs, A. M. 57
Jagacinski, R. J. 80
James, M. 155
James, W. 232–233
Jamieson, G. A. 177
Jamson, A. H. 135, 137, 143, 222, 225, 238
Janssen, C. P. 218
Jarosch, O. 218
Jentsch, F. 57
Jerison, H. J. 89, 92–93
Jian, J. Y. 178
Johansen-Berg, H. 236
Johnson, M. 36, 51, 60, 191, 198–199, **202**,
 205–206
Johnston, W. A. 238
Jones, D. G. 156, 161–*162*, **164**
Jones, K. S. 234
Jones, L. 173
Jones, R. E. 283
Jonker, C. M. 36

Author Index

Jorna, P. G. 158
Judge, C. L. 112

K

Kaber, D. B. 20, 157–158, 191–192, 195–196, 205, 249
Kahneman, D. 3–5, 20–21, 72
Kam, J. W. Y. 240
Kamzanova, A. T. 88, 94, 130,
Kanki, B. G. 119
Kant, I. 275
Kantowitz, B. H. 4, 226
Kao, C. S. 132
Kasneci, E. 233
Kauffman, S. A. 277–278, 281
Keizer, G. 1
Keller, D. 30–31
Kemp, S. 13
Kennedy, R. S. 173
Kidwell, B. 53
Kielbasa, L. 221
Kim, J. 177
Kim, S. Y. 56, 177
Kircher, K. 134, 242
Kiris, E. O. 108, 151–152, 156, 161, 217
Kirlik, A. 198–199, **202**, 209
Klauer, S. G. 237
Klein, G. A. 50, 53, 56, 59, 72, 75, 77, 80
Klimmt, C. 237
Knight, J. L. 4
Kochan, J. A. 35
Kockelman, K. 213, 232
Koestler, A. 270, 276–277, 280, 284, 287
Koltko-Rivera, M. E. 57
Koonce, J. M. 172, 184
Koopman, P. 53
Körber, M. 134, 221
Kring, J. A, 172
Krishnan, A. 259
Krueger, G. P. 172
Kühn, M. 135, 221
Kun, A. L. 218
Kuorelahti, J. 233
Kurtz, P. 279
Kurzweil, R. 260, 303
Kusano, K. D. 241
Kustubayeva, A. M. 88, 130

L

LaChapell, J. 142, 177
Lai, F. C. H. 135, 222, 238
Lakhmani, S. G. 156
Lambert, F. 243
Landauer, R. 272
Landsbergis, P. A. 237

Langheim, L. K. 132, 134
Langner, R. 143
Lanzetta, T. M. 89–90
Lathin, D. 237
Latorella, K. 270
Lauber, J. K. 118
Laurie-Rose, C. 88, 91, 93
Layton, C. 28
Lazarus, R. 130
Le Blaye, P. 52
Leahey, T. H. 234
Lee, D. 142, 177
Lee, G. 5
Lee, H. 177
Lee, J. D. xii, 10, 26, 105, 137, 152, 172–177, 183, 199, **202**, 216–218, 221–222, 224, 226, 237–238
Lee, S. E. 177, 237
Lees, M. N. 177
Lepers, R. 8
Lerner, N. 237
Leveson, N. x, 81
Lewis, B. 240
Li, D. 177
Li, H. 20, 172, 197
Li, Y. 50, 177
Liang, Y. 218
Liao, H. 242
Lin, J. 130, 132–133, 136, 143
Lin, M. K. 26
Lindblom, C. E. 70, 77, 79
Liu, J. 256
Liu, Y. 10, 217
Llaneras, R. E. 238
Lleras, A. 235
Lopez-Sanchez, J. I. 304
Lorenz, L. **225**, 235, 242
Louw, T. 222
Lozano-Perez, T. 232
Lüdtke, A. 34
Lyell, D. 23–24, 30, 32
Lyons, J. B. 29

M

Mackworth, N. H. 234
Madhavan, P. 177
Magrabi, F. 24
Malin, J. 110, 114, 116
Malthus, T. R. 272
Mangos, P. 180
Manning, V. 8
Manzey, D. H. x, 20–21, 24–26, 28–29, 32, 34–35, 51, 137–138, 141, 143, 152, 172, 176–177, 197, 234
Maran, N. 140
Marcora, S. M. 8

Marin-Lamellet, C. 241
Marks, M. A. 140
Mars, F. 243
Martens, M. H. 133
Maslow. A. H. 290, 295
Mata, J. 7
Mathan, S. 140
Mathieu, J. E. 140
Matthews, G. x, 85–88, 91–93, 127–135, 137, 139–140, 143–144, 177, 221, 234–235
Mattia, D. 130
May, J. F. 144
Mayer, K. 141
McAnulty, H. 222
McBride, S. E. 26, 177
McCarley, J. S. 30
McClumpha, A. 155
McConica, J. 295
McCoy, C. E. 28
McDonnell, L. 33, 112
McEwen, T. 80
McGarry, K. 25
McGehee, D. V. 177
McGuirl, J. M. 29, 35
McKibben, W. 274, 285
McKibbon, K. A. 23
McNaughton, L. 7
McNeese, M. 50
Mehler, B. 222–223
Mehta, R. 30
Meir, M. 215
Meng, F. 238
Merat, N. 135, 222, **225**, 238
Merrifield, C. 130
Merritt, S. M. 142, 174, 177
Methot, L. L. 92
Metzger, U. 23, 152
Meyer, J. 21
Michon, J. A. 214
Midden, C. J. 177
Middlemiss, A. 93
Mikulka, P. J. 158, 165
Militello, L. G. 50
Miller, C. A. ix, 53, 140, 199–200, **202**, 209
Miller, G. A. 70
Miller, L. C. 90
Möbus, C. 34
Molloy, R. 26–27, 104–107, 111, 156, 172, 175, 177–178, 184
Moore, S. E. 272, 303
Moravec, H. 272, 286–287, 293,
Moray, N. xi–xii, 31, 33, 105, 152, 173–174, 176, 267, 293
More, T. 262n1
Morgan, J. F. 214
Morris, N. M. 107, 111, 117,
Morrison, J. G. 103, 106, 108–109, 112–*113*

Mosier, K. L. 19–22, 26–27, 33–36,
Mouloua, M. ix–x, 89–90, 93, 104, 106, 132, 156, 169–170, 172–174, 177, 184, 249, 260, 279, 288, 301, 306n1
Muir, B. M. 105, 152, 173, 176
Mulder, M. 79
Murphy, A. Z. 93
Murphy, L. B. 89
Murphy, R. 49
Murray, C. 53
Muzet, A. 221

N

Nagel, T. 292
Naikar, N. 79, 200, **203**, 205
NASA Autonomous Systems Capability Leadership Team (NASA ASCLT), 191
National Highway Traffic Safety Administration (NHTSA), 213, *254*, xviii, 231
Naujoks, F. 218
Nehme, C. 171
Neigel, A. R. 86, 91
Neisser, U. 221
Nelson, W. T. 131
Neubauer, C. 129–132
Neuschwander, A. 141
Newton, D. C. 134, 276
Neyt, X. 234
Nickerson, R. S. 267, 272
Nieva, V. F. 115
Nogami, G. 119
Norman, D. A. 117
Nourbakhsh, I. xi, 253, 303–304

O

Oakley, K. 303, 306n6
Office of Technical Intelligence (OTI), 63
Olson, J. R. 117
Onnasch, L. 20, 24, 29, 51–52, 138, 143, 152, 172, 177, 197
Onuska, L. 3
Orlady, H. M. 257
Orlady, H. W. 118, 257
Oron-Gilad, T. 281
Osborne, M. A. 250
Oser, R. L. 131
Ouimet, M. C. 237
Oulasvirta, A. 233
Oyman, K. 30

P

Pageaux, B. 8
Paivio, A. 4

Pak, R. 36, 177
Palmer, E. A. 22
Palmer, M. T. 119
Panganiban, A. R. 130, 177
Panou, M. C. 226
Parasuraman, R. ix-x, xii, 20–21, 23, 25–27, 31, 34–35, 49–51, 53–54, 86–87, 89–94, 103–109, 111, 127, 130, 133, 137, 139–141, 143, 152, 156–158, 172–173, 175–178, 183–184, 196, 200–201, 221, 224, **225**, 234–235, 239, 249, 260, 270, 288, 301, 306n1
Paris, J.-C. 241
Parks, D. L. 218–219
Parsons, H. M. 103
Parsons, K. S. 95, 131
Pashler, H. 5, 235
Patterson, V. T. 7
Pattyn, N. 234
Pavlas, D. 172
Peavler, W. S. 3, 4
Peirce, C. S. 70, 76
Perez, M. A. 107, 232
Perrow, C. 277
Petrides, Y. 177
Piaget, J. 76
Pickett, R. M. 89
Piggott, J. 222
Pitichat, T. 1
Pollock, E. 226
Pomranky, R. A. 172
Pope, A. T. 158
Posey, S. M. 79
Posner, M. I. 235
Poviakalo, A. A. 24
Povyakalo, A. 24
Pribram, K. H. 70
Price, M. 36, 177
Priest, H. A. 141
Prinzel, L. J. III, 26, 158
Pritchett, A. R. 56, 253

R

Racine, D. P. 242
Radlmayr, J. 218
Rasmussen, J. 54, 73, 81, 197, **202**, 214
Rau, P. P. 177
Raup, D. M. 277, *278*
Regan, M. 214
Reichenbach, J. 24, 138, 143, 152, 177
Reimer, B. 223
Reinecke, M. 116
Reinerman, L. E. 134
Reising, J. M. 116
Revesman, M. E. 110
Rice, S. 30–31, 176

Rieck, A. M. 115
Rieskamp, J. 30
Riggs, S. L. 91
Riley, J. 158
Riley, M. 131
Riley, V. x, 173, 175–176, 183–184, 224, 284
Robert, G. 137
Roberts, D. M. 237
Rochlin, G. I. 77
Rogé, J. 221
Rogers, W. A. 26, 177
Rosa, R. R. 89
Rose, C. L. 89
Rosenblatt, B. 33
Rosenstiel, W. 233
Ross, H. A. 87, 94–95
Roth, E. M. 59, 253
Roto, V. 233
Roudsari, A. 23
Rouse, S. H. 107
Rouse, W. B. 103, 104, 105, 107–112, 117, 120
Rovira, E. 23, 25, 27–28,
Ruff, H. A. 53, 132
Ruscio, D. 235
Rushworth, M. F. S. 236
Rusnock, C. F. 137
Russell, P. N. 6, 7, 8, 87, 91, 93, 95

S

Sagan, C. 280, 282, 304
Salas, E. 114, 117–118, 131, 141
Salinger, J. 238
Salmon, P. 303, 304
Samuel, S. 52, 291
Sanders, T. L. 304
Santamaria, A. 24
Sarter, N. B. 23, 28–29, 35, 57, 105, 108, 111, 118, 223, 251
Sauer, J. 24, 35, 132, 137
Sawers, A. 93
Sawin, D. A. 87, 90, 94
Sawyer, B. D. 5, 92
Saxby, D. J. 129, 131–133, 135, 137, 140, 143
Scallen, S. F. 52, 106–107, 253,
Scerbo, M. W. 26–27, 87, 89–90, 93–95, 114–116, 130, 158, 172, 184, 236
Schaefer, K. E. 169, 173, 177–178, 257, 304,
Schickedantz, B. 89
Schneider, S. 136
Schneider, W. 118
Schoenfeld, V. S. 93–94
Schoenmakers, T. M. 237
Schooler, J. W. 221
Schreckenghost, D. 110, 114
Schroeder, B. 23, 28, 29

Schumsky, D. A. 90
Schwartz, B. 7, 77
Scialfa, C. T. 177
Scott, L. A. 158
Scott, R. 59
Searle, J. 289
Sebok, A. L. 25, 137
Sedgwick, J. 267
See, J. E. 26, 86, 90–92,
See, K. A. 26, 137, 152, 172, 176, 183, 226
Selker, T. 224
Selkowitz, A. R. 156
Sellers, B. C. 57
Senders, J. W. xi–xii, 171, 214, 279
Seong, Y. 34
Seppelt, B. D. 223, 224
Shappell, S. A. 170
Sharkey, N. 259
Sharples, S. 29
Shattuck, L. G. 79, 80
Shaw, T. H. 91–93, 129, 131–132, 282
Sheridan, T. B. x, 20, 29, 35, 43–46, 49–51, 64, 139, 156–157, 170, 196, 199, 201, **225**, 249, 253–*254*
Shields, J. 49
Shiffrin, R. M. 118
Shinar, D. 215
Shladover, S. E. 54, **225**
Sidney, P. 293
Sierhuis, M. 36
Sikdar, S. 92
Silbermann, L. 177
Simon, H. A. 72, 74, 79
Simons, D. J. 21, 237
Simons-Morton, B. G. 237
Simpson, H. M. 4
Sims, D. 141
Sims, V. 5
Sinek, S. 79
Singh, A. L. 89, 94
Singh, I. L. 26–27, 89, 94, 104–107, 172, 175, 178, 184
Sisa, L. 87, 91
Skitka, L. J. 20–22, 26–27, 33–34
Slovic, P. 21
Slusher, A. L. 7
Small, R. L. 112, 114, 116–117
Smith, K. 173
Smith, M. R. 8
Smith, P. J. 28, 59
Smither, J. 172–173
Snow, C. P. 294, 304
Society of Automotive Engineers (SAE) International, 135, 158, 196, 215, 232–*233*
Soetens, E. 234
Solis-Marcos, I. 242

Sonnentag, S. 140
Sparrow, B. 256
Spence, C. 238
Spörrle, M. 177
Spriggs, S. 53
Staiano, W. 8
Stammers, R. B. 140
Stanciulescu, M. 240
Stanton, N. A. 127, 131, 133–134, 222, 224, 242
Staveland, L. E. 85, 132, 134
Stedmon, A. W. 177
Steg, L. 240
Steinke, F. 177
Stevenson, H. 8
Stewart, J. xi, 253, 303–304
Stolzmann, W. 233
Strayer, D. L. 238
Streufert, S. G. 119
Strigini, L. 24
Sturre, L. 36, 177
Suchman, L. 259
Sullivan, S. 224
Sullivan, T. E. 93
Sutcliffe, A. 242
Svenson, O. 30
Sweller, J. 226
Szalma, J. L. ix, xi, 86–91, 94–95, 129–130, 133, 144, 173, 177, 235, 281

T

Tamminen, S. 233
Tannenbaum, S. I. 114
Taylor, F. 80
Taylor, G. S. 177
Taylor, R. M. 116
Temple, J. G. 71, 88, 92, 234
Teo, G. 88, 90–91
Terrence, P. I. 177
Thompson, J. M. 104
Thoreau, H. D. 265
Thornburg, K. M. 130
Tildesley, H. 240
Tiwari, T. 89, 94
Todd, P. M. 20, 33, 72
Toffetti, A. **225**
Tolkkinen, K. 268
Touliou, A. A. 226
Treisman, A. 93
Tripp, L. D. 130
Trumbo, M. 242
Tsai, T. L. 23
Tsao, H.-S. J. **225**
Tucci, J. 89
Tung, F. W. 177
Tunstall, C. 30
Tversky, A. 21

Author Index

U

Ünal, A. B. 240
United States Air Force, xi, 154, 162, 169

V

Valdes-Dapena, P. 216
Vallee, J. 272, 294
van Arem, B. 134, **225**
Van De Mheen, D. 237
van den Broek, E. L. 177
Van Den Eijnden, R. J. 237
van der Linden, D. 140, 141
van der Meulen, H. 218
van Driel, C. J. G. **225**
Van Dyne, L. 141
van Paassen, M. M. 79
Vecchiato, G. 130
Verberne, F. M. 177
Vermulst, A. A. 237
Verplank, W. L. 44–46, 64, 156, 196, 199, 201, 253
Ververs, P. M. 140
Vicente, K. J. 73, 79, 80,
Victor, T. W. 218, 222, 237
Vieane, A. Z. 25, 137
Vincenzi, D. A. 172, 173, 184
Vogelpohl, T. 135–136, 221
Vohs, K. D. 13
Vollrath, M. 135, 221
Voorhorst, F. A. 304

W

Wacker, G. J. 103
Waddington, C. H. 283
Walker, R. A. 175
Wallach, W. 262n5
Wang, D. 1
Wang, L. 177
Ward, M. 5, 177
Warm, J. S. ix-xi, 85–94, 97, 104, 106, 127, 129–131, 133–135, 143, 170, 172, 221, 234–235, 239, 255
Wastell, D. G. 24, 132, 137
Waytz, A. 177
Wegner, D. M. 256
Weil, L. 93
Weinbeer, V. 135
Weiss, L. G. 139
Welpe, I. M. 177
Wen, J. 11, 12, 13
Wertheimer, M. 71–72
Westbrook, J. I. 23
Westerink, J. H. 177
Westerman, S. J. 140
Westrum, R. 269, 297
Wheatley, D. 177
Whinnery, J. E. 111
Whitlow, S. D. 140
Wickens, C. D. x, 4,5, 10, 20, 24–25, 49, 51, 85–87, 104–105, 117, 120, 127, 137, 139, 157, 171–172, 183, 196–197, 201, **203**, 205, 209, 217, 235, 249
Wiczorek, R. 28–29, 35
Wiegmann, D. A. 170
Wiener, E. L. 21, 103–105, 111, 118, 120, 151, 155–156, 251, 257
Wiener, N. 259
Wierwille, W. A. 214
Williams, P. S. 234
Wilson, J. R. 29
Wiltse, C. M. 7
Winner, L. 250
Wohleber, R. W. 129–130, 132, 136, 138, 163–164
Wojciechowski, J. Q. 172
Wong, D. 106, 111
Woodham, A. 5–6
Woods, D. D. x, 35, 45, 54, 56–57, 63–64, 81, 103–105, 108, 110–111, 114, 117–119, 191, 223, 251
Wright, E. J. 132
Wright, P. 4
Wright, T. J. 52
Wyatt, J. C. 23

Y

Yadav, A. K. 89
Yates, J. F. 50
Yeh, Y. Y. 86–87
Yoo, H. S. 52, 57
Young, K. 214

Z

Zaccaro, S. J. 115, 140
Zeuner, F. E. 306n6
Zhang, Y. 177
Zilberstein, S. 52
Zimmermann, M. 134, 221
Zuckerman, M. 237

Subject Index

Note: Page numbers in **boldface** represent tables and those in *italics* represent figures.

A

Abduction, 70, 76–77
Adaptive Function Allocation for Intelligent Cockpits (AFAIC), 112–113, 116
Adaptive systems, 35, 103, 117–120
Advanced driving assistance systems (ADAS), 231–232, 237–240, 243
Affording, 69–70, 79
Alarms *see also* alerts, 32, 141
Alerts *see also* alarms, 24, 29
Allocation of function *see* function allocation
Arousal, 3–4, 85, 129–130, 134
Attention: capacity, 3–4, 233; divided, 3–4, 237–238; filter model, 2–3; multiple resource model, 5; restoration strategies, 239–240; sustained, 93, 95, 129, 233–235
Augmented technology, 12–13
Authority, 110–111, 142
Automated decision support systems (DSSs), 19–20; in healthcare, 23
Automation: benefits, 104; bias, 21–36; costs 104–105; degrees of, 20, 29, 51–52, 197, 202–**203**, 205, 254; levels of, 20, 43–45, 48, 54–55, 59–60, 105–108, 132–133, 156–**157**, 192, 206, 223, *233*, 240, *254*; reliability, 25–27, 105, 152, 165, 175, 183–184

C

Coactive design, 36, 198, **202**, 205–206
Cognitive maps, 12–13
Cognitive work analysis (CWA), 197, 200, **203**, 205, 209
Complex movement tasks, 6, 13
Compliance, 21, 31–33, 137, 174–176, 182
Computer-aided detection aids (CADs), 24
Concepts of operation, 152–**153**

D

Decision-making, 139, 141–142, 159–160, 162, 199, 201, 209
Displays *see also* interfaces, 2, 79–80, 91, 93, 242–243
Distraction, 135–136, 213, 231, 237–240

E

Ecological interface design (EID), 73–76
Effort-regulation, 135–137, 139
Engagement, 129–133, 156, **181**, 216, *220*, 227, 237
Event rate, 89–90, 92–94, 132–133, 234–235
Expertise, 54, 71–73, 197, **202**

F

False alarms, 31, 171
Fitts' List, 43, 49, 54, 57–58, 61–62
Function allocation, 52–54, 56, 64, 196–197, 205–206
Fundamental attribution error (FAE), 261

G

Global Positioning System (GPS), 9–10, 19–20, 35, 137, 159, 162, 256–257

H

Head mounted wearable displays (HMDs), 2, 5, 9–10
Human automation interaction (HAI) modeling, 196–198, **202–203**, 204, 209
Human autonomous system oversight (HASO) model, 152, 154–*155*, 158–159, 161, 164–165, 197–198, **202**, 205

I

Intelligent autonomous systems (IASs), 127, 139–144
Interfaces *see also* displays, 13, 58, 69, 71, 73–74, 79, 112, 119, 155–156, 163–165, 285–286

M

Mental models, 161
Mind wandering, 97, 129–130, 235, 237, 239–240
Muddling, 70, 76–80, 82

N

NASA Task Load Index (TLX), 85, 96, 132, 134–135

319

O

Omission errors, 21–22, 24, 27–29, 31–33

P

Performance-workload associations, 86–90, 92–97
Pilot's Associate (PA) program, 112, 116

R

Reliance, 13–14, 20–21, 32, 137, 174–177, 182–183

S

Satisfying, 69–71, 79
Specifying, 69, 79
"Stages" of human-automation activity, 50–51, 253
Supervisory control, 152–153, 158–159, 170–171

T

Take-over scenarios, 135, 174, 208, 214, 217–219, 223–224, 226

Time pressure, 30–32, 35–36
Traffic collision avoidance system (TCAS), 20, 35
Transactional model of stress, 130–131
Trust: calibration, 136–138, 176–177, 182–183; and individual differences, 26, 173–174; measurement, 178, overtrust, 26, 152, 175–176, 223, 234; undertrust, 175; *see also* reliance

U

Unmanned aerial systems (UASs), 131–133, 169; fatigue, 132–133; multi-UAS operation, 136, 138, 144, 171; operator training and selection, 178–182; and trust, 177–178; *see also* unmanned aerial vehicles (UAVs)
Unmanned aerial vehicles (UAVs), 191, 193, *see also* unmanned aerial systems (UASs)

W

Workload, 133; overload, 131, 171–172, 214, 217–219; underload, 214, 217–228